Tourism and Hospitality Studies

İrfan Yazıcıoğlu / Özgür Yayla / Alper Işın /
Turhan Çetin (eds.)

Tourism and Hospitality Studies

PETER LANG

Bibliographic Information published by the Deutsche Nationalbibliothek
The Deutsche Nationalbibliothek lists this publication in the Deutsche Nationalbibliografie; detailed bibliographic data is available online at http://dnb.d-nb.de.

Library of Congress Cataloging-in-Publication Data
A CIP catalog record for this book has been applied for at the Library of Congress.

Cover illustration: © 3D_generator/istockphoto.com

ISBN 978-3-631-83886-0 (Print)
E-ISBN 978-3-631-84009-2 (E-PDF)
E-ISBN 978-3-631-84010-8 (EPUB)
E-ISBN 978-3-631-84011-5 (MOBI)
DOI 10.3726/b17779

Peter Lang – Berlin · Bern · Bruxelles · New York · Oxford · Warszawa · Wien

This publication has been peer reviewed.

www.peterlang.com

Preface

This book has been prepared, by pursuing a scientific goal in the field of tourism and hotel management, in order to create a resource for the academicians and sector representatives who conduct researches on the subject. The book consists of 28 chapters written by 38 different authors. The current issues in the book have been determined by taking into consideration the lack of resources in the field of tourism and hotel management and the latest research areas in the international literature.

Within this context, the book **'Tourism and Hospitality Studies'**, considering the tourism and hotel management terminology, is expected to be a source book for the theoretical and practical scientific studies in the fields with which tourism is in close relationship such as gastronomy, recreation and marketing.

The book is an outcome of teamwork. We'd like to express our endless gratitude primarily to our valuable colleagues, who are the authors of the chapters, for their contribution and dedication to the chapters and to all our lecturers, whose names we couldn't include in here, for their strenuous efforts.

September, 2020
Editors

Prof. Dr. İrfan YAZICIOĞLU
Asst. Prof. Dr. Özgür YAYLA
PhD. Alper IŞIN
Prof. Dr. Turhan ÇETİN

Contents

List of Contributors

Melda AKBABA

Asst. Prof. Dr., Kilis 7 Aralık University, Vocational School of Tourism and Hotel Management, Hotel, Restaurant and Catering Services Department, mharbalioglu@kilis.edu.tr, https://orcid.org/0000-0001-8701-017X

Burcu Ayşenur AKBULUT

Asst. Prof. Dr., Recep Tayyip Erdoğan University, Ardeşen Faculty of Tourism, Department of Gastronomy and Culinary Arts, burcuaysenur.akbulut@erdogan.edu.tr, https://orcid.org/0000-0002-0679-3692

Emin ARSLAN

Asst. Prof. Dr., Tokat Gaziosmanpaşa University, Zile Dinçerler Tourism and Hotel Management College, Department of Gastronomy and Culinary Arts, Tokat, Turkey, emin.arslan@gop.edu.tr, https://orcid.org/0000-0003-1592-8162

Fatmanur Kübra AYLAN

Asst. Prof. Dr., Selçuk University, Tourism Faculty, Department of Recreation Management, kubraceliloglu@selcuk.edu.tr, https://orcid.org/0000-0002-6703-2934

Serkan AYLAN

Asst. Prof. Dr., Selçuk University, Tourism Faculty, Department of Tourism Guidance, serkan.aylan@selcuk.edu.tr, https://orcid.org/0000-0002-7262-4440

Hilal AYVALI

Eskişehir Osmangazi University, Institute of Social Sciences, hkayvalı@gmail.com, https://orcid.org/0000-0002-1127-0001

Fuat BAYRAM

Ph.D., Bolu Abant İzzet Baysal University Mengen Vocational School, bayram_f@ibu.edu.tr, https://orcid.org/0000-0003-0513-7674

Muhammet Cenk BİRİNCİ

Asst. Prof. Dr., Erzincan Binali Yıldırım University School of Physical Education and Sports, Department of Recreation E-mail: cenk.birinci@erzincan.edu.tr, https://orcid.org/0000-0002-0309-8256

Koray ÇAMLICA

Ph.D., Nevşehir Hacı Bektaş Veli University, Tourism Faculty, Department of Tourism Management, koraycamlica@nevsehir.edu.tr, https://orcid.org/0000-0003-0746-285X

Sinan ÇAVUŞOĞLU

Phd., Bingol University, Vocational School of Social Science, Management and Organization Services, Bingol, Turkey, 12000, sinankys42@gmail.com, https://orcid.org/0000-0001-9365-8677

Halil CODA

Eskişehir Osmangazi University, Institute of Social Sciences, coda.106@hotmail.com, https://orcid.org/0000-0002-5197-2446

Barış DEMİRCİ

Asst. Prof. Dr., Eskişehir Osmangazi University, Faculty of Tourism, Department of Gastronomy and Culinary Arts, bdemirci@ogu.edu.tr, https://orcid.org/0000-0002-1247-1841

Sait DOĞAN

Asst. Prof. Dr., İskenderun Technical University, Faculty of Tourism, Department of Gastronomy and Culinary Arts, Central Campus, 31200, İskenderun/Hatay, Turkey, Tel: 0.326 613 56 00, sait.dogan@iste.edu.tr, https://orcid.org/0000-0003-4486-8699

Olca SEZEN DOĞANCILI

Asst. Prof. Dr., Sinop University, School of Tourism Management and Hotel Management, Department of Tourism Guidance, o.dogancili@sinop.edu.tr, https://orcid.org/0000-0002-7070-4076

Ebru GÖZEN

Asst. Prof. Dr., Akdeniz University, Manavgat Tourism Faculty, Department of Recreation Management, ebrugozen@akdeniz.edu.tr, https://orcid.org/0000-0002-3476-5023

Ediz GÜRİPEK

Asst. Prof. Dr., Tokat Gaziosmanpasa University, Zile Dincerler Tourism and Hotel Management College, Department of Tourism and Hotel Management, Tokat, Turkey. email:ediz.guripek@gop.edu.tr, https://orcid.org/0000-0001-9573-2674

Alper IŞIN

Ph.D., Ankara Hacı Bayram Veli University, Faculty of Tourism, Department of Gastronomy and Culinary Arts, alper.isin@hbv.edu.tr, https://orcid.org/0000-0003-2232-0606

Ali İSKENDER

Asst. Prof. Dr., Atatürk University, Tourism Faculty, Department of Tourism Guidance, ali.iskender@atauni.edu.tr, https://orcid.org/0000-0003-2753-1242

Ercan KARAÇAR

Asst. Prof. Dr., Sinop University, School of Tourism Management and Hotel Management, Department of Recreation Management, ekaracar@sinop edu.tr, https://orcid.org/0000-0002-1124-9667

Yusuf KARAKUŞ

PhD., Nevşehir Hacı Bektaş Veli University Faculty of Tourism Department of Tourism Management E-mail:ykarakus@nevsehir.edu.tr, https://orcid.org/0000-0002-4878-3134

Ali KELEŞ

PhD. Tourism Entrepreneur, lindaalik@hotmail.com, https://orcid.org/0000-0002-0807-937X

Hüseyin KELEŞ

Lecturer, Akdeniz University, Manavgat Tourism Faculty, hkeles@akdeniz.edu.tr, https://orcid.org/0000-0003-1469-0017

Üzeyir KEMENT

Assoc. Prof. Dr., Bingöl University, ukement@bingol.edu.tr, https://orcid.org/0000-0002-3190-9079

Hakan KENDİR

Asst. Prof. Dr., Tokat Gaziosmanpaşa University, Zile Dinçerler Tourism and Hotel Management College, Tourism and Hotel Management Department, Tokat, Turkey, hakan.kendir@gop.edu.tr, https://orcid.org/0000-0002-1356-1339

Emrah KESKİN

Asst. Prof., Nevsehir Hacı Bektaş Veli University, Tourism Faculty, Department of Gastronomy and Culinary Arts, ekeskin@nevsehir.edu.tr, https://orcid.org/0000-0001-6565-4265

Handan ÖZÇELİK BOZKURT

Asst. Prof. Dr., Sinop University, School of Tourism and Hotel Management., Department of Gastronomy and Culinary Arts, handanozcelikbozkurt@gmail.com, https://orcid.org/0000-0002-0035-3362

Mehmet Mert PASLI

Asst. Prof. Dr., Giresun University, Bulancak School of Applied Sciences, Tourism and Hotel Management, mert.pasli@giresun.edu.tr, https://orcid.org/0000-0003-0323-0461

Esra ŞAHİN

PhD., Akdeniz University, Manavgat Tourism Faculty, Department of Gastronomy and Culinary Arts, e-mail:esrasahin@akdeniz.edu.tr, https://orcid.org/0000-0001-6399-0177

Ayşe SÜNNETÇİOĞLU

Asst. Prof. Dr., Canakkale Onsekiz Mart University, Gelibolu Piri Reis Vocational School, aysesunnetcioglu@comu.edu.tr, https://orcid.org/0000-0002-4295-9776

Tufan SÜREN

Asst. Prof. Dr., Ankara Hacı Bayram Veli University, Faculty of Tourism, Gastronomy and Culinary Arts, tufan.suren@hbv.edu.tr, https://orcid.org/0000-0002-7877-8256

Tuba TÜRKMENDAĞ

PhD., Atatürk University, Tourism Faculty, Recreation Management, E-mail: tuba.gezen@hotmail.com, https://orcid.org/0000-0003-1528-2883

Zafer TÜRKMENDAĞ

PhD., Atatürk University, Tourism Faculty, Department of Tourism Guidance, zafer.turkmendag@gmail.com, https://orcid.org/0000-0002-7712-1500

Eray TURPCU

Asst. Prof. Dr., Giresun University, Bulancak School of Applied Sciences, Tourism and Hotel Management, eray.turpcu@giresun.edu.tr, https://orcid.org/0000-0003-2732-6222

Ahmet ÜNVER

Eskişehir Osmangazi University, Institute of Social Sciences, ahmetu20@gmail.com, https://orcid.org/0000-0003-3733-6547

Özgür YAYLA

Asst. Prof. Dr., Akdeniz University, Manavgat Tourism Faculty, Department of Recreation Management, ozguryayla@akdeniz.edu.tr, https://orcid.org/0000-0001-7124-9311

Mehmet YEŞİLTAŞ

Prof. Dr., Cyprus International University, Faculty of Economics and Administrative Sciences, CYPRUS, myesiltas@ciu.edu.tr, https://orcid.org/0000-0003-0110-9684

Vedat YİĞİTOĞLU

Asst. Prof. Dr., Akdeniz University, Manavgat Tourism Faculty, Department of Gastronomy and Culinary Arts, viyitoglu@akdeniz.edu.tr, https://orcid.org/0000-0003-2215-1961

Bülent YORGANCI

Doctoral Student, Nevsehir Hacı Bektaş Veli University, Department of Gastronomy and Culinary Arts, bulentyor@yahoo.com, https://orcid.org/0000-0002-6598-2733

Zafer TÜRKMENDAĞ

1 Information Technology Studies in Tourism and Hospitality

1 Introduction

The way of competing has changed in tourism with the increment in innovative solutions and the development of information technologies. In this context, the impact of technological developments on tourism has emerged as an important field of study. Developments such as the internet of things, big data, artificial intelligence, and robots have also created the need to restructure the service and management processes in tourism. Information technology (IT) is widely used in the tourism sector but managements still have worries when making strategic decisions. Unfortunately, many managers have communication problems with IT specialists because of the different points of view on the context. Managers mainly focused on marketing and strategies while IT specialists are interested in systems, knowledge, and computers (Law et al., 2014). As technology grows every day, this cause tourism market requirements and customer needs evolve. As this happens, the connection between stakeholders of destination comprising customers and all processes within firms or between firms and customers has entered into a new era. So, a new competitive environment includes networks, new interfaces with clients and stakeholders, renovations, and innovations in products (Buhalis & Law, 2008). Likewise, destinations need more information to attract tourists. Big data composed of the traveler's profiles, patterns, and desires, may be a piece of strategic information for destinations (d'Amore et al., 2015).

2 Smart Tourism

Tourism has a complex structure with multinational and SMEs thus understanding the different types of underlying information may support tourism professionals to make innovations (Aldebert et al., n.d.). This gives tourism firms the ability to create new valuable products and services which gives them also the strength to use IT in their processes and efficiency. This occurs in every part of the firm including employees, mid-level, and top-level managers, and in their way of thinking (Buhalis & Law, 2008). Even though data is very important in the IT field, understanding this data by making it comprehensible is an occupation of tourism workers. Because not data but knowledge has a fundamental status in

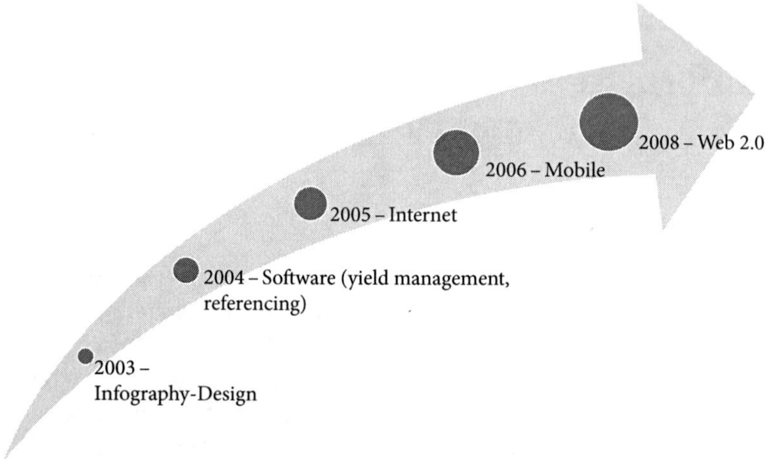

Fig. 1: Main technologies by year (Aldebert et al., n.d.).

Source: Aldebert, B., Dang, R. J., & Longhi, C. (2011). Innovation in the tourism industry: The case of Tourism@. Tourism management, 32(5), pp: 1212.

making innovations and thus they are particular to the relative sector's technological developments. Because tourism is dynamic and always evolving, has an exclusive state comparing to others (Aldebert et al., n.d.). This relation is not unidirectional, that is to say, data or knowledge also changes the environment and structure of travel and hospitality (Buhalis & Law, 2008). The ecosystem between intelligent systems and consumers can help firms to understand the interactive relation. Thus, sociological perspectives may have problems understanding the phenomenon but swarm theory is a featured one to understand these social actions which empower software for intelligent systems (Gretzel, 2011).

If we want to get down in the history of smart tourism (see Fig. 1), it can be seen that the western countries tried to apply technology to their marketing needs and per contra, China-centered smartness on tourism thereby switched to a modern service industry. This can be called as digitalization of tourism processes while creating more information. Smart tourism provides much information to tourists in the traveling phase. It is logically constituted in four levels (Li et al., 2017):

1. Smart tourism is tour information,
2. The smart tour is a ubiquitous tour information service,
3. Smart tourism information is supplied to a person, not group,

4. Smart tourism is a revolutionary tour information service which is developed in tour facilities.

Productivity in marketing coupled with web 2.0 features needs to pay attention to the incremental numbers in online purchasing and reviews. As positive online word of mouth increases hotels online room sales must focus on their reviews (Ye et al., 2011). Because the internet has become the main source before traveling with its particular channels like OTAs, social media, search engines, portals, websites of destinations, etc. Therefore, tourism industry players must include these channels to their marketing strategies and operations to be able to communicate and sell online better (Xiang et al., 2015).

3 Websites

The internet provides information and facilitates business processes for consumers with websites and internet applications. Consumers generally look at the website of the company to see who is the object of it. The website is an identity for the company and allows the consumer for having an idea in mind. Thus, website design and quality are critical to attracting consumers and have an impact on satisfaction (Bai et al., 2008). Besides convenience and financial security are crucial in consumer assessments of online satisfaction (Szymanski & Hise, 2000). One of the novelties brought by web 2.0 is the hotel rating list which gives customers the ability to quickly find the best-rated hotels on the top of the list. These lists are important and asymmetric due to their effect on purchase behavior thus hotels seek the ways to be chart-topping. The hotel rating list is very confidential and helpful when it is created by a known website (Casaló et al., 2015). Besides the quality of the website is determined by its usability, functionality, and privacy. The user interface design of the website and ease of use give more trust and providing more detailed information gives more performance. All these facts plus the recognition of users' personal information being in a safe place improve their trust level (L. Wang et al., 2015).

Electronic word of mouth (E-WOM) lets people share their bad experiences, and %48 of them share it with ten others. To decrease the amount of negative E-WOM firms should improve their quality of service. Besides they may have customer relationship management systems to have a better relationship with customers and privatize services for them. Therefore, hotels that use intermediaries for reviews, tend to respond honestly to negative reviews to improve their sincerity (Tsao et al., 2015). With technological developments, experiences also have been shared which revealed the experience economy.

These experiences have been created as tourists co-create their own experiences and use technology to co-create better experiences. Destinations have physical and virtual environments for customers before and after a trip which creates multiple settings (Neuhofer et al., 2012).

4 Social Media & Mobile Technologies

Social media, as one of the important developments of the 21st century, affected tourism mobility. Travelers post experiences or comments on social media. Besides, the follower can follow social media and get information about new places. The most important counter of consumer approval in social media is a pleasure. Enjoyable and bearer media sharing enables them to build a better human-computer and should be hedonic. Marketers who want to use social media in the tourism sector should also buy visited media. It is really difficult to establish such platforms because of the structure of the service in tourism and its excessive word of mouth (Ayeh et al., 2013).

Businesses are using new technology with market pressures, but they cannot adopt much in the virtual environment. Social media is used extensively by consumers as a part of the virtual environment. Similarly, tourists benefit from reliable information on social media. This news is seen as a strategic tool for social media businesses, especially in a promotion (Leung et al., 2013). Tourism businesses should ensure information security and make it more enjoyable to increase the use of social media. Due to the complex structure of tourism knowledge, businesses can provide a pleasant and enjoyable environment by communicating quickly with more simple and understandable information (Chung & Koo, 2015).

Tourists tend to use their mobile phones during their trip to unfamiliar places and when they need more information about the place. Mobile innovation is a great and immediate helper for travel challenges (Gretzel, 2010). Mobile technology is a perfect compatible solution for tourists as it can be taken during the travel. By the reason of this technology, travelers stay always in connection with each other also when they are out of their place of residence. Moreover, tourists have the opportunity to take pictures and share them instantly with their family or friends. While some tourists don't want to use their phone to be disconnected from people and live the place visited (Dickinson et al., 2016). Mobile phone usage is driven by cognitive beliefs of individual and especially contextual factors. (D. Wang et al., 2014). With the mobile phone, the use of NFC also increases and support service quality improvements and marketing efforts. With NFC tags tourists can share and get information simultaneously without touching. It is widely used for mobile payments in markets and other shops (Pesonen & Horster, 2012).

5 Virtual & Augmented Reality

Entertainment and design in virtual reality (VR) is widely used in tourism. It is expected that the use of VR in tourism will increase in parallel with the increase in its use in society. For this reason, tourism enterprises should follow this technology and be able to evaluate it (Guttentag, 2010). For example, museums get the opportunity to market themselves better with augmented reality (AR), but they are uneasy for investment. Also, user interface and user experience design for AR in museums affect the efficiency of their use of this technology (tom Dieck & Jung, 2017).

VR should not just think of the image. It can be used with wearables, nanotechnology, and sensor technologies that support it. By adding moisture feel through brain signals in a VR environment, it can be made to feel like in a rainy area or this can be experienced with wearable technology. Thus, with VR and AR, people can experience different emotions and their emotions can be controlled. AR applications are used in tourism businesses as an assistant. Thanks to such assistants, the participation of virtual tourists in travel are increased and they can be provided together for touristic products. This also helps businesses to innovate. VR and AR develop themselves in four areas, especially in (Loureiro et al., 2020):

1. Physical and sensory stimulation,
2. Longitudinal virtual experiences,
3. Prosperity development,
4. Use of artificial intelligence.

Customization can be made in AR and reveals a service quality concept within itself. Service quality in AR is related to the content it has and the ability of visitors to obtain customized information (Jung et al., 2015).

6 Artificial Intelligence & Robotics

Robotics and artificial intelligence are competing for the cost of labor in the tourism industry. Due to its structure and information complexity, the use of artificial intelligence in tourism finds more places (Gursoy, 2018). While digital transformation is changing the way of working and communication in the service sector, businesses need to use artificial intelligence to develop new strategies and personalization for consumers (Özdemir Akgül, 2019). Using artificial intelligence is a new norm in every level which gives many advantages but also the complexity. To invest in artificial intelligence businesses should have financial and technical infrastructure which is very expensive. Besides the increment in the use of such technologies get human interaction to be decreased in tourism (Samala et al., 2020).

Studies on robots take place especially in the field of tourism. Robots can be used in many areas of tourism due to their service structure and information-intensive features. The adoption of robots in enterprises and the expansion of their use affects their efficiency (Ivanov et al., 2019). In the next period, we will be talking about the cloud robotics which will share information and have a common big data. So, the robots will have more ability to serve in a customized way (Tung & Law, 2017).

7 Conclusion & Suggestions

As technology affects all fields changes also tourism and hospitality due to its abstract and information-intensive nature. Information technologies continue to show itself rapidly in the tourism sector. This change takes place with different techniques in every area from customers to businesses. With the new approaches in internet technology and the development of infrastructures, tourism information is produced and shared every second that passes. Websites and social media accelerate this flow of information and make changes not only in terms of marketing but also in the lives of people on a larger scale. Developments in website designs and content and media quality in social media stand out as determining consumer behavior. Besides, mobile technologies are the most suitable for tourism. Thanks to mobile technologies that can also be used while traveling or away from home, the tourist can be online at any time and record or share their experiences simultaneously. With the introduction of internet technology into smartphones with technologies such as 4G, 4.5G, and 5G, a new communication network has been established in the world and has turned into a communication tool that provides instant information transfer to tourists. Tourists can obtain the required information or find solutions to their troubles immediately thanks to the mobile phone, messaging, mapping, and suggesting applications or websites. Besides, tourists can determine the restaurant or the hotel he wants to go by looking at the suggestions on his smartphone at the location visited. Mobile technologies also provide great benefits to businesses through instant communication, efficient management of processes, and reporting. Moreover, virtual reality, artificial intelligence, and robots also change the service processes and quality of businesses. These technologies, which take the competition to a different level, can affect the behavior of tourists positively or negatively. Tourists can meet these technologies at all stages of their travel and when managed effectively, can provide a more competitive advantage to businesses.

References

Aldebert, B., Dang, R. J., & Longhi, C. (2011). Innovation in the tourism industry: The case of Tourism@. Tourism Management, 32(5), 1204–1213.

Ayeh, J. K., Au, N., & Law, R. (2013). Predicting the intention to use consumer-generated media for travel planning. Tourism Management. https://doi.org/10.1016/j.tourman.2012.06.010

Bai, B., Law, R., & Wen, I. (2008). The impact of website quality on customer satisfaction and purchase intentions: Evidence from Chinese online visitors. International Journal of Hospitality Management. https://doi.org/10.1016/j.ijhm.2007.10.008

Buhalis, D., & Law, R. (2008). Progress in information technology and tourism management: 20 years on and 10 years after the Internet-The state of eTourism research. Tourism Management, 29(4), 609–623. https://doi.org/10.1016/j.tourman.2008.01.005

Casaló, L. V., Flavián, C., Guinalíu, M., & Ekinci, Y. (2015). Do online hotel rating schemes influence booking behaviors? International Journal of Hospitality Management. https://doi.org/10.1016/j.ijhm.2015.05.005

Chung, N., & Koo, C. (2015). The use of social media in travel information search. Telematics and Informatics. https://doi.org/10.1016/j.tele.2014.08.005

d'Amore, M., Baggio, R., & Valdani, E. (2015). A practical approach to big data in tourism: A low cost raspberry pi cluster. In Information and Communication Technologies in Tourism 2015. https://doi.org/10.1007/978-3-319-14343-9_13

Dickinson, J. E., Hibbert, J. F., & Filimonau, V. (2016). Mobile technology and the tourist experience: (Dis)connection at the campsite. Tourism Management. https://doi.org/10.1016/j.tourman.2016.06.005

Gretzel, U. (2010). Travel in the network: Redirected gazes, ubiquitous connections and new frontiers. In Post-Global Network and Everyday Life, 41–58.

Gretzel, U. (2011). Intelligent systems in tourism. Annals of Tourism Research. https://doi.org/10.1016/j.annals.2011.04.014

Gursoy, D. (2018). Future of hospitality marketing and management research. Tourism Management Perspectives. https://doi.org/10.1016/j.tmp.2017.11.008

Guttentag, D. A. (2010). Virtual reality: Applications and implications for tourism. Tourism Management. https://doi.org/10.1016/j.tourman.2009.07.003

Ivanov, S., Gretzel, U., Berezina, K., Sigala, M., & Webster, C. (2019). Progress on robotics in hospitality and tourism: A review of the literature. In Journal of Hospitality and Tourism Technology. https://doi.org/10.1108/JHTT-08-2018-0087

Jung, T., Chung, N., & Leue, M. C. (2015). The determinants of recommendations to use augmented reality technologies: The case of a Korean theme park. Tourism Management. https://doi.org/10.1016/j.tourman.2015.02.013

Law, R., Buhalis, D., & Cobanoglu, C. (2014). Progress on information and communication technologies in hospitality and tourism. International Journal of Contemporary Hospitality Management. https://doi.org/10.1108/IJCHM-08-2013-0367

Leung, D., Law, R., van Hoof, H., & Buhalis, D. (2013). Social media in tourism and hospitality: A literature review. In Journal of Travel and Tourism Marketing. https://doi.org/10.1080/10548408.2013.750919

Li, Y., Hu, C., Huang, C., & Duan, L. (2017). The concept of smart tourism in the context of tourism information services. Tourism Management. https://doi.org/10.1016/j.tourman.2016.03.014

Loureiro, S. M. C., Guerreiro, J., & Ali, F. (2020). 20 years of research on virtual reality and augmented reality in tourism context: A text-mining approach. In Tourism Management. https://doi.org/10.1016/j.tourman.2019.104028

Neuhofer, B., Buhalis, D., & Ladkin, A. (2012). Conceptualising technology enhanced destination experiences. Journal of Destination Marketing and Management. https://doi.org/10.1016/j.jdmm.2012.08.001

Özdemir Akgül, S. (2019). Turizm 4.0 ve Yapay Zeka Uygulamaları. In M. Sezgin, S. Özdemir Akgül, & A. Atar (Eds.), Turizm 4.0 (pp. 142–153). Detay Yayıncılık.

Pesonen, J., & Horster, E. (2012). Near field communication technology in tourism. Tourism Management Perspectives. https://doi.org/10.1016/j.tmp.2012.04.001

Samala, N., Katkam, B. S., Bellamkonda, R. S., & Rodriguez, R. V. (2020). Impact of AI and robotics in the tourism sector: a critical insight. Journal of Tourism Futures. https://doi.org/10.1108/JTF-07-2019-0065

Szymanski, D. M., & Hise, R. T. (2000). E-satisfaction: An initial examination. Journal of Retailing. https://doi.org/10.1016/S0022-4359(00)00035-X

tom Dieck, M. C., & Jung, T. H. (2017). Value of augmented reality at cultural heritage sites: A stakeholder approach. Journal of Destination Marketing and Management. https://doi.org/10.1016/j.jdmm.2017.03.002

Tsao, W. C., Hsieh, M. T., Shih, L. W., & Lin, T. M. Y. (2015). Compliance with eWOM: The influence of hotel reviews on booking intention from the perspective of consumer conformity. International Journal of Hospitality Management. https://doi.org/10.1016/j.ijhm.2015.01.008

Tung, V. W. S., & Law, R. (2017). The potential for tourism and hospitality experience research in human-robot interactions. In International Journal of Contemporary Hospitality Management. https://doi.org/10.1108/IJCHM-09-2016-0520

Wang, D., Xiang, Z., & Fesenmaier, D. R. (2014). Adapting to the mobile world: A model of smartphone use. Annals of Tourism Research. https://doi.org/10.1016/j.annals.2014.04.008

Wang, L., Law, R., Guillet, B. D., Hung, K., & Fong, D. K. C. (2015). Impact of hotel website quality on online booking intentions: ETrust as a mediator. International Journal of Hospitality Management. https://doi.org/10.1016/j.ijhm.2015.03.012

Xiang, Z., Magnini, V. P., & Fesenmaier, D. R. (2015). Information technology and consumer behavior in travel and tourism: Insights from travel planning using the internet. Journal of Retailing and Consumer Services. https://doi.org/10.1016/j.jretconser.2014.08.005

Ye, Q., Law, R., Gu, B., & Chen, W. (2011). The influence of user-generated content on traveler behavior: An empirical investigation on the effects of e-word-of-mouth to hotel online bookings. Computers in Human Behavior. https://doi.org/10.1016/j.chb.2010.04.014

Eray TURPCU

2 Innovation in Tourism Businesses

1 Introduction

With the development of information technology, multimedia and communication networks in the 21st century, businesses that trying to keep up with the changing and complex new world order have had to adapt to innovative movements in order to survive and continue their activities in the competitive environment. Businesses have taken radical decisions in accordance with innovation in order to harmonize their structures, management strategies, business processes, traditions and cultures, private self and competencies they used in the past in order to keep up with the times and the developing world order. In doing so, in order to survive against its competitors in the competitive environment and to move ahead (Kodama, 2003: 235), they have acquired many opportunities such as to make productivity efficient, to establish a quality standard, to have the ability to reduce costs, to ensure price competition, to create new markets and to meet expectations (Çetin and Gedik, 2017: 173). In order to acquire these opportunities, businesses have to make an effort to provide a flexible and sustainable working environment suitable for the changing and developing business life (Kodama, 2003: 235). Because the unique way businesses can continue to survive in changing competitive conditions is through innovation (Korucuk et al, 2020: 158).

Innovation activities vary greatly from business to business by their nature. While some companies participate in well-defined innovation projects such as the development and launch of a new product, others primarily make continuous improvements to their products, processes and operations (Oslo Manual, 2005: 47). Because regardless of the sector and size, it is inevitable for all companies to innovate, but this also pertinent for the tourism sectors. For example, a restaurant can switch to a computer-controlled ordering and billing system. A tourism agency can start serving its customers with a reservation and information service through its website (Elçi, 2007: 29). Considering the abstraction of their products in innovation activities in tourism enterprises, the tourist may be unstable about the quality of the services provided. Despite or because of this abstraction, the value of tourism products can be increased by adding feelings (design, smell, light, color, emotional attachment) (Weiermair, 2006: 58). Researches indicate that innovations can be funny for tourists, and tourists expect to experience new and different things. Nowadays, tourists can also travel for their own spiritual renewal. It is also emphasized in the researches that the search for innovation is an important motivation in the tourism sector. However, it can be ignored where and how innovation is sought, what should be new, and at what level the innovation should be (Kanca, 2019: 15).

In this part of the book, theoretical information about the concept of innovation is given, innovation processes are explained, innovation-related types are mentioned, information about innovation in tourism enterprises is evaluated with studies carried out in tourism enterprises in general, and the place and importance of innovation in tourism is revealed with the conclusion section.

2 The Concept and Process of Innovation

The concept of innovation, which derived from the Latin word "innovare", in its broadest sense means "doing something new" (Flynn et al., 2003: 420). The concept of innovation is named with the words *"yenileşim"* and *"inovasyon"* in the Turkish Language Association's Current Turkish Dictionary and it is defined as *"the introduction of new methods in social, cultural and administrative environments to adapt to changing conditions"* (TDK, 2020). It was interpreted as "the driving force of development" by the economist and political scientist Joseph Schumpeter for the first time in the literature (Elci et al., 2008: 25) and he saw innovation as the heart of economic change. Schumpeter (1997) evaluated the definition of innovation within the definition of manufacturing a new product, making changes in the nature of the existing, going to innovate in a certain period of time, opening up to new markets, making changes in the industrial organization and finding new resources for the organization (as cited in Weiermair, 2006: 55).

Over the years, the concept of innovation has been a research subject for many authors and scientists in different disciplines, especially in the service sector, and it has been tried to be defined with different approaches. Some definitions made by the researchers regarding the concept of innovation are shown in Tab. 1.

When the below definitions and common points regarding the concept of innovation are examined, it can be thought that the concept of innovation has gradually expanded to include concepts such as invention, change, discovery and technology, and the main purpose of the concept is creating economic or social value (Gunay and Calik, 2019: 10). Because concept "innovation" contrary to common belief, it covers optimization, development activities with greater steps as a consequence of radical ideas never experienced and developed and with distinct variation not only for the product but also in the process, organization

Tab. 1: Definitions of Innovation Concept

Source	Definition
Drucker (1998: 3)	Innovation is defined as the creation of new resources by an entrepreneur, either in a public institution or in an enterprise owned by him/her, or increasing the existing potential by improving the existing resources.
Tang (2006: 68)	Innovation is defined as an increase in economy and productivity with the learning and application process by discovering new technologies and techniques in the activities of an enterprise.
Rogers (1983: 11)	Innovation is ideas, practices and objects perceived as new by a person or institution.
Kim et al. (2011: 1190)	Innovation is a set of results that progresses by addressing innovation and social value creation practices.
Damanpour (1996: 694)	Innovation covers the process of adoption, implementation and development of new ideas and behaviors by the organization.
Halligan (2009: 17)	Innovation is defined as creating an economic value by developing and changing the current way of working of an enterprise, creating a new or improved product and service.
Goldhar (1980: 284)	Innovation is a succession of individual and organizational behaviors from the creation of new ideas to their commercialization.
Udwadia (1990: 65)	Innovation is defined as the successful creation, development and delivery of new goods, processes and services.
Elci (2007: 2)	Innovation is the transformation of knowledge into economic and social benefit.

Source: Compiled by the author

and marketing with huge breakthrough and also the optimization and development activities those can be created with step-by-step and consecutive processes. Different from other concepts, it is necessary to create values to get an innovation for changes and novelties (Memis and Korucuk, 2019: 1110–1111). Since the concept of innovation deals with creative talent, creative idea, intuition and research skills, it exists with entrepreneurship and includes all human talents. Innovation can be described as seeing things that everyone looks at but cannot see in the existing system, although conceptually it includes positive change and development. At the beginning level of the innovation concept, many ideas are presented and the most important of these ideas is separated from the others to be used and after use, a commercial value is obtained for the business (Gunay and

Calik, 2019: 10). In the Innovation and R & D Management Guide for Clusters created by the Turkish Ministry of Economy, General Directorate of Exports, the innovation process consists of elements such as idea, concept, prototype, production and put on marketing (Fig. 1).

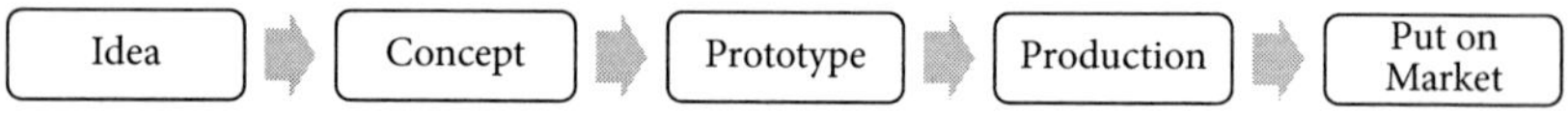

Fig. 1: Innovation process.
Source: www.smenetworking.gov.tr

In the innovation process, when an innovation will be made regarding a product, researches on the properties of that product have a great importance. Generating innovative ideas about the features of the product is seen as the first phase of the innovation process. In businesses, usually brainstorming is done during the idea creation phase, and many ideas are created and evaluated. The important point here is which of the innovative ideas put forward will be put into practice, both inside and outside the enterprise. For this, the studies that will be carried out at the concept stage are supported by feasibility studies and the most suitable idea for the business is evaluated. The concept phase becomes vital for the business in this respect. After the innovative choice to be made is determined, the ideas that are put on paper before starting production are brought to the fore in the prototype stage, together with the good and bad sides of the product. Although the prototype stage can be considered as the last stage before production, there is no perception that every product deemed appropriate at the prototype stage will go into mass production. In the prototype stage, the stage of putting the product to the market, which is deemed appropriate and desired to be put into mass production, is considered to be a way of presenting the product to the consumer. Here, when a newly produced product or an innovative product is put on the market, it has a vital importance that it is put on the market through systems such as research, advertisement, promotion, sales development and public relations (www.smenetworking.gov.tr).

Since businesses that succeed in the innovation process can be considered as productive, profitable and sustainable enterprises of which have competitive advantage and can insure their future, it is vital for businesses to carry out their innovation processes in accordance with the purpose. However, since countries with innovative businesses in their economies have developed by gaining competitive advantage on a global scale, it will be beneficial for businesses to have a management idea that aims to develop their innovation structures as a solution to the multi-dimensional crisis experienced on a global scale (Toraman et al., 2009: 115).

3 Innovation Types

The concept of innovation has continued to evolve, especially in the last three decades, as the driving force of the development, technological advancement and business success of nations, and today it is seen as it ticks all boxes not only for the "creation of something new" but also for solving a wide variety of problems. The term "innovation" has been combined with words such as metaphor, political promise, slogan or fashion rather than a strict scientific concept by politicians, marketing experts, advertising professionals and management consultants, and they have been used more and more frequently in business (Kotsemir et al., 2013: 3). Because it is thought that the development of all economic processes and activities owned by enterprises with the entry of innovation into business life are closely related to new technologies and innovations. The type of innovations of which will be used in the production of business-related production, scientific discoveries, project activities, creation of new high-tech products, provision of services and execution of businesses is regarded as an important issue for businesses to survive in the competitive environment (Kogabayev and Maziliauskas, 2017: 60).

Innovation types are seen as a key concept in the innovation literature. Regarding this subject in the literature, types of innovation are discussed in four ways in the Oslo Manual (2005) as product innovation, process innovation, marketing innovation and organizational innovation, and different types of innovation such as technological innovation and managerial innovation were also considered as research topics by different authors and disciplines.

Product Innovation: Product innovation is defined as the development of new products that differ significantly from existing products. Product improvement is defined as a significant improvement in the technologically existing product. Product innovation is a radical change for a business in proportion to product improvement (Lee and Kang, 2007: 344). Product innovation contributes to the reduction of production costs and the production process, as well as an increase in return on investment and production efficiency. It also contributes to improving product quality and makes products more competitive in foreign markets. Product innovation, which is an important driving factor in economic growth and efficiency, ensures the continuity of customer loyalty, creates new product patterns, meets the needs of customers with new features and offers new opportunities to use new resources by producing solutions to production problems (Reguia, 2014: 148).

Process Innovation: Oke et al. (2007: 738) define process innovation as tools and information that enable the development of methods to produce existing products with higher quality and less cost, or the transformation of new inputs into outputs for a sector, organization or unit. Although process innovation is a type of innovation that can be done alone or can be sustained, it maintains its connection and can be associated with other types of innovation. Customer understanding and the level of change in knowledge, which are requirements of the competitive environment and one of the most important elements of innovation, ensure that process innovation also affects other types of innovation and plays an important role. In this context, process innovation can also be perceived as a radical improvement of basic business processes through the use of new tools and business designs, revealing the process view of enterprises (Gules and Bulbul, 2004: 43).

Organizational Innovation: Organizational innovation refers to new and different structures that will ensure that material and human resources are brought together in the most effective way (Yavuz, 2010: 147). Organizational innovation is concerned with all parts of the organization and all aspects of its functioning; therefore it includes all kinds of innovations. Moreover, the adoption of innovation is designed to encompass the initiation, development and implementation of new ideas or behaviors. Organizations can either develop or purchase innovation. However, in both cases, the adopted innovation is intended to contribute to the performance or effectiveness of the organization (Damanpour, 1992: 376). The difference of organizational innovation from other organizational changes is the fact that innovation is implemented for the first time as a result of the decision of the management that has not been done/tried before. In this context, the first use of a method that the company has not used in itself or in its external relations can be considered as organizational innovation. The commercial value of these innovations made within the business or in foreign relations is also an important issue in deciding whether the innovation made can be defined as organizational innovation (Gunay, 2007: 17). An organizational innovation occurs in the business's business practices, workplace organization, or external relations (Soylu and Gol, 2010: 117).

Marketing Innovation: In the market economy, there are innovations in the marketing of products as well as innovations in products and production processes. The development of new marketing tools and methods plays an important role in the development of industries (Chen, 2006: 101). Marketing innovation is defined as the search for creative and new solutions to problems and needs. In order for businesses to become more competitive and improve their performance, they must constantly develop new products and strategies (Ungerman et al., 2018: 132). In such innovations, it is aimed to increase the sales and market share by changing the condition of the product. New sales channels are also included in marketing innovations. If the company implements a marketing method of which has never been used before, and there is a change from the marketing tools it uses, it means that the company has achieved a marketing innovation (Oğuztürk, 2011: 50). During the marketing innovation phase, serious changes are made in the design, packaging, placement, promotion or pricing of the product (Elçi, 2007: 12).

Technological Innovation: Technological innovation is a process that leads to changes in companies' business activities and technologies (Ersoy and Sengul, 2008: 61). It is the most emphasized type of innovation because it is more controlled in terms of technological innovation, formation, research and observation. The return of technological innovation is quite high in providing economic returns and employment for countries and companies (Halac et al., 2014: 167). Technological innovation or innovations are important for the direction the future of the companies. Management levels of strong businesses express that making use of technological developments to increase and develop the quality of products and services offered in the market, and in order to ensure consumer satisfaction and loyalty can contribute to businesses (Sagir and Turkeri, 2016: 277).

Managerial Innovation: Managerial innovation is the event of making effective and appropriate decisions in choosing human resources and evaluating alternatives (Cannarella and Piccioni, 2003: 377). Managerial Innovation appears to have an unmatched power to create dramatic and lasting changes in competitive advantage compared to other types of innovation. So much so that long-term times may be necessary for competitors to perceive and react to innovations introduced. Because it is very difficult for competitors to copy systemic advantages, especially covering the innovation network that spans many management processes (Hamel and Bren, 2007: 69). A new management system involves the initiation of the administrative process or staff development program. Managerial innovations include automated personnel records and skills systems, formalized strategic planning process, management by objectives, zero-based budgeting, continuous training programs for staff, job rotation, flexible time, coordination, incentives or reward systems for staff (Damanpour and Evan, 1984: 400).

4 Innovation and Tourism Relationship

Nowadays, with the increase of leisure time and income, developments in transportation systems and the high-quality comfort of accommodation businesses have caused an increase in tourism activities. Due to these conditions, the increase in the number of tourists and the growth of the economic pie has made tourism businesses one of the fastest growing sectors in the world economy. In terms of tourism enterprises operating as a service sector in the market where there are many manufacturers and many consumers, and the demands and needs of the customer are in the first place. Today's consumers have the opportunity to choose from a wide variety of options whenever they demand any good or service. Like all industries, businesses in the tourism sector also offer product bundles of different options to consumers in order to maintain their market share or survive in a competitive environment. As an extension of this situation, consumers shape their demands by holding the great power to choose options with innovative features (Yagci, 2008: 409). Tourism enterprises constantly change and develop depending on the innovative patterns in terms of touristic products and processes offered in transportation, entertainment and hotel services, depending on the competition for preference by tourists. For this reason, it is important to see innovation systemically or as an integral part of the tourism system as a whole (Hall and Williams, 2008: 3).

Innovation management in the tourism sector is a kind of cultural, economic and commercial activity of which tourism businesses create on the basis of innovative processes to achieve their goals. Innovation management includes various functions such as planning, marketing, organization, control and supervision within tourism businesses. The specificity of innovation in the tourism sector is determined not only by the characteristics of tourism services, but also by the characteristics of consumers and producers. As can be seen, tourist activities are constantly looking for new ways, solutions and therefore innovations. In order to develop and implement innovative ideas, it is necessary to create a flexible, adaptable and sustainable management system of which will meet current market requirements (Sandybayev, 2016: 9). Strategic innovations and changes that will be developed in tourism enterprises are important for the business to renew itself. Competitiveness can be supported by integrating technology into internal or external applications. In addition, it has an effect on reducing the costs of the products offered in the enterprises. In this context, increasing business efficiency and responding to consumer demands contribute to the adoption of innovative ideas in businesses (Ayazlar, 2013: 122).

The tourism industry has a number of specific characteristics that can affect the innovation process both positively and negatively. The failure of the innovation that will be realized in the tourism sector should be regarded as a natural situation. It cannot be thought that every innovation move that will be made in organizations will definitely produce positive results. For example, the reasons can be listed among the factors that may cause innovation activities to fail such as; producing an idea by managers personally in tourism businesses without conducting market research, the wrong positioning of the innovative activity that will be carried out in the market, conducting non-effective advertisement and promotion in order to get the attraction of the potential customers to the innovation activities, wrong timing, bureaucracy obstacles and cannot reaching the technology of rival enterprises. However, the ideas of which will be realized regarding the innovation process may be innovative, but it may also be disappointing for the business if the market of the tourism business is too small or the costs are too high with the innovative activity that will be carried out. Therefore, tourism enterprises need to develop new services in order to cope with such negativities. However, in this case too, there is a high risk of failure, and there is also limited knowledge on how to create successful innovations (Ottenbacher, 2007: 432). Eliminating or minimizing the negative effects listed for tourism enterprises of which decide to innovate will directly affect the success of decisions and practices. It is clearly seen that removing these negativities, each of which is called obstacle, will enable tourism enterprises to successfully manage innovation (Yagci, 2008: 415).

It seems that the concept of innovation has become a buzzword for anything new in most cases. The theories, concepts and methodologies that tourism researchers have known, used and applied in other sectors for years are supported by the studies they use in the concept of innovation and tourism today. However, when the tourism innovation literature is examined, it is seen that it has reached a level that is comparable with the studies in other economic sectors with the latest researches. Many tourism researchers deal with the wide-ranging issues that fall under the scope of innovation in tourism and carry out their work by expanding the methodological scope (Hjalager, 2010: 1). In this context, some of the researches conducted by tourism researchers on innovation in the tourism sector, which has increased especially after the 2000s, are given and summarized below in the following section.

Petrick (2002), conducted a research to test the reliability and validity of the innovation scale and in order to evaluate the benefit of golf tourists to the tourism market. As a result of the research, it is concluded that the scale used is useful for evaluating the effects of tourists coming for golf on the tourism market. However, it is concluded that young golf tourists are more prone to innovation than older golf tourists, golf tourists who play more often avoid innovation more than golf tourists who play less, and finally, golf tourists who come for holiday purposes believe that they get quality service for their money. Babaita et al. (2010), conducted a study to examine the relationship between leadership styles and innovation on 168 employees of 3 and 4-star hotel enterprises in Timisoara, Romania. As a result of the study, a meaningful conclusion was reached between leadership style and innovation, and it was concluded that leaders should behave close to their employees and leaders should control it when employees perform an action. In their research, Najda-Janoszka and Kopera (2014), conducted a research with the aim of determination of the factors of which hinder the innovative activities of micro, small and medium-sized tourism enterprises. While the research was carried out with tourists in the Southern region of Poland, innovation barriers were determined in three different ways: organizational, environmental and innovation process. As a main result, it has been concluded that the organizational, environmental and innovation process barriers to innovation in the tourism sector are closely interrelated. Isik et al. (2015), conducted a research in Ortahisar, Urgup and Goreme in the Cappadocia region with the aim of determination of the qualified workforce deficit of accommodation enterprises and in order to develop innovative suggestions for lower-middle and upper-level employees. Within the scope of the research, questionnaire forms were distributed to 95 people and 30 department managers and their opinions on the subject were obtained. In the study, it was concluded that the managers complained about the insufficient number of qualified workforce and the employees performed qualified work despite not being had innovative training and suggestions were made on the subject. Esitti and Erdem (2017), conducted a research in food and beverage enterprises which are operating in Istanbul with tourism certificates in order to investigate the effect of employees' compliance with the organization on organizational innovation. The opinions of the participants were obtained by applying a questionnaire on a total of 344 employees. As a result of the research, it was seen that positive and significant relationships were reached between organizational innovation and its sub-dimensions and individual-organization harmony. Zengin and Dursun (2017), conducted a research to determine the contribution of innovative activities carried out in 4 and 5 star accommodation establishments in East Marmara to competitive advantage. As a result of the research, it was revealed that the most innovative activities related to the architectural structure were carried out in hotel enterprises by evaluating the consumer demands and needs.

When the above studies are examined, the effects of the concept of innovation on alternative tourism types such as golf tourism, health tourism, different accommodation types and different destinations are investigated and its contributions to the tourism sector are revealed. In general, the results obtained have provided an environment where innovation has positive effects on the tourism sector and businesses can adapt to today's rapidly changing technology and provide a competitive environment in a competitive environment. In addition, the number of studies on the concept of innovation, which has social, economic, technological and cultural effects in business life and the lives of employees, has been increasing from the past to the present and innovative activities gain importance in many branches of the service sector.

5 Conclusion

Today's world is changing rapidly. Movement of technology with a high acceleration, social changes, much easier of access to information and communication compared to the past, causing differentiation in consumer needs and behaviors. Now the rules of the game have changed, dynamism has replaced stagnation, and the old established rules are no longer valid (Unal and Kilinc, 2016: 130). Nowadays, the adaptation of enterprises to these rapidly developing developments/changes and innovations is decisive in competition. Globalizing trade, universal communication, information-based technologies seriously affect the activities of businesses. Growth in world trade brings along growing markets with high competition and increasing demand. As a natural result of this transformation, technological investments are increasing. Therefore, the importance of information is increasing day by day in competition. In a technology-intensive knowledge-based economy, businesses must keep up with today's changes to become a global player. As a requirement of the era, a constant need for change and renewal arises in businesses (Cetin and Gedik, 2017: 171).

Businesses need innovation both at the stage of generating the idea and as a result of the actual application of technology. Innovation plays an important role in gaining a place in the market and ensuring competitiveness with competitors as a result of innovative studies in the efforts to produce and market the product (Zengin and Dursun, 2017: 43). When the subject is evaluated in terms of tourism enterprises, it is seen that tourism enterprises differentiate their products and services in this intense competitive environment in order to maintain their existence and dominate the market. The way to realize this differentiation will be possible by following scientific and technological innovations. Therefore, today, where the knowledge economy is dominant, the use of innovative activities by tourism enterprises is seen as a necessity rather than a need (Gursel and Su Ersoz, 2017: 127). Staying behind technology and innovation for tourism businesses that find a place for themselves in the monopolistic competition market, which is a market with many consumers and many producers meet, will cause negativities in terms of struggling with competitors in this market environment. Because the customer has the right to choose in this type of market and always wants to see the value of his/her money in the service sectors, which indicates that he/she will prefer the service that will give the most advantage in different ways. The fact that the supplying enterprises produce goods and services in line with the demands and needs of the demanding customers and the use of innovations that can be described as different in response to the demand in tourism enterprises makes it easier to stand out in the competitive environment. For example, for a customer who will choose between two hotels with the same price, if one hotel does not have wi-fi while the other one has, the customer will prefer the hotel with wi-fi, which provides a plus service. Therefore, it does not seem possible for the accommodation businesses involved in the service sector to avoid innovative activities and keep up with technology in a competitive environment.

As a result, the concept of innovation is inevitable for all businesses in the service sector, whether in tourism businesses or in different businesses. The important point here is that, considering an innovation movement related to the enterprise, both the employees, the managers and the structure of the enterprise should be suitable for this. On the other hand, marketing of the innovations that will be made as a promotional activity for the potential and existing customers can be seen as another important issue. If this happens, innovation activities will contribute positively to the organization in terms of achieving the goals of tourism businesses and the success of the innovation process and they will be factors that increase success of the business.

References

Ayazlar, G. (2013). *Otel işletmelerinde yenilikçilik performansı.* In Mesci, M. & Bozkurt, Ö. (Ed.), *Otel işletmelerinde yenilik yönetimi.* İzmir: Sidas Yayınları.

Babaita, C., Sipos, G., Ispas, A. & Nagy, A. (2010). Leadership style and culture for innovation in hotel industry. In *Proceedings of the 5th WSEAS International Conference on Economy and Management Transformation*, Vol. 2, pp. 526–531.

Cannarella, C. & Piccioni, V. (2003). Innovation transfer and rural SMEs. *Journal of Central European Agriculture, 4*(4), 371–388.

Chen, Y. (2006). Marketing innovation. *Journal of Economics & Management Strategy, 15*(1), 101–123.

Çetin, K. & Gedik, H. (2017). Karaman'daki işletmelerde inovasyon süreci ve inovasyon pazar yapısı ilişkisi. *Route Educational and Social Science Journal, 4*(8), 170–182.

Damanpour, F. (1992). Organizational size and innovation. *Organization Studies, 13*(3), 375–402.

Damanpour, F. (1996). Organizational complexity and innovation: Developing and testing multiple contingency models. *Management Science, 42*(5), 693–716.

Damanpour, F. & Evan, W. M. (1984). Organizational innovation and performance: The problem of organizational lag. *Administrative Science Quarterly, 29*(3), 392–409.

Drucker, P. F. (1998). The discipline of innovation. *Harvard Business Review, 76*(6), 3–8.

Elçi, Ş. (2007). *İnovasyon: Kalkınmanın ve rekabetin anahtarı.* 13. Baskı. Ankara: Technopolis.

Elçi, Ş., Karataylı, İ. & Karaata, S. (2008). Bölgesel inovasyon merkezleri: Türkiye için bir model önerisi. TÜSİAD-T/2008–12/477, İstanbul, 1–208.

Ersoy, B. A. & Şengül, C. M. (2008). Yenilikçiliğe yönelik devlet uygulamaları ve AB karşılaştırması. *Yönetim ve Ekonomi Dergisi, 15*(1), 59–74.

Eşitti, B. & Erdem, S. H. (2017). Birey-örgüt uyumunun örgütsel inovasyon üzerindeki etkileri: Yiyecek-içecek işletmeleri üzerine bir araştırma. *MANAS Sosyal Araştırmalar Dergisi, 6*(4), 475–489.

Flynn, M., Dooley, L., O'Sullivan, D. & Cormican, K. (2003). Idea generation for organisational innovation. *Journal of Innovation Management, 7*(4), 417–442.

Goldhar, J. L. (1980). Some Modest Conclusions. In Dean, B. V. & Goldhar, J. L. (Eds.), *Management of Research and Innovation* (pp. 283–284). Amsterdam: North Holland Publishing Company.

Güleş, H. K. & Bülbül, H. (2004). *Yenilikçilik işletmeler için stratejik rekabet aracı.* Ankara: Nobel Yayınları.

Günay, D. & Çalık, A. (2019). İnovasyon, icat, teknoloji ve bilim kavramları üzerine. *Üniversite Araştırmaları Dergisi, 2*(1), 1–11.

Günay, Ö. (2007). *KOBİ'lerde yenilik türlerinin analizi ve yenilik engellerinin belirlenmesine yönelik bir araştırma.* (Yayımlanmamış Yüksek Lisans Tezi). İstanbul Üniversitesi, İstanbul, Türkiye.

Gürsel, Ü. & Su Eröz, S. (2017). Örgütsel bağlılık ve yenilik yönetimi ilişkisi: Mutfak personeli üzerine bir araştırma. *International Journal of Academic Value Studies (Javstudies), 3*(13), 109–132.

Halaç, D. S., Eren, H. & Bulut, Ç. (2014). Sosyal yenilikçilik: Bir ölçek geliştirme çalışması. *Hacettepe Üniversitesi İktisadi ve İdari Bilimler Fakültesi Dergisi, 32*(1), 165–190.

Hall, C. M. & Williams, A. M. (2008). *Tourism and innovation.* London: Routledge

Halligan, U. (2009). Skills in Creativity, Design and Innovation, Forfas, http://www.skillsireland.ie/media/egfsn-091104-cdi.pdf (Accessed:17.08.2020).

Hamel, G. & Bren B. (2007). *The future of management*. Boston: Harvard Business School Press.

Hjalager, A. M. (2010). A review of innovation research in tourism. *Tourism Management, 31*(1), 1–12.

Işık, C., Keskin, G. & Serçeoğlu, N. (2015). Turizm sektöründe nitelikli işgücü açığına yenilikçi yaklaşım: Kapadokya örneği. *Girişimcilik ve İnovasyon Yönetimi Dergisi, 4*(4), 1–21.

Kanca, B. (2019). Turizmde Yenilik Arayışı. In Kanca, B. ve Ertaş, Ç. (Eds.), *Turizmin Geleceği: Yeni Arayışlar* (s. 1–22). Ankara: Detay Yayıncılık.

Kim, C., Song, J. & Nerkar, A. (2011). Learning and innovation: Exploitation and exploration tarde-offs. *Journal of Business Research,* 65(8), 1189–1194.

Kodama, M. (2003). Strategic innovation in traditional big business: Case studies of two Japanese companies. *Organization Studies, 24*(2), 235–268.

Kogabayev, T. & Maziliauskas, A. (2017). The definition and classification of innovation. *HOLISTICA–Journal of Business and Public Administration, 8*(1), 59–72.

Korucuk, S., Memiş, S. & Ergün, M. (2020). Kobi'lerde inovasyon stratejilerinin önceliklendirilmesine yönelik bir çalışma: Giresun ili örneği. *Karadeniz Teknik Üniversitesi İletişim Araştırmaları Dergisi, 2020*(1), 156–168.

Kotsemir, M., Abroskin, A. & Meissner, D. (2013). Innovation concepts and typology–an evolutionary discussion. *Higher School of Economics Research Paper No. WP BRP, 5.*

Kümeler için İnovasyon ve AR-GE Yönetimi Kılavuzu, T.C. Ekonomi Bakanlığı İhracat Genel Müdürlüğü. http://www.smenetworking.gov.tr/userfiles/pdf/belgeler/ekonomiBakanligi/8_inovasyon.pdf (Accessed: 25.07.2020).

Lee, K. & Kang, S. M. (2007). Innovation types and productivity growth: Evidence from Korean manufacturing firms. *Global Economic Review, 36*(4), 343–359.

Memiş, S. & Korucuk, S. (2019). Dematel ve vikor bütünleşik yaklaşımı ile gıda işletmelerinde inovasyon çeşitlerinin önceliklendirilmesi ve en ideal firma seçimi. *Manas Sosyal Araştırmalar Dergisi,* 8(1/2), 1099–1113.

Najda-Janoszka, M. & Kopera, S. (2014). Exploring barriers to innovation in tourism industry–the case of southern region of Poland. *Procedia-Social and Behavioral Sciences, 110*, 190–201.

OECD & Eurostat (2005). Oslo Manual: Guidelines For Collecting and Interpreting Innovation Data, Third Ed., Paris: OECD Publishing.

Oğuztürk, S. B. (2011). Güney Kore'nin kalkınmasında inovasyonun rolü. *Süleyman Demirel Üniversitesi Vizyoner Dergisi, 3*(5), 48–53.

Oke, A., Burke, G. & Myers, A. (2007). Innovation types and performance in growing UK SMEs. *International Journal of Operations & Production Management, 27*(7), 735–753.

Ottenbacher, M. C. (2007). Innovation management in the hospitality industry: different strategies for achieving success. *Journal of Hospitality & Tourism Research, 31*(4), 431–454.

Petrick, J. F. (2002). An examination of golf vacationers' novelty. *Annals of Tourism Research, 29*(2), 384–400.

Reguia, C. (2014). Product innovation and the competitive advantage. *European Scientific Journal, 1*, 140–157.

Rogers, M. E. (1983). *Diffusion of Innovations.* 3rd Edition, New York: The Free Press.

Sağır, M. & Türkeri, İ. (2016). Turizmde İnovasyon Çalışmaları. E. Akgöz (Ed.). Turizm Paradigmaları İçinde. Ankara: Detay Yayıncılık.

Sandybayev, A. (2016). Strategic innovation in tourism. A conceptual and review approach. *International Journal of Research in Tourism and Hospitality (IJRTH), 2*(4), 5–10.

Soylu, A. & Göl, Ö. M. (2010). Yönetim inovasyonu. *Sosyoekonomi Dergisi, 1*, 113–130.

Tang, J. (2006). Competition and innovation behaviour. *Research Policy, 35*(1), 68–82.

TDK (2020). Türk Dil Kurumu, Güncel Türkçe Sözlük. https://sozluk.gov.tr/ (Accessed: 04.08.2020).

Toraman, C., Abdioğlu, H. & İşgüden, B. (2009). İşletmelerde inovasyon sürecinde entelektüel sermaye ve yönetim muhasebesi kapsamında değerlendirilmesi. *Afyon Kocatepe Üniversitesi İktisadi ve İdari Bilimler Fakültesi Dergisi, 11*(1), 91–120.

Udwadia, F. E. (1990). Creativity and innovation in organizations: Two models and managerial implications. *Technological Forecasting and Social Change,* 38, 65–80.

Ungerman, O., Dedkova, J. & Gurinova, K. (2018). The impact of marketing innovation on the competitiveness of enterprises in the context of industry 4.0. *Journal of Competitiveness, 10*(2), 132–148.

Ünal, A. & Kılınç, İ. (2016). İnovasyon Yönetimi. In Çatı, K. (Ed.), *Girişimcilik ve İnovasyon Yönetimi* (s. 99–134). Nobel Akademik Yayıncılık.

Weiermair, K. (2006). Product improvement or innovation: What is the key to success in tourism? In OECD, *Innovation and Growth in Tourism* (pp. 53–69). Paris: OECD Publishing. http://dx.doi.org/10.1787/9789264025028-5-en.

Yağcı, Ö. (2008). Turizm İşletmelerinde Yenilik Yönetimi. In Okumuş, F. & Avcı, U. (Eds.), *Turizm İşletmelerinde Çağdaş Yönetim Teknikleri* (s. 401–425). Ankara: Detay Yayıncılık.

Yavuz, Ç. (2010). İşletmelerde inovasyon-performans ilişkisinin incelenmesine dönük bir çalışma. *Girişimcilik ve Kalkınma Dergisi*, *5*(2), 143–173.

Zengin, B. & Dursun, C. (2017). Otel işletmelerinde uygulanan inovatif faaliyetlerin rekabet avantajı üzerine etkileri: Doğu Marmara örneği. *Turar Turizm & Araştırma Dergisi*, *6*(2), 38–58.

http://www.smenetworking.gov.tr/userfiles/pdf/belgeler/ekonomiBakanligi/8_inovasyon.pdf (Accessed: 23.07.2020).

https://www.egiad.org.tr/wp-content/uploads/arastirma-raporlari/inovasyon-raporu.pdf (Accessed: 25.07.2020).

Sinan ÇAVUŞOĞLU

3 Eco-Label Applications in Tourism Industry

1 Introduction

Tourism generally sees the characteristics of "flueless industry" (having a low environmental impact) in providing important opportunities such as employment and foreign exchange income for the host country (Kızılırmak, 2011: 2). For the majority of people who have the financial means to participate in recreational travel, tourism is an activity where little conscious thought is likely given beyond just enjoying the last vacation and deciding where to go for the next vacation (Holden, 2007: 3). The longer a tourist spends in a particular country or region, the more financial benefits it provides to the country's economy. Although tourism creates economic benefits for local companies, it cannot be denied that it still has a significant impact on the natural environment. These impacts indicate environmental problems such as increased consumption of water, energy and raw materials, as well as increased greenhouse gas emissions (e.g. carbon dioxide) and waste reduction. Due to the ever-increasing concerns about environmental problems such as pollution, depletion of non-renewable resources, global warming, consumers reflect their interest in the environment and turn to green consumerism by changing their purchasing habits (Çavuşoğlu and Durmaz, 2020: 131; Demirağ ve Durmaz, 2020).

In recent years, a number of environmental and social initiatives have been implemented by environmental organizations and associations. Sustainable practices have been encouraged because of the industry's significant impact on the environment and its reliance on maintaining a more livable environment. Tourism can have a detrimental effect on the environment, which is the main resource of the tourism industry (Holden, 2007). Therefore, it is important for tourism services providers to adopt environmentally sustainable practices. Many tourism organizations adhere to more sustainable forms of tourism aimed at preserving natural and cultural heritage (Stabler and Goodall, 1997). World Environment and Development Commission declared that sustainable development is "development that meets the needs of the present without compromising the ability of future generations to meet their own needs" (Erdogan, 2018: 21). In the United Nations publication, it is stated that one of the most important goals of sustainable tourism is environmental cleaning (UNEP, 2005: 18).

Voluntary initiatives such as eco-labels, awards, environmental management systems, codes of conduct, individual commitments and declarations play an increasing role in the regulation of tourism operations (Erdogan, 2018: 22). The most common method among these attempts is the eco-labeling system. In this section, practices created within the scope of eco-labeling in the tourism sector will be discussed and examined in a comprehensive way.

2 Eco-Labeling Applications in Tourism

As tourism becomes more important to communities around the world, the need to develop tourism sustainably becomes a major problem. Sustainability is defined as the ability to sustain a certain ecosystem without interruption, degradation, destruction and without overloading resources (Sezgin and Karaman, 2008: 429). People have become increasingly environmentally friendly in recent years, recognizing the seriousness of ecological problems (Yayla and Güven, 2019). This ecological awareness has led an increasing number of people to engage in environmentally friendly behaviors in their daily lives (Kalafatis et al., 1999). With the effect of this situation, businesses started to adopt a culture that takes into account the awareness of social responsibility. Because the customers have taken on a structure that is questioning, searching and difficult to satisfy. This structure of customers no longer accepts a business model that harms the environment. Responsibilities toward nature have led businesses to different strategies. The most basic of these strategies is the eco-labeling system known as environmental labeling (Çavuşoğlu and Durmaz, 2020: 119). Eco-labeling (green labeling, environmental label) is a certificate showing that a product is environmentally sensitive and friendly (Kınacı et al., 2011: 98).

Eco-labels have been used to encourage less harmful, more environmentally friendly businesses as well as the development of potential tools to educate and influence customer behavior. Ecological labeling started in 1985 with the Blue Flag award given by the Foundation for Environmental Education (FEE) for beaches and marinas where environmental protection is a high priority (UNWTO, 2020). Most of the standards focus on accommodation and especially hotels, but there are also very specific standards covering for example beaches, marinas, travel agency, tour operators, tour guides, protected areas (Erdogan, 2018: 22).

Eco-labeling means providing consumers with information that a service is more environmentally friendly than other services in the same category (Nunes and Riyanto, 2005). However, it was originally developed to communicate to consumers that a service is truly as environmentally friendly as the business claims. Eco-labels are available globally, but they are mostly common in developed countries such as North America and Europe. Many businesses have changed certain aspects of their products to meet environmental requirements such as using different materials, changing manufacturing processes or using renewable energy (Clark and Russell, 2005: 114). Labels help consumers understand which services are better for the environment at the point of purchase. It also provides a competitive advantage for businesses because consumers are more likely to believe in environmental claims reviewed by an organization that certifies eco-labeled products (Zohbi, 2012: 4).

With labeling, it is aimed to encourage producers to make practices and innovations that are less harmful to the environment and to encourage consumers to buy more environmentally responsible habits (McKerlie, 2002: 1). Today, environmental concerns have gained momentum and these concerns have begun to change the basic habits of consumers (Stisser, 1994). Many studies have shown that green consumers are willing to change their consumption behaviors to protect the environment (Stern, 2000; Kement and Çavuşoğlu, 2017; Çavuşoğlu and Durmaz, 2019; Kement, 2019). For example, as the popularity of environmental labeling is seen as a marketing tool and product differentiation to position the image of the product, its use by businesses has increased in recent years. Studies show that sales increase when an eco-label is acquired, but in general there is no statistical data showing the market power an eco-label can provide on a product (OECD, 1997).

The eco-labeling system has an informative feature (Kınacı et al., 2011: 98). Essentially, the presence of an eco-label indicates that an enterprise or service complies with the sustainability standard. In this way, eco-labels function as a sign of trust between producers and consumers (van der Ven, 2016: 5). In tourism, eco-labels are generally given in the form of a star or flag with a certificate, award or logo (Kınacı et al., 2011: 98). If a product or business is considered ecologically safe, it is rewarded with an eco-label (Alagöz, 2007: 2). Eco-labeling can be based on first-party verification in which social or environmental characteristics are introduced, or third-party verification that an independent source labels according to certain social or environmental criteria (DeAngelis, 2013: 8). Non-governmental organizations, public institutions, buyers, marketing groups or other relevant third parties can develop certification standards. While governments set mandatory standards in the form of regulations, voluntary standards are established by NGOs, industry associations and public-private sector partnerships and are usually verified by third-party auditing (Giovannucci and Ponte, 2005). Eco-label certification criteria are typically voluntary standards (DeAngelis, 2013: 8).

3 Purposes of Eco-Label

Eco-labeling has some major purposes. Acting in accordance with these objectives contributes significantly to the success of the eco-labeling program at the international level. Eco-labeling application has three main purposes in general. These purposes;

✓ Protection of nature,
✓ Incentiving of activities,
✓ Raising the awareness of consumers.

3.1 Protection of Nature

Owing to eco-labeling programs, it is aimed that consumers prefer environmentally friendly products and thus keep the damage to nature at the lowest level. States, non-governmental organizations and other relevant third parties support the emergence of a nature-sensitive production and consumption by guiding the purchasing decision of consumers with these activities. In this respect, eco-labels are an economic activity that leads the development of applications beneficial to nature (Kından, 2006: 14).

3.2 Incentiving of Activities

Within the scope of eco-labeling programs, it aims to reward the services that cause the least harm to nature and to incentive the enterprises to improve their environmental activities (Taşdemir, 2014: 59). In this way, pro-natural ideas and inventions become incentived. Businesses that engage in environmental activities help their consumer masses to positively affect their purchasing process and gain a positive image. In this respect, businesses are interested in eco-label programs in order to gain competitive advantage (Çavuşoğlu and Durmaz, 2020: 135). The eco-labeling program is generally based on rewarding activities that meet certain criteria. In this respect, programs enable production and service enterprises to continue increasing their quality in terms of incentiving environmental activities and creating environmentally friendly products (Kından, 2006: 15).

3.3 Raising the Awareness of Consumers

Another purpose of eco-labeling is to make consumers adopt sustainability and environmental awareness. While eco-labeling raises the awareness of consumers about the environment, it also guides their preferences. Studies show that the vast majority of consumers have more confidence in a product with a green label and use their purchasing preferences for these services (Kalafatis et al., 1999; Stern, 2000; Çavuşoğlu and Durmaz, 2019). In order to incentive the choice of products with green labels, it is important to inform the consumers about the effects of the products on the market through eco-labels in countries with high consumer awareness and sensitivity (Taşdemir, 2014: 59).

4 Benefits of Eco-Label Application

Eco-labels are a communication system that tries to take environmental concerns more into account and also to influence consumer behavior. Eco-labeled businesses can use these certificates for their own benefit to inform consumers that their services minimize negative environmental impacts (Buckley, 2002). Displaying eco labels can be used as a differentiator compared to other offers (Burgin and Hardiman, 2010). Forsyth (1995) found that businesses try to gain competitive advantage by offering environmental sustainability to consumers instead of using ecological labels as a price war.

Word Tourism Organization has advised the public sector to establish or support national certification systems for sustainable tourism and mentioned the benefits of these certificates for society, environment and businesses (WTO, 2003: 16–17). In this context, it is useful to examine the benefits of eco-label applications in terms of society, environment and tourism enterprises.

4.1 Social Benefits of Eco Label Applications

Social advantages of eco-labeling can be stated as follows (Kından, 2006: 14):

- ✓ Businesses with a certificate as an eco label are committed to minimizing their damage to the environment, while also providing social and economic benefits. Thus, host societies will benefit environmentally, socially and economically.
- ✓ The more tourism businesses participate in these label programs, the stronger the awareness of sustainability issues in host communities will be.

4.2 Environmental Benefits of Eco Label Applications

The advantages of eco-labeling against the environment can be stated as follows (Çavuşoğlu and Durmaz, 2020: 119):

- ✓ It leaves a suitable space for the next generations by ensuring the efficient use of resources.
- ✓ It facilitates reuse and recycling by reducing wastes.

- ✓ It provides protection of ecological order and diversity in nature.
- ✓ It helps to reduce the use of chemicals in production and to manage them correctly.

4.3 Benefits of Eco Label Applications for Tourism Businesses

The eco-labeling application provides great contributions to businesses as well as to the environment (Kınacı et al., 2011: 99):

- ✓ The business increases the corporate image by making the necessary advertisements to the consumers with the label it has.
- ✓ It increases the competitive power by proving that it is different from other businesses with the document it has.
- ✓ A business with an eco-label increases its preference level by offering healthier environments to its consumers and employees.
- ✓ By reducing the use of natural resources such as energy and water, businesses will provide a more efficient view both to the ecological system and within its own structure.
- ✓ Following strong environmental management strategies will provide significant cost savings for the company.

5 Eco-Label Examples Emerging in the Tourism Sector

In the world, eco-labeling practice is applied under different names. The number of applications in the tourism industry is not known exactly (Çavuşoğlu and Durmaz, 2020). Some of the best known and common labels in the world are as follows; Blue Flag, Green Globe, Greening Hotels, Green Key.

5.1 Blue Flag

Blue Flag; It is an eco-label given for beaches, marinas and yachts. The label is a system that emerged in 1985 to mark places on the beaches of France that are perfect for the beach and the environment. This idea was introduced by the Foundation for Environmental Education (FEE) in 1987 and developed internationally (Blazquez et al., 2002). Blue Flag guarantees that industrial or urban wastes will not affect water quality, ensuring that adequate waste bins are placed and complies with a specific policy, ensuring that the selected beach is perfect for swimming (Mir-Gual et al., 2015: 108).

The award is given for 1 year. Generally determined criteria are grouped under four main headings (Çavuşoğlu and Durmaz, 2020: 122):

- ✓ Quality of water used for swimming
- ✓ Directing environmental education and information activities
- ✓ Ensuring beach layout security
- ✓ Environmental management

All of these criteria must be met for the Blue Flag award. As of 2020, 4664 Blue Flag awards (for beaches, marinas and yachts) have been awarded in 46 countries around the world. This figure belongs mostly to Spain. In Spain, the number of Blue Flag awarded beaches is 590, the number of marinas is 94 and the number of yachts is 5 (689 in total). Spain respectively, Greece (517) and Turkey (508) are followed (Blue Flag, 2020).

5.2 Green Globe

The Green Globe certification is a structured assessment of the sustainability performance of travel and tourism businesses and supply chain partners. The Green Globe standard includes 4 basic criteria supported by more than 380 compliance indicators divided into four main sections; sustainable management, social economy, environment and cultural heritage. With members in more than 90 countries around the world, the Green Globe represents a true international membership that supports the highest sustainability standards. There are 3 different certification levels in the Green Globe. These; certified membership (certified member), gold membership (gold member) and platinum membership (platinum member).

Certified Member status is granted to Green Globe members certified according to all mandatory criteria within the Green Globe Travel and Tourism Standard. The certification is approved annually when more than 50 % of the indicators related to the criteria are reached. Gold Membership is awarded to certified members for 5 years in a row. This membership is only possible when a member meets all the required criteria under the Green Globe Standard for Travel and Tourism and completes its independent and mandatory audits in varying years. Platinum Member status is granted to members who have been certified for 10 years in a row. This is the highest level certification offered by the Green Globe and is possible when it meets all the required criteria and completes its independent and mandatory audits in varying years (Green Globe, 2020).

5.3 Greening Hotels

Greening Hotels certification project incentives tourism facilities to become greener for sustainable tourism. This eco-label aims to increase the number of green accommodation facilities, increase the sustainability awareness of the tourism sector and increase the motivation of accommodation facilities to be environmentally sensitive. To obtain Greening Hotels certification, candidate businesses are evaluated on four main areas:

- Energy Management
- Water Management
- Improving Indoor Air Quality
- Waste Reduction and Recycling

Based on the minimum success status of the total criteria in the Checklist, the accommodation facilities are entitled to receive "Greening Hotels" Certificate, respectively Bronze (60 % success), Silver (75 % success) and Gold (90 % success). The validity period of the certificate is two years, and hotels that want to use the existing Greening Hotel certificate after this certificate can continue to use their certificates by receiving a half-day renewal audit (Greening Hotels, 2020).

5.4 Green Key

Green Key is an important eco-label awarded in the field of environmental responsibility and sustainability in the tourism industry. This prestigious eco-label represents a commitment to meeting the criteria set by the Foundation for Environmental Education (FEE). Green Key is known as the first international environmental label developed for hotels and hostels, campsites and holiday parks, small accommodations, conference centers, attractions and restaurants, and is a voluntary eco-label awarded to more than 3,200 hotels and other establishments in 65 countries.

The goals determined by the Green Key tag in general can be listed as follows (Green Key, 2020):

- ✓ Increasing the use of environmentally friendly and sustainable working and technology methods in businesses and thus reducing the general use of resources.
- ✓ To create awareness and behavioral changes in the guests, employees and suppliers of individual tourism businesses.
- ✓ To raise awareness and use of environmentally friendly and sustainable methods to create behavioral changes in the hospitality and tourism industry in general.

Green Key has 13 criteria; environmental management, guest information, staff participation, water saving, cleaning and washing, waste, energy saving, food, indoor climate, management, green activities and social responsibility and cooperation. The criteria are set internationally and are common to all countries. The label is re-evaluated every year to protect the certificate and meet the environmental requirements of the Foundation for Environmental Education (Erdogan, 2018: 24).

6 Conclusion

As tourism has become important to communities around the world, the need to develop tourism in a sustainable way has begun to change. For the tourism industry to move toward sustainability, practices need to be changed in all areas of the industry. As sustainability has become an important global phenomenon, the environment has started to be prioritized along with its social and economic aspects. In order to talk about a sustainable development, it is necessary to protect the environment from deterioration and to minimize the effects on the environment.

Increasing the environmental awareness of people contributes to the improvement of environmental performance. Today, with the increasing environmental problems, the characteristics of the preferred and used services have started to gain importance. While the determining measure of environmental sensitivity is the reduction of the environmental damage caused by the products to the minimum level, the applied eco-labeling programs direct environmental desires. Eco-labels, which are awarded by certain organizations and are an award, show that the service meets certain environmental standards and the damage it causes to nature is less than others.

The society benefits more from the eco-labeling system in social, economic and environmental aspects, which encompass three aspects of sustainability. Businesses with an environmental label try to create benefits by reducing the negative effects of these three aspects. Eco-labels will make the contribution of tourism activities to the sustainable development of host communities more distinct, more measurable and more accountable. If the majority of tourism companies and/or destinations are certified, the level of awareness of sustainability issues will be further strengthened in the host community. Eco-labels can benefit consumers by providing more information and guidance on travel choices decisions and service quality assurances. The eco-labeling system will also facilitate governments. This system will provide an effective alternative to direct regulation, which can be more difficult and time consuming to implement. It will also allow governments to take a flexible approach to monitoring the tourism industry. Governments, instead of controlling and supervising tourism companies with eco-labeling, will provide businesses with opportunities specific to their circumstances, ensuring a comprehensive environmental and social development, thereby helping to reduce the financial burden. In general, eco-labels emerge as an important form of certification in terms of increasing the recognition of tourism in the country, its competitive power in national markets and its image in international markets.

References

Alagöz, S. B. (2007). Yeşil Pazarlama ve Eko Etiketleme. *Akademik Bakış*, 10, 1–12.

Blazquez, M., Murray, I. & Garau, J. M. (2002). In: Muntaner, Lleonard (Ed.), *El Tercer Boom. Indicadors de Sostenibilitat Del Turisme de Les Illes Balears 1989–1999.* Centre d'Investigacions Tecnologiques i Turístiques de les Illes Balears, p. 427.

Blue Flag (2020). https://www.blueflag.global/all-bf-sites (Accessed: 07.08.2020).

Buckley, R. (2002). Tourism ecolabels. *Annals of Tourism Research*, 29(1), 183–208.

Burgin, S. & Hardiman, N. (2010). Eco-accreditation: Win-win for the environment and small business? *International Journal of Business Studies*, 18(1), 23–38.

Clark, C. D. & Russell, C. S. (2005). Public information provision as a tool of environmental policy. In S. Krarup & C. S. Russell (Eds.), *Environment, information and consumer behaviour* (pp. 111–140). Cheltenham, England: Edward Elgar.

Çavuşoğlu, S. & Durmaz, Y. (2019). Yeşil Davranışlara Karşı Tutumun Ziyaret Niyetine Etkisinde Yeşil İmajın Düzenleyicilik Rolü: Yeşil Oteller Örneği. *Fırat Üniversitesi Sosyal Bilimler Dergisi*, 29(2), 303–315.

Çavuşoğlu, S. & Durmaz, Y. (2020). *Deneyimsel Pazarlama ve Yeşil Davranış*. (1. Basım). İstanbul: Hiper Yayın.

DeAngelis, N. (2013). Rhode Island Aquaculturists' Perceptions of Eco-labeling Certification Programs (Master's Thesis), University of Rhode Island, USA.

Demirağ, B. & Durmaz, Y. (2020). *Marka Yönetimi*. (1 Basım). İstanbul: Hiper Yayın.

Erdogan, N. (2018). Sustainability in tourism: Ecolabel and certification programs at hotels in Turkey. *Yearbook-Faculty of Tourism and Business Logistics*, 3(2), 20–27.

Forsyth, T. (1995). Business attitudes to sustainable tourism: Self-regulation in the UK outgoing tourism industry. *Journal of Sustainable Tourism*, 3(4), 210–231.

Giovannucci, D. & Ponte, S. (2005). Standards as a new form of social contract? Sustainability initiatives in the coffee industry. *Food Policy*, 30(3), 284–301.

Green Globe (2020). https://greenglobe.com/ (Accessed: 07.08.2020).

Green Key (2020). https://www.greenkey.global/ (Accessed: 07.08.2020).

Greening Hotels (2020). https://www.yesillenenoteller.com/ (Accessed: 07.08.2020).

Holden, A. (2007). *Environment and Tourism*. Florence: Routledge.

Kalafatis, S. P. Pollard, M. East, R. & Tsogas, M. H. (1999). Green Marketing and Ajzen's Theory of Planned Behavior: A Cross-Market Examination. *Journal of Consumer Marketing*, 16(5), 441–460.

Kement, Ü. (2019). Yeşil Tutumunun Davranışsal Niyet Türlerine Etkisi: Yeşil İmajın Aracılık Rolü. *OPUS Uluslararası Toplum Araştırmaları Dergisi*, 10(17), 370–394.

Kement, Ü. & Çavuşoğlu, S. (2017). Hafızaya Yönelik Müşteri Deneyimlerinin Müşteri Sadakatine Etkisi: Yeşil Oteller Örneği. *Uluslararası Sosyal ve Eğitim Bilimleri Dergisi*, 4(8), 172–194.

Kınacı, B., Pehlivan, N. A. & Seyhan, G. (2011). *Turizm ve Çevre (Çevre Koruma)*. Ankara: Pegem Akademi.

Kından, A. (2006). *Bir Eko-Etiket Olarak Mavi Bayrak'ın Türkiye Kıyı Turizminde Bir Pazarlama Unsuru Olabilirliğinin Araştırılması*. Yayımlanmamış Yüksek Lisans Tezi, Ankara Ünversitesi Sosyal Bilimler Enstitüsü Sosyal Çevre Bilimleri Anabilim Dalı, Ankara.

Kızılırmak, İ. (2011). Dünya'da ve Türkiye'de ki Turizm İşletmelerinde Çevre Korumaya Yönelik Uygulamalar: Amacı ve Önemi. *Sosyal Bilimler Dergisi*, 2, 1–12.

McKerlie, K. (2002). Implications of an Eco-Certification Program for Alberta; Discussion Paper C3–04, *Climate Change Central*, July. pp. 1–57.

Mir-Gual, M., Pons, G. X., Martín-Prieto, J. A. & Rodríguez-Perea, A. (2015). A critical view of the Blue Flag Beaches in Spain using environmental variables. *Ocean & Coastal Management*, 105, 106–115.

Nunes, P. A. & Riyanto, Y. E. (2005). Information as a regulatory instrument to price biodiversity benefits: Certification and ecolabeling policy practices. *Biodiversity & Conservation*, 14(8), 2009–2027.

OECD (1997). Eco-labelling: Actual Effects of Selected Programmes. *Paris: Organization for Economic Cooperation and Development.*

Sezgin, M. & Karaman, A. (2008). Turistik Destinasyon Çerçevesinde Sürdürülebilir Turizm Yönetimi ve Pazarlaması. *Selçuk Üniversitesi Sosyal Bilimler Enstitüsü Dergisi*, 19, 429–438.

Stabler, M. J. & Goodall, B. (1997). Environmental awareness, action and performance in the Guernsey Hospitality Sector. *Tourism Management*, 18(1), 19–33.

Stern, P. C. (2000). New environmental theories: Toward a coherent theory of environmentally significant behavior. *Journal of Social Issues*, 56(3), 407–424.

Stisser, P. (1994). A deeper shade of green. *American Demographics*, 16(3), 24–29.

Taşdemir, G. (2014). *Yeşil Pazarlamada Ambalaj Kararının Önemi ve İçme Suyu Sektöründe Bir Araştırma* (Yüksek Lisans Tezi), Galatasaray Üniversitesi Sosyal Bilimler Enstitüsü, İstanbul.

United Nations Environment Programme (UNEP), Division of Technology (2005). *Making Tourism More Sustainable: A Guide for Policy Makers*. World Tourism Organization Publications.

UNWTO, World Tourism Organization (2020). *Voluntary initiatives in tourism*, https://www.e-unwto.org (Accessed: 12.08.2020).

van der Ven, H. (2016). *Eco-labeling and the conditions for rigorous and credible transnational governance* (Doctoral dissertation). Department of Political Science University of Toronto.

WTO, Word Tourism Organization (2003). *Wto Asia-Pacific Conference on Sustainability Certification of Tourism Activities-Final Report*. Kuala Lumpur, Malaysia, 11–13 December.

Yayla, Ö. & Güven, Y. (2019). *Rekreasyon ve Ekoloji*. Editör (Yayla, Ö. Ve Karaçar, E.). Rekreasyon Yönetimi Güncel Animasyon Uygulamaları. Detay Yayıncılık: Ankara.

Zohbi, H. (2012). Meeting environmental certification in design: A toolkit facilitating the process of eco-labelling through LCA for eletronic products (Doctoral dissertation), Carleton University.

Sait DOĞAN & Mehmet YEŞİLTAŞ

4 A Research on Performance Appraisal and Performance Appraisal Form Development Process in Hotel Enterprises[1]

1 Introduction

Rapid technological changes and developments in our age have both globalized and increased competition. The most basic foundation of organizations wanting to be successful in such an environment is human resources. Because the efficient and economical (optimal) use of other resources in line with the goals and objectives of the organization without wasting depends highly on human and the knowledge, skills and equipment of human resources. However, no matter how equipped the employees are, if they are not well motivated and the expected benefit of them is not clearly stated, the expected benefit cannot be fully achieved. In other words, their performance cannot be properly exploited. For this reason, in order to motivate employees and provide the expected benefit, first of all, their performance levels should be measured and evaluated and the results should be reflected to individuals within the fairness and justice measures (Tunçer, 2013).

Hotel businesses are also labour-intensive businesses, and the main foundation of these businesses is human resources. For this reason, the workforce has an important effect in these enterprises, and at the same time, it is an important issue to ensure that the qualified workforce remains in the business for a long time in hotel enterprises. Performance appraisal is important in terms of determining the extent of this activity, meeting the expectations of the individual and rewarding the successful individuals (Demir et al., 2005: 129; Gavcar et al., 2006). While performance appraisal studies are carried out, it is not a correct method to apply a single performance appraisal form to all employees in hotel businesses where many different jobs are performed. There are many jobs that have their

1 This book section is derived from the author's doctoral thesis. For more information and comparison, see: Doğan, S. (2017), "Otel İşletmelerindeki İşgören ve Yönetici Adaylarına Yönelik Performans Değerleme Formu Geliştirilmesi: Ankara İlinde Bir Uygulama", (Yayımlanmamış doktora tezi), Gazi Üniversitesi/Sosyal Bilimler Enstitüsü, Ankara."

own features in a hotel business, and it is essential here that the employees are evaluated within the scope of the features and characteristics of the job.

In this context, due to the fact that hotel businesses are service-intensive businesses and due to their unique features, instead of using the performance appraisal forms that exist in the market as a uniform type, it is important for hotel enterprises to come up with a new performance appraisal form by analyzing the specific characteristics of hotel enterprises and their departments that provide different services. In this book chapter, partial results of the doctoral thesis study conducted with the process of developing a performance appraisal form in hotel enterprises are included, and the processes of performance appraisal, job analysis, job descriptions and job requirements in hotel enterprises are mentioned. Within the scope of the limitations of the study, 4 departments (Reception, Service (Restaurant), Kitchen, Housekeeping) with high human resources were examined in this study, and among the employees working in these departments, Receptionist (Receptionist), Service Attendant (Waiting Staff), Cook and Housekeeper were included in the research.

2 Performance Appraisal and Its Importance in Hotel Enterprises

Hotel businesses provide various services and perform these services through their employees in different departments. The fact that the services are different has made the division of labor necessary in hotel enterprises and within the scope of this division of labor, departments have been created in order to perform the services effectively. The working area and functions performed by each department within the hotel business differ. Considering these differences, it is not right in terms of equity and fairness to expect that both the personality and vocational knowledge and skills of employees in various departments of hotel businesses will be the same in terms of performance appraisal. On the other hand, it will not be correct to evaluate all hotel employees with a single performance appraisal form, and it will not be correct and sufficient to create a performance appraisal form specific to each department. Many employees work in the same department, and each employee here has a different position (duty). For this reason, the job duties and definitions of the employee whose performance will be evaluated must be taken into consideration. In the development of performance appraisal forms, it will be appropriate to determine the criteria in accordance with the job/job description in terms of efficiency.

Although the performance appraisal system is important in hotel businesses, employee motivation and productivity can only be increased by using it effectively. On the other hand, performance appraisal is important for the employee as well as for the organization. The successful employee wants to see the reward of his work. As a result of an employee performance appraisal that is not based on valid data, the morale of the employee who sees that an employee who does less effort or does not work well is valued deteriorates and the enthusiasm for work decreases over time (Tunçer, 2013). On the other hand, employee turnover is an important problem in hotel businesses and employee turnover rate is also high. While high employee turnover is an important problem for all enterprises, it has become an even more important problem in the service sector (Avcı & Küçükusta, 2009). In hotel enterprises, together with the existence of high employee turnover rate due to the structure of the sector, the increasing need for qualified workforce increases the importance of Human Resources Management (HRM) and Human Resources Planning (HRP) in terms of trying to bring quality to the forefront. Since the essence of hotel management is based on manpower, hotel enterprises are a sector that has to employ qualified workforce. Therefore, the defining characteristic of the hotel workforce is the basic knowledge and skills that enable workers to provide quality service to customers. In other words, one of the important components of service quality and quality in hotel businesses, even the first one, is the quality of the employee (Olalı & Korzay, 1993: 10–11). For this reason, following the process that starts with the recruitment of suitable personnel, the studies carried out within the scope of the orientation, training, and protection of the rights and the performance of the employees are very important. In addition, finding and recruiting suitable employees in hotel businesses is more costly than retaining the existing staff. The fact that performance appraisal is not used completely in some hotel enterprises, filling the forms merely and the employee's not taking an active role in this process, affect the expected result/benefit from performance appraisal negatively. Managers' monitoring of the deficiencies identified through performance appraisal also increases the performance potential of the employee (Tarlan & Tütüncü, 2001). According to Boadu et al. (2014), in order for hotels to be successful and to cope with the difficulties of global competition, the performance of hotel employees should be continuously evaluated.

3 Job Analysis and Job Descriptions in Departments/Units of Hotel Enterprises

Job analysis is considered the cornerstone of all HRM practices. Job analysis also provides a basis for subsequent HRM programs (e.g. performance reviews, staff selection (recruitment) systems, promotion criteria, and training courses). The effectiveness of performance appraisal studies to be carried out in businesses depends largely on pre-determined performance standards. Usually these standards are desired to be numerical, but this is very difficult; however, the fact that the standards are realistic, understandable, precise, specific and relevant to the work done is as important as their being numerical. When setting standards with these characteristics, the most powerful resources are job descriptions and job analysis. Therefore, before starting a performance appraisal study, it is necessary to examine whether the job descriptions are sufficient or not, and if there are deficiencies, job analyses should be made before the appraisal study and job descriptions should be created (Köksal, 2005: 106–107).

In order to create job descriptions, a job analysis must be created in advance (Kozak, 2001: 5). Job analysis is a technique that collects information and examines the quality, quantity, requirements, responsibilities and working conditions of each job in the organization with scientific methods in order to evaluate the jobs in an accurate, effective and healthy manner." (Bingöl, 2010: 82).

In terms of employee selection, job analysis can be defined as systematically examining the details of the work to be done. In job analysis, the answers to the questions such as what the requirements of the job are, what the job expresses to fulfil, how, when, where and why the job will be done are sought. The job analysis form which is prepared is also the basis for the preparation of job descriptions and job specifications. (Erdoğan, 1991: 15). The information obtained through job analysis is summarized as job description and job requirements. This information affects all functions of human resources management and forms the basis of these functions. Job descriptions and job requirements in human resources management are used for;

- Human resources planning,
- Personnel procurement and selection,
- Job appraisal and remuneration,
- Training and development,
- Performance appraisal,
- Development of working relationships,
- Basic functions such as ensuring job security (Can et al., 2009: 82; Bingöl, 2010: 81).

When job description is examined, although there is no standard in writing the job description; job description in most definitions include: job identity, job summary, responsibilities and areas of duty, duty authorization, performance standards, working conditions, job specifications (Dessler, 2013: 144). According to Kozak (2001: 5), the job description is to put the duties, responsibilities and working conditions in writing within the scope of any job.

Job requirements, on the other hand, contrary to the job description, contain descriptive information that reveals the qualities that an employee must have in order to perform a job in the best way. In addition, the higher the requirements of the jobs, the more effort is required to find employees. If the job description and job requirements are briefly stated; while job description is the profile of a job, job requirements are a profile of the qualifications required by the employee who can perform the job (Tortop et al., 2013: 75).

When considered from the point of view of tourism and hotel management organizations, doing "job analysis" in these organizations brings along some risks, as well. According to Kaufman et al. (2009), recent theories suggested that a job is a series of tasks that an individual evaluates and performs to fulfil his responsibilities. In tourism, many jobs/duties do not fit this definition easily. Since the work done in tourism enterprises consists of interrelated activities, the boundaries of the jobs are somewhat uncertain (Baum & Odgers, 2001; Gilbert et al., 2001; Kyriakidou & Gore, 2005), and the cooperation in tourism organizations is increasing gradually (Hu et al., 2009). Due to the nature of the service produced in hotel businesses, although the human factor is very important, hotel services also require close cooperation between employees (Aras, 1993: 5). On the other hand, managing an organization includes adapting both work-related tasks and the changing environment (Agut et al., 2003). The excessive risks in the external environment of tourism organizations (for example: increasing competition, innovations in technology, rapidly developing markets and economic limitations) cause constant change in the quantity and quality of the knowledge, skills and skills required by human resources. As a result, the demand for multi-skilled employees is also growing (Foster et al., 2010). In organizations, there is evidence of increasing employee skill levels through the teaching of new and wider different job skills, with combinations of skills within many business and vocational groups (Baum & Odgers, 2001). This diversity of knowledge, skills and abilities required in human resources makes tourism organizations able to fight against risks in the external environment (Saad, 2013). As a result of an experimental study, Zehrer (2011) found that many tourism organizations had to accept some human resources practices such as cross-training, job rotation and project teams within/between departments in order to gain or enrich the knowledge, skills and skills required for employees. According to Saad (2013), due to these risks in tourism and hotel enterprises:

- The critical boundaries of jobs in tourism are questionable, as it is still difficult to define specific tasks that are limited to each of the jobs in tourism.
- There is no obligation for the knowledge, skills and abilities of the employees in tourism and hotel businesses to fully match the knowledge, skills and abilities listed in job descriptions.
- Identifying the skills and potential needed in tourism jobs is not easy, especially for jobs involving customer communication. While the attitudes and behaviors of the employee are effective on customer satisfaction, the success of service delivery essentially depends on the attitude and behavior of the employee. However, identifying the behavioral skills and standards needed to perform a job is a matter of research.

Before determining the performance appraisal criteria to be made in this direction, it is important to specify and determine the department-based job analysis (job descriptions and job requirements) of hotel enterprises, in order to fulfil the performance appraisal function.

The job descriptions and requirements to be defined within this scope have been taken from the National Occupational Standards (Tourism, Hotel, Food and Beverage Services) archive and the relevant literature on the Vocational Qualifications Authority website. The job descriptions, working environment and conditions and job requirements of the relevant departments and units are given below.

3.1 Reception

According to the National Occupational Standards, looking at the job description of a Reception (Front Office) Officer working in the Front Office department, (Position of the Vocation in International Classification Systems: ISCO 08: 4224 (Hotel Receptionists)):

> In private or public accommodation establishments, Front Office Attendant is a person who has the knowledge and skills to carry out daily room operations and check-in and check-out operations, to fulfil reservation and switchboard services, to carry out safe and night reports, to inform guests about in-hotel and out-of-hotel activities, and to deal with guest requests and their complaints (Vocational Qualifications Authority, 2010a: 8).

Mısırlı (2003: 48) defined the Receptionist as a qualified person who has the knowledge and skills to inform the guests before arrival to the facility, to resolve their problems, to perform the check-out and post-check-out procedures on his own and within a certain time in the accommodation establishments. The main duties and operations he fulfilled are listed as follows:

- Organizing work,
- Preparing for work,
- Making reservations,
- Performing guest check-in procedures,
- Carrying out the procedures during the guest's stay,
- Carrying out cash procedures,
- Performing guest check-out procedures,
- Making end of day procedures,
- Carrying out activities related to vocational development.

3.2 Service Unit

Within the scope of the research constraint, the job definition, working environment and conditions of the waiting staff, which has a large number of service unit employees, were examined. Koçak (2006: 39) defines the waiting staff as the personnel responsible for all the services of a post consisting of 4–8 tables, while Sökmen (2008: 45) defines him/her as a member of the team and who communicates well with the customer. He defines him as the person who provides the best service, meets the requests and needs and represents his restaurant in the best way. According to the definition made by the Turkish Language Association (2015a), waiting staffs is a person who serves customers in places such as restaurants, patisseries and coffee houses.

In this context, the waiting staff has a number of personality traits (honesty, smiling, good communication, etc.), s/he makes the necessary preparations before the service, takes the orders without delay in line with the requests of the guests coming to the restaurant, follows them, makes an effort for the guests to have a pleasant time and can be defined as a service unit employee who welcomes the guest and contributes their leaving the enterprise happily. On the other hand, he must behave in accordance with business and operating rules while performing these processes.

A service attendant (waiting staff) for restaurant operations is someone who avoids waste, productive, cares about safety and sanitation, provides a quality service to guests, and follows the business rules and regulations set by the management in a team approach. In addition, a qualified waiting staff should be reliable, collaborative, pleasant in personal appearance, healthy, clean and tidy, knowledgeable, persuasive and attentive (as cited in Dahmer and Kahl: Wen & Chen,

2006). According to the National Occupational Standards, the job description of a waiting staff working in the service unit of the food and beverage department (Position of the Vocation in International Classification Systems: ISCO 08: 5131 (Waiting Staff)):

> Service Attendant (Level 4), in the accommodation, entertainment and catering businesses, is defined as a person who, in line with the business policies and principles, arranges the hall and table, welcomes guests, makes reservations, takes food and beverage orders, prepares and serves various food and beverages using special techniques, makes the coordination between the service stations, bars and the kitchen. S/he is also responsible for assisting the Service Manager (Level 5) in conducting activities related to the service area (such as weddings, banquets, cocktails), performing account taking and closing operations (Vocational Qualifications Authority, 2010b: 7)

Lillicrap and Cousins (1996) stated that a service attendant should have fourteen qualifications and obligations to reach certain standards. These include personal hygiene, food and beverage knowledge, punctuality, local (regional) knowledge, personality, guest attitude, strong memory, honesty, management and conduct, pleasant personal appearance, sales ability, understanding of urgency, giving importance to customer satisfaction and being solution oriented. When different sources about the duties of a waiting staff are analyzed, they are listed as follows: (Koçak, 2006: 39; Sökmen, 2008: 45–46):

- Ensuring that the mail is ready, supervising the work of pages and training them,
- Being aware of reservations and making appropriate preparations,
- Attending the daily meetings, being informed about changes and exchanging ideas with their superiors,
- Controlling the tables he is responsible for,
- Having necessary spare equipment in his/her station and service,
- Ensuring that the trolleys and equipment to be used are clean and ready,
- Stocktaking and having them approved,
- Conveying the problems he cannot cope with,
- Cleaning his/her tables or having it done,
- Tranche and flambé,
- Collecting the customer bills and transmitting it to the cashier,
- Listening to customer complaints and conveying them to their superiors,
- Performing other duties assigned by their superiors,
- Assisting the customer in the selection of food and beverage.

3.3 Kitchen Unit

The job description of a Cook working in the kitchen unit of the food and beverage department according to the National Occupational Standards (Place of the Vocation in International Classification Systems: ISCO 08: 5120 (Cooks)):

> The cook (Level 4) is a person who has the knowledge and skills to prepare standard recipes and menus, local, national and international soups and cannons, hot and cold sauces, pastry dishes, meat and seafood dishes, dishes made from vegetables and legumes, olive oil dishes, rice and pastas, hot and cold appetizers, salads, garnishes and desserts in accordance with the recipe/menus and hygiene rules and make them ready for service. S/he assists the supervisor in preparing the budget for the department, conducting the department, organizing and providing trainings (Vocational Qualifications Authority, 2010c: 8).

The staffs working in the kitchen are generally called cooks. The cook is the person who prepares the foods that need to be prepared during the day, within a certain plan, using his knowledge and skills, paying attention to the cleaning rules, and making the food ready for service (Küçükaslan, 2006: 99). According to the definition made by the Turkish Language Association (2015b); a cook is a person who has a vocation in cooking or someone who cooks and sells food. The duties that a cook must fulfil are as follows (Küçükaslan, 2006: 99–100; İşkur Food Aid Directive, 2013):

- Working in line with the business plan of the chef,
- Keeping the preservation of the given material in the most economical and sanitary way without any loss,
- Ensuring that meals are prepared in a timely and delicious manner in accordance with health rules,
- Ensuring the general cleanliness of the kitchen,
- Ensuring the maintenance, good use and cleanliness of the machines and fixtures in the area where s/he works,
- Keeping the containers and equipment used in the kitchen clean,
- Following the cleanliness and work of the staff working in the kitchen,
- Ensuring that meals are served on time,
- Fulfilling other duties assigned by their superiors.

3.4 Housekeeping Department

According to the National Occupational Standards, the job description of a Room Cleaning and Organizing Staff working in the housekeeping department

(Place of the Vocation in International Classification Systems: ISCO 08: 9112, ISCED 97: 811, NACE: 55.10.):

> Room Cleaning and Organizing Staff (Room Attendant) (Level 2) in accommodation establishments, is a person with the knowledge and skills who works in jobs related to cleaning, order, maintenance and protection of guest rooms, floor corridors, stairs, elevators and offices, within a certain period of time and abiding by the hygiene and sanitation rules and operations (Vocational Competence Institution, 2009: 7).

The staff who are defined as "Room Cleaning and Organizing Staff" in the vocational competence institution are defined as "Housekeepers" by TDK (Turkish Language Association). The Housekeeper is the person responsible for the cleaning of the rooms and floors and the supply of the necessary materials, according to the definition made by TDK (2015c). Kozak (2001: 14), like TDK, named this vocation as a housekeeper; however, unlike TDK, named it as a Female Housekeeper (Maid, Fam or Famme), Male Floor Attendant (Vale). According to Kozak, in this department where women are generally employed in cleaning the room, a housekeeper is a female employee responsible for room cleaning (room, bathroom cleaning, dusting, and bedding).

4 Conclusion and Suggestions

The differences in job descriptions and job requirements make it important to appraise each employee to be audited with different criteria. However, different criteria come to the fore between departments in terms of performance appraisal. Therefore, determining different performance standards in each department and unit is important in terms of accurate appraisal and success. According to the definition in Republic of Turkey Council of Ministers' Law No. 2005/8948, Regulations Regarding Certification and Qualifications of Tourism Facilities Article 19, hotels are facilities whose main function is to provide accommodation for the customers, besides this service, they can also have auxiliary and complementary units for their food, beverage, sports and entertainment needs. When the definition of hotel is examined, it is seen that besides the accommodation service, which is their main function, hotel enterprises include food and beverage, sports, entertainment and other different services. For this reason, for each different service provided, there is a need for employees specialized in that service. In addition to performing the main function in hotel businesses, the most important target is customer satisfaction. In order to achieve this in hotel businesses, each department and unit has some duties and responsibilities to fulfil. The people who will fulfil these duties and responsibilities are the employees of the relevant department and unit. This situation varies according to the department or unit in some jobs, duties and responsibilities. For example; "With the performance standards, it is aimed to make cleaning at a certain quality and increase the efficiency in the housekeeping department" (Kozak: 2001: 9). Therefore, the most important task for the housekeeping department is to clean the rooms at a certain quality and time. From the point of view of food and beverage services, food and beverage businesses are subjected to economic, social and disciplined vocations, with their structure, technical equipment, comfort and care conditions, material, social value and service quality of and qualitative staff. (Sökmen, 2008: 1). When the definitions are examined, the most important task for food and beverage services is to meet the nutritional needs of the people, and to receive a quality service through experienced staff. In this direction, while the production of a quality and delicious meal is the most important task for the kitchen unit; for the service unit, the timely and quality presentation of the service comes to the fore. From the perspective of reception-guest relations, the most important tasks for the reception unit are welcoming the guest well, performing the entrance procedures in the most urgent way and dealing with the guests during the stay and monitoring their accounts, as well as timely reporting and faultless communication with other departments.

As is seen, different services are fulfilled in every unit of hotel enterprises. Therefore, it cannot be expected that all staff working in different units of hotel businesses and in different job positions have the same vocational knowledge and skills. Due to the different functioning and job descriptions of each department, it is necessary to differentiate the performance appraisal of the employees according to the department and the position they are in. Within the scope of this study, these differences were considered and tried to be determined by focusing on them.

When summarized as a result of the research, the most important 5 personality traits for the relevant vocations are as follows:

For the receptionists: "*Compliance with the working time, (coming to and leaving work on time), Working Spirit (Willingness), Being Solution Oriented, Being Honest and Trustworthy, Being Ethical.*"

For the housekeeping attendants: "*Being Moral, Being Honest and Trustworthy, Working Spirit (Willingness), Being Careful and Elaborate (meticulous), Having Personal Appearance and Hygiene.*"

For the waiting staff: "*Being Smiling and Polite, Compliance with the working time, Honesty and Reliability, Being Ethical, Working Spirit (Willingness).*"

For the cook: "*Being Ethical, Compliance with Working Time, Personal Appearance and Hygiene, Being Open to Change and Innovation, Being Suitable for Flexible Working Hours.*"

Other criteria that are not listed here for the relevant vocations are also accepted as important or very important criteria. The results show that the employee or employee candidates who work or want to work in all four occupational groups should be moral, honest-reliable, disciplined, willing to work (have high motivation to work) and give importance to personal appearance-hygiene, within the scope of "Personality Traits". One of the three performance appraisal dimensions, the "Personality Traits" dimension is the most important and weighted performance appraisal dimension for the "Housekeeping Attendant" with 39 %, it is the second most important dimension for the "Receptionist" with 28 % and for the "Waiting staff" with 26 %, and it was determined as the least important dimension for the cook with 24 %.

Within the scope of vocational knowledge and skills, as a result of research on four vocations, the most important 5 vocational competencies for the relevant vocations are as follows:

For the receptionist: *"Hotel Rooms Information, Hotel Information, Guest Check-in-Check-out Information/Skill, Knowledge/Skill for Executing and Following Guest Stays and Having Information on Shift Transactions."*

For the housekeeping attendants: *"Having the Knowledge of Hygiene, Knowledge of Cleaning Materials, Knowledge/Skill for Making a Bed, Job Description Knowledge and Having the Knowledge and Skills of Room Cleaning."*

For the waiting staff: *"Table Set-Up (Service Preparation) Knowledge, Menu Mastery and Advisory Skills, Follow-up and Control Principle Knowledge, Hygiene Knowledge and Preliminary Preparation (Mise en Place) Knowledge/Skill."*

For the cook: *"Hygiene Knowledge, Soup Knowledge/Skill, Hand Skill, Substitute Product Knowledge and to Have Quality and Control Knowledge."*

Other criteria that are not listed here for the relevant vocations are also accepted as important or very important criteria. The above criteria list the 5 most important skills that people who want to work in these vocations must have in a professional sense.

While the "Vocational Knowledge and Skills" dimension, which is one of the three performance appraisal dimensions in employee performance appraisal, is the most important and weighted performance appraisal for "Cook" with 50 %, "Waiter" with 48 % and "Receptionist" with 41 %, it is determined as a secondary dimension for the "Housekeeping Officer" with 33 %. As a result of the research, it was concluded that vocational competencies (knowledge and skills) have at least 33 % effect on the total performance level. This rate is around 50 % for the "Waiting Staff" and "Cook". While vocational knowledge and skills are more prominent in waiting and cookery vocations, it is seen that personality traits and behaviors are more prominent for reception and housekeeping staff. Although vocational knowledge and skill is the most important dimension for the receptionist, the personality traits dimension, which is the second important one, emerged as an important main criterion with 31 %. In recruitment processes, it may be suggested that HRM and relevant decision-makers prioritize vocational knowledge and skills at the job interview stage, especially for cooks and waiting staff.

All of the "Personality Traits" sub-criteria (6 common criteria), another of the research results determined for the related vocations, were seen as very important criteria in terms of performance appraisal. All of the criteria are scored with very high scores (min. average: 9.33) between 1 and 10. In line with the results, all four vocations (Receptionist, Waiter, Cook and Housekeeping Officer) are expected to be loyal employees of their companies by following the rules and laws in their work processes and to speak well of their businesses on different platforms. However, in their business processes, it is important that they do not forget their responsibilities in terms of behavior toward their superiors, subordinates and colleagues, especially guests, and that they organize their business relations in the workplace in a good way. The rate of the "Personality Traits" dimension, one of the employee performance appraisal dimensions as been determined, respectively, as: 31 % for the "Receptionist", 28 % for "Housekeeping Attendant", 26 % for "Waiting Staff" and "Cook". Managers emphasized that in addition to the "Personality Traits" and "Vocational Knowledge and Skills", the behavior of the staff is also very important.

As a result of the study, a draft of a scored and rated performance appraisal form for each determined job position (Waiting Staff, Receptionist, Cook and Housekeeping Attendant) was presented. This study is limited to the relevant vocations (Waiting Staff, Receptionist, Cook and Housekeeping Attendant), and it is important for the researchers who will conduct research on this subject in the future to develop assessment models for employees in other units of the hotel or different sectors, in order to contribute to both the literature and the sector. One of the limitations of the study is that only a certain part of the thesis can be included within the scope of the book chapter. This study was produced from the PhD thesis titled "Developing performance appraisal forms for employees and manager candidates at hotels (A case study in Ankara)" completed by Doğan in 2017.

References

Agut, S., Grau, R. & Peiro, J. M. (2003). Competency needs among managers from Spanish hotels and restaurants and their training demands. *International Journal of Hospitality Management,* 22(3), pp. 281–295.

Aras, H. (1993). *Konaklama İşletmeciliğinde Yiyecek-İçecek Maliyet Kontrolü*, Ankara: Turizm Bakanlığı Yayınları.

Avcı, N. & Küçükusta, D. (2009). Konaklama İşletmelerinde Örgütsel Öğrenme, Örgütsel Bağlılık ve İşten Ayrılma Eğilimi Arasındaki İlişki, *Anatolia: Turizm Araştırmaları Dergisi*, Cilt 20, Sayı 1, Bahar: 33–44.

Baum, T. & Odgers, P. (2001). Benchmarking best practice in hotel front office. *Journal of Quality Assurance in Hospitality and Tourism*, 2(3/4), pp. 93–109.

Bingöl, D. (2010). *İnsan Kaynakları Yönetimi*, 7. Baskı, İstanbul: Beta Basım Yayın Dağıtım.

Boadu, F., Dwomo-Fokuo, E., Boakye, J. K. & Frimpong, A. O. (2014). Employee appraisal and performance in the hospitality industry, *Research in Business and Management*, ISSN 2330–8362, 1(2), pp. 146–165.

Can, H., Kavuncubaşı, Ş. & Yıldırım, S. (2009). *Kamu ve Özel Kesimde İnsan Kaynakları Yönetimi.* 6. Baskı, Ankara: Siyasal Yayın-Dağıtım.

Dahmer, S. J. & Kahl, K. W. (1996). *The Waiter and Waitress Training Manual.* 4th edition, New York: VNR.

Demir. C. (Editör). (2005). *Konaklama İşletmelerinde İnsan Kaynakları Yönetimi: İlkeler ve Uygulamalar.* Ankara. Nobel Yayın Dağıtım.

Dessler, G. (2013). *Human Resources Management.* 13th edition, England: Pearson Education Ltd.

Erdoğan. İ. (1991). *İşletmelerde Personel Seçimi ve Başarı Değerleme Teknikleri*, İstanbul: İşletme Fakültesi Yayını.

Foster, C., McCabe, S. & Dewhurst, H. (2010). Management development skills in the hospitality and tourism sector: Needs and issues from a regional perspective. *Tourism and Hospitality Planning and Development*, 7(4), pp. 429–445.

Gavcar, E., Bulut, Z. A. & Engin, K. (2006). Konaklama İşletmelerinde Uygulanan PerformansDeğerleme Sistemleri ve Uygulama Alanları (Muğla İli Örneği), *Celal Bayar Üniversitesi İ.İ.B.F. Yönetim ve Ekonomi Dergisi*, Cilt:13 Sayı:2, s. 31–45.

Gilbert, D., Child, D. & Bennett, M. (2001). A qualitative study of the current practices of 'no-frills' airlines operating in the UK. *Journal of Vacation Marketing*, 7(4), pp. 302–315.

Hu, M., Horng, J. & Sun, Y. (2009). Hospitality teams: Knowledge sharing and service innovation performance. *Tourism Management*, 30, pp. 41–50.

İnternet: İşkur Food Aid Directive/İşkur Yiyecek Yardımı Yönergesi, (2013), Türkiye İş Kurumu Yiyecek Yardımı Yönergesi, Web: http:www.iskur.gov.tr/.../Türkiye%20İş%20Kurumu%20Genel%20Müdürlüğü%20... adresinden 09 Eylül 2015 tarihinde alınmıştır.

İnternet: Regulations Regarding Certification and Qualifications of Tourism Facilities/Turizm Tesislerinin Belgelendirilmesine ve Niteliklerine İlişkin Yönetmelik (2005), 8948 Sayılı Yönetmelik, Web: http://www.resmigazete.gov.tr/eskiler/2005/06/20050621-11.htm adresinden 08 Ağustos 2016 tarihinde alınmıştır.

İnternet: Turkish Language Association, TDK(2015a), Garson, Web: http://www.tdk.gov.tr/index.php?option=com_gts&arama=gts&guid=TDK.GTS.55f02e36d8e579.75947797 adresinden 09 Eylül 2015 tarihinde alınmıştır.

İnternet: Turkish Language Association, TDK (2015b), Aşçı, Web: http://www.tdk.gov.tr/index.php?option=com_gts&arama=gts&guid=TDK.GTS.55f027feb664c0.55131560 adresinden 09 Eylül 2015 tarihinde alınmıştır.

İnternet: Turkish Language Association, TDK (2015c), Kat Görevlisi, Web: http://tdk.gov.tr/index.php?option=com_gts&arama=gts&guid=TDK.GTS.55f58b9fa725d7.26925497 adresinden 13 Eylül 2015 tarihinde alınmıştır.

Kaufman-Scarborough, C. & Forsythe, S. (2009). Current issues in retailing: Relationships and emerging opportunities. *Journal of Business Research,* 62(5), pp. 517–520.

Koçak, N. (2006). *Yiyecek İçecek Hizmetleri Yönetimi,* 3. Baskı, Ankara: Detay Yayıncılık.

Kozak, M. A. (2001). *Otel İşletmelerinde Kat Hizmetleri Yönetimi,* 3. Baskı, Ankara: Detay Yayıncılık.

Köksal, M. (2005). *İşletmelerde Performans Değerlendirme,* İstanbul, Avcı Ofset.

Küçükaslan, N. (2006). *Yiyecek-İçecek İşletmelerinde Mutfak Hizmetleri Yönetimi,* 1. Baskı, Ankara: Detay Yayıncılık.

Kyriakidou, O. & Gore, J. (2005). Learning by example: Benchmarking organizational culture in hospitality, tourism and leisure SMEs. *Benchmarking: An International Journal,* 12(3), pp. 192–206.

Lillicrap, D. R. & Cousins, J. A. (1993). *Food and Beverage Service,* 3rd edition. Chapter 2, London: Hodder and Stoughton https://books.google.co.uk/books?id=2EJ1BAAAQBAJ&pg=PT45&hl=tr&source=gbs_toc_r&cad=3#v=onepage&q&f=false, accessed March 05, 2016.

Mısırlı, İ. (2003). *Konaklama İşletmelerinde Önbüro Teknikleri ve Uygulamaları,* 2. Basım, Ankara: Detay Yayıncılık.

Olalı, H. & Korzay, M. (1993). *Otel İşletmeciliği,* İstanbul, Beta Basım Yayım Dağıtım A.Ş.

Saad, S. K. (2013). Contemporary challenges of human resource planning in tourism and hospitality organizations: A conceptual model, *Journal of Human Resources in Hospitality and Tourism*, 12, pp. 333–354.

Sökmen, A. (2008). *Yiyecek İçecek Hizmetleri Yönetimi ve İşletmeciliği*, 4. Baskı, Ankara: Detay Yayıncılık.

Tarlan, D. & Tütüncü, Ö. (2001). Konaklama İşletmelerinde Başarım Değerlemesi ve İş Doyumu Analizi, *Dokuz Eylül Üniversitesi, Sosyal Bilimler Enstitüsü Dergisi*, Cilt 3, Sayı:2, s. 141–163.

Tortop, N., Aykaç, B., Yayman, H. & Özer, M. A. (2013). *İnsan Kaynakları Yönetimi*, 5. Basım, Ankara: Nobel Akademik Yayıncılık.

Tunçer, P. (2013). Örgütlerde Performans Değerlendirme ve Motivasyon, *Sayıştay Dergisi*, Sayı:88, Ocak - Mart, s. 87–108.

Vocational Qualifications Authority/Mesleki Yeterlilik Kurumu (2009). Ulusal Meslek Standardı, Oda Temizleme Ve Düzenleme Görevlisi, Seviye 2, Referans Kodu/09ums0024–2, Resmi Gazete Tarih-Sayı/26/11/2009 - 27418.

Vocational Qualifications Authority/Mesleki Yeterlilik Kurumu (2010a). Ulusal Meslek Standardı, Ön Büro Görevlisi, Seviye 4, Referans Kodu/10UMS0067–4, Resmi Gazete Tarih-Sayı/05/07/2010–27632 (Mükerrer).

Vocational Qualifications Authority/Mesleki Yeterlilik Kurumu (2010b). Ulusal Meslek Standardı, Servis Görevlisi, Seviye 4, Referans Kodu/10ums0049-4, Resmi Gazete Tarih-Sayı/11/03/2010-27518 (Mükerrer).

Vocational Qualifications Authority/Mesleki Yeterlilik Kurumu (2010c). Ulusal Meslek Standardı, Aşçı Seviye 4 Referans Kodu/10UMS0110-4 Resmi Gazete Tarih-Sayı/29.12.2010 - 27800 (2.MÜKERRER).

Wen, M. L. Y. & Chen, J. H. Y. (2006). How effectively do vocational junior college food and beverage students assess themselves in server's competencies in Taiwan. *International Journal of Technology and Engineering Education*, Winter, 3(2), pp. 62–70.

Zehrer, A. (2011). Knowledge management in tourism—The application of Grant's knowledge management model to Austrian tourism organizations. *Tourism Review*, 66(3), pp. 50–64.

Tuba TÜRKMENDAĞ

5 Career Development in Tourism Sector

1 Introduction

Individuals want to choose a profession that suits their abilities and pursue a career in their profession throughout their working life. Therefore, promotion and development opportunity is one of the most powerful motivation tools for people to join and work in organizations. Individuals want to get more power and new opportunities during the process of career development (Ozdasli, Kanten and Kanten, 2009).

The concept of career, which deals with the current and future working life of the individual, refers to the arrangement of the series of things to be done in time. Although the number of jobs in the tourism sector is traditionally high, the development of these jobs within the concept of career and carrying out tourism specific studies related to this development is a relatively new trend (Ayres, 2006). Career is the employee's experience and development process through job in the organization (Baruch and Rosenstein, 1992).

Both recruitment practices and career development in the tourism sector require various skill requirements and educational qualifications. This situation led to the occurrence of career development as a subject that covers individuals and human resources managers making career plans in the sector. Additionally, the tourism sector is a continuously growing sector that offers direct, indirect, and stimulated employment opportunities. For this reason, career development in tourism will be discussed in the context of education, employment and working life.

1.1 Concept of Career

Career is a process related to "developing order of a person's work experience over time" (Arthur, Hall and Lawrence, 2010) and jobs organized in a prestige hierarchy where people act in a predictable order (Stewart ve Knowles, 1999). Traditional career understanding based on hierarchical and relatively rigid structures focused on linear progress (Rosenbaum, 1979). In this respect, a traditional career understanding is addressed in terms of progress in organizational position and professional employment (Gunz and Heslin, 2005). Traditionally, the career path has emphasized upward movement in an organization, and many career theories are based on this concept (Arthur and Rousseau, 1996). In the modern sense, career is defined as the experience and development process of the employee in one or more organizations (Weber ve Ladkin, 2008). In a modern sense, career is multidimensional and develops beyond the boundaries of a single organization or professional environment (Arthur, 1994; Collin and Young, 2000).

Career is closely related to structural opportunities such as the size of the industry, organizational structure, and information needs, and human talent and desire. Structural opportunities in a sector set the framework for a profession, while individual talent and desire determine how people choose within structural opportunities (Gutridge, Leibowitz and Shore, 1993; Ladkin and Weber, 2009). Career is also a dynamic process involving the individual's self-related, likes, dislikes, strengths and weaknesses, and work-related experiences (Ayres, 2006).

Although there are many models of career development, few of them have been effective and permanent (Yarnall, 1998). According to Super's "career phase theory" which is a fundamental in the area, persons characteristically go through different career ladders throughout their lives. The stages in this career theory are; establishment, advancement, maintenance, and withdrawal stages (Ayres, 2006, p. 17).

The other example of career studies is the "career anchors theory" which put forward by Schein. These anchors include five value dimensions that affect individuals' career decisions. These are; general managerial competence, technical/functional competence, security/stability, creativity, and autonomy/independence (Schein, 1974). Then three more dimensions; sense of service, pure challenge, and lifestyle, were added to career anchors (Schein, 1990). Schein also suggested that individuals establish a dominant career link that will play an important role in future career choices and decisions (Stewart and Knowles, 1999).

Another study on career research is Derr's "career success map". These proposed models are used as a basis for career choice. Moreover, the career studies carried out by Schein and Derr provide an advantage by accepting the need for a balance between the individual and the organization, rather than purely individual focus (Yarnall, 1998). However, the concept of modern career cannot be defined with a single organization and alternative career concepts arise with various titles. Within the scope of modern career studies, DeFillippi and Arthur (1994) introduced the concept of "boundaryless career" which emphasizes the blurry of career-related boundaries inside and beyond. Peiperl and Baruch (1997) emphasize the general change of the system with the concept of "post-corporate career". The new system proposes a combination of both traditional hierarchies and innovative initiatives. One of these concepts is the concept of "protean career". (Hall, 1996; Hall and Moss, 1998). Protean career is a process managed by the person, not the organization. It consists of individual's experiences such as education, working in various organizations, and professional changes. In a protean career, individual's personal career choices and self-fulfillment seeking are integrative elements in his/her life. The criterion of success is not external, it is internal (Hall and Moss, 1998). In the "protean career", while the employee does not blindly trust the institution related to his/her career, the organization does not fall under an unacceptable obligation. This task-oriented relationship is not only healthier for the individual and the organization, also provides the necessary diversity for the future (Hall, 1996). At the same time, individuals define career success according to traditional income, promotion, and status criterias as well as criterias such as self-satisfaction, life balance, autonomy, and freedom. Accordingly, individuals can adopt non-traditional approaches such as lateral movement and change of direction to achieve success in their careers (Weber and Ladkin, 2008).

1.2 Career Development

Career development requires management of growth and progress in a person's career. Career development is expressed as a lifelong process that involves the management of learning, work, leisure, and transitions to move toward a personally determined and preferred future (Gyansah and Guantai, 2018, p. 40). According to another definition, career development is the process of a series of stages characterized by unique concerns, themes, and tasks (O'neil and Bilimoria, 2005). Lifelong career development is considered necessary for individuals to make choices, express themselves and adapt to their environment. Career development of individuals is adapting to a range of career transitions from school to work, job to job and profession to profession that achieved with the goal of human-environment integration. Career transitions from school to work or from one job to another affect the individual's career adjustment in an environment of uncertainty (Tuna, Kanten, Yeşiltaş, Kanten and Alparslan, 2014, p. 142).

Career development can be seen in a particular profession or lifelong jobs (Hall, 2002). Career development of individuals is important not only for hiring and keeping for themselves or for organizations, but also for the development of human capital in the labor market. Career development related to outcomes such as career advancement, career mobility, career ladders, career barriers and career compromise. (Weber and Ladkin, 2008).

Career development responsibility does not belong to either the individual or the organization. Structurally, only a part of career development is related to individual preferences (Pazy, 1988, Orpen, 1994). Different stakeholders such as bosses, human resources managers, professional associations, and life partners inside and outside the organization affect the career development of the individual (Stewart and Knowles, 1999). It can be said that common responsibility is the best way for a successful career development (Kong, Cheung and Zhang, 2010; Lewis and Arnold, 2012). In addition to the goals for the organization and the individual, there are differences in the core of the career process. When it comes to targets, representatives of the organization have an expectation to increase the compatibility between the needs of the organization and the qualifications of the employees. The individual tries to meet his/her needs. Career development enables the individual to reach high-grade needs through the search for recognition and promotion (Baruch and Rosenstein, 1992).

Career development in the organizational context related to; (1) how individuals manage their careers within organizational structures, (2) how organizations organize their careers. In addition to how personal career development manages an individual's career, it is about how he/she sees and guides in terms of his/her goals, values, beliefs, and expectations (Ladkin and Kichuk, 2017, p. 72). In this context, there is a connection between personality and career choice (Stewart and Knowles, 1999). Career development will vary from person to person. This difference depends on one's vision and efforts on this path (Özdaşlı et al., 2009). In order to increase the awareness of the person collecting career information is important both in vocational guidance and in the career decision making process. While making this conscious attempt to reduce the complexity, risk and uncertainty involved in the process, the individual seeks relevant career information from both primary and secondary sources (Zopiatis, Theocharous and Constanti, 2016).

Depending on individuals, career development is affected by age, race, ethnicity, locus of control, socio-economic status, disengagement, and gender. The complex interaction of these factors affects the degree to which individuals are prepared to specialize in tasks suitable for various stages of career development (Kerka, 1998). For an organization, career development can be an important advantage for achieving high productivity from a trained and skilled workforce. Organization's human resources manager will increase competitiveness by using this advantage, with a well-planned career development program that includes planning and succession organization-centered career planning, talent management, performance evaluation, development activities, transfer, and promotion opportunities (Gyansah and Guantai, 2018, p. 41).

1.3 Career in Tourism Sector

Tourism is a very dynamic sector worldwide and future expectations remain strong even in mature tourism destinations where growth slows down. With the expected long-term growth of the global tourism market, competition for labor and skills is growing rapidly, requiring a larger workforce. Moreover, tourism is in competition with other sectors of the economy, which are seen as more attractive in the labor market (OECD Tourism Trends and Policies, 2012, p. 88). In the estimations made by the World Travel and Tourism Council (2018), it is predicted that approximately 413 million people will be employed in travel and tourism jobs in the last decade. It can be stated that the employment of tourism and entertainment includes a wide range of activities, arrangements and collaboration. The diversity of sectors provides a wide range of job opportunities that are difficult to classify. As a result, choosing the career path and the opportunities it contains become complicated (Ladkin and Kichuk, 2017). Accordingly, both employment and building career in tourism industry includes many personal talents and school training. These topics concern the improving career fields in the sector, the individuals getting ready to enter these career fields, and everyone involved in human resources management (Ayres, 2006). The advantage of this is the employment opportunity in many fields related to tourism and hospitality (Ladkin and Kichuk, 2017).

Despite the diversity in the tourism and hospitality sector, it can be stated that employment in these sectors has certain characteristics. These broad features are often identified and discussed (Janta, Ladkin, Brown and Lugosi, 2011; Duncan, Scott and Baum, 2013). Of these features, those that have an impact on career development can be summarized as follows (O'Leary and Deegan, 2005; OECD Tourism Trends and Policies, 2012; Brunton, 2016; Ladkin and Kichuk, 2017):

- Tourism-related occupations differ widely in skill requirements. This can be a tool to develop skills throughout a career journey.
- Entry barriers are low in most professions. Unskilled or low-skilled jobs provide entry points to jobs in the industry.
- Contract arrangements vary. The flexibility of contracts can be perceived positively to suit different individual needs, but negatively, since work safety is not always achieved.
- Lack of clear career paths in some professions may result in the sector's weak outlook, which may have a negative impact on recruitment.

- One of the universal characteristics of the sector felt in most countries is that it has a higher turnover rate compared to professions in other sectors.
- The sector is a sector where immigrants are employed due to the ease of finding a job. Low skill and training requirements, short-term contracts expand the range of employees to be employed in most jobs.
- Consumer demand patterns in tourism often require working conditions, which are described as non-social, shift hours, weekend changes, night changes, or holiday periods, and irregular working hours.
- Providing better working conditions and eliminating fluctuations in demand requires employees to have more skills and show compliance and flexibility.
- Routine jobs with extremely low pay, less benefits and opportunities for promotion and career development are limited.
- In addition to the high staff turnover rate, trained and experienced staff in tourism cannot be kept. Exploring career advancement in the industry is emphasized as a critical need to secure and retain a professional and skilled workforce to meet current and emerging needs.

Basic working conditions have remained largely unchanged. The negative of these features constitutes a rather weak outlook for career development in the sector. Few countries actively promote career development in the industry, and occupations are not considered a prestigious employment option (OECD Tourism Trends and Policies, 2012).

Within the framework of the mentioned features, it can be accepted that most of the tourism related jobs require low and medium level technical skills. As a matter of fact, it is stated that approximately two thirds of the jobs in the tourism sector are low-skilled (Holman and McClelland, 2011). However, it has been demonstrated that studies in countries such as Korea and Italy show that the jobs requiring high skill are at a level that cannot be underestimated, and the most demanded workforce is the workforce with medium skills (OECD Tourism Trends and Policies, 2012, p. 89). However, tourism enterprises must accept individual and occupational advancement problems in order to keep up the flow of highly skilled employees to the sector. Keeping and hiring good employees is a huge challenge. However, there are opportunities to develop these employees with career development tools (Ayres, 2006). In the tourism sector, there is a need for consistency in developing open career paths that students and employees can use to determine their careers. It is stated that it is difficult for young people to define and express their current career in the sector without ensuring consistency (Brunton, 2016). A career-oriented employee is seen as an employee who moves quickly to keep up with change, dedicated to continuous learning, has career management, and committed to the success of the business (Iles, 1997). However, businesses need to develop a more diverse and dynamic career path to retain high-quality managers (Ayres, 2006).

The most common approach in predicting the career choice of individuals after graduation is the interaction between the two elements which are specified as "individual elements (personality, tendency, interest, attitude)" and "contextual factors of the structure and culture that develop or build the individual's social world" (Barron and Ali-Knight, 2017). Individuals stand for reaching to knowledge on career improvement paths and training competences to pursue a career in tourism. Individuals must be able to comprehend the different and flexible nature of the tourism industry. Additionally, in order to meet the sector's need for highly skilled human resources, in-depth studies on the tourism career are needed (Ayres, 2006).

Career studies provide information about the career development of individuals to both internal and external target audiences. In addition to studies play an important role in contemporary higher education through assessment, accountability and market-oriented research and also by specifying the level of preparation of graduates for the labor market (Rowley and Purcel, 2001; Cabrera, Weerts and Zulick, 2005). Generally, businesses and especially the tourism and hospitality sector, are recruiting from nearby higher education institutions and new graduates to attract young talents. The factors that shape the effectiveness of such approaches can be evaluated as attitudes that affect career development (Chang and Tse, 2015). Therefore, studies related to career goals have vital importance for higher education institutions that educate students to the sector and subjected to economic and political pressure (Cabrera vd., 2005). These pressures are closely related to external factors that shape the higher education environment and include marketing, consumerism, higher accountability of higher education institutions, as well as the diversity of students, diversification of lessons (Nabi and Bagley, 1998; Brennan, Williams and Woodley, 2005).

The promotion of government, industry, and trade unions for the adoption of formal education and desired qualifications and the recognition that the sector needs high-skilled workers increases the number and quality of higher education institutions that provide tourism education. In addition, tourism and hospitality courses have attracted attention and the supply of such courses has increased due to this demand. However, it is an undeniable fact that tourism employers often employ non-tourism graduates who demonstrate the general skills required for jobs in the field of tourism (Dale and Robinson, 2001). Therefore, students concerned with long-term career goals in the tourism sector are concerned about personal issues, the nature of tourism jobs, human factors, educational factors, and some issues related to management factors (Jiang and Tribe, 2009). In addition, many employers in the sector do not require their employees to receive formal training. Less formal forms of education, such as in house or vocational training, are considered within the framework of career development. Such trainings are seen as an important tool for gaining and maintaining the skills necessary to increase employability (Ayres, 2006).

Besides the employment and career and education relationship in technology, technology can be considered as another subject affecting career development in the sector. As in other economic sectors, technology has had a significant impact on tourism-related employment both in support and to change the way services are provided. On the one hand technology has replaced many repetitive jobs, on the other hand reduced the need for certain skills. In addition to these technical or difficult skills, many of which are learned during work or professional programs, soft skills that technology cannot be used or affect to a lesser extent gain importance. What is important in the service industry in general, and especially in tourism, is not the task-based aspect of the business, but its relational aspect. Therefore, the skills required to approach in accordance with the job requirements and customer expectations are also important. These qualifications are among the qualities that require the employee to have a certain pre-education and cultural understanding (OECD Tourism Trends and Policies, 2012, 89).

2 Conclusion

It is understood that the importance of career choice and career development in the individual's life is closely related to the definition of the concept of career. The profession chosen by the individual is a whole with personality traits, as it is related to his/her talent, expectation, future plan and effort. At the same time, organizational compliance, expectations of the organization, efforts to retain the employee and contributions to career development are also considered as important variables for career development along the career path. Therefore, career is expressed as a process that predicts continuous development for both the individual and the organization. Especially in the modern sense, the concept of career has revealed that the organization as well as the individual must make efforts to support the career development of the individual. It has been concluded that it is not possible for the individual to make a career plan alone and the organization cannot achieve success without supporting the career development of the individual.

Studies on the concept of career are seen in the literature as a subject that do not go long way back but they continue to be relevant. It is thought that especially technological, legal, political, socio-cultural, and economic developments cause important changes in business life. These developments ensure that human rights issues become more important globally and improvement of employee rights and opportunities, as well as bringing new obligations to employees. It can be concluded that this situation creates a competition in business life for both the organization and the individual. The way to survive in business life, to provide social, cultural, and economic multi-dimensional satisfaction from the work done is through career development. The individual who makes career development can be a happy individual in both business and social life and has the opportunity to take himself/herself one step further in business life. Organizations that can incorporate an individual who has completed their career development or are in a career effort and support their career development can also provide competitive advantage while contributing to their own development.

It is possible to state that a number of positive and negative characteristics arising from the structure of the tourism sector can affect individuals' loyalty to the sector and career development in the same direction. The fact that the works to be carried out in this context are generally seen as jobs requiring a low level of qualification also causes individuals who have not received tourism education to be employed in the sector. The sector stands out with its features such as heavy workload and long and irregular working hours. Some features (such as foreign language, general culture, communication skills) required by the tourism sector cannot be gained to the individual through education or are inadequate. The individual's efforts in this regard may also be insufficient. Often, businesses do not support the career development of their employees, as they see investment in education as a burden on time and cost. Political arrangements and businesses may fail to improve working conditions in the sector. Therefore, it has been emphasized that education, education institutions, technology, government policy and the direct or indirect activities of the companies in the sector should be analyzed, understood and should be in harmony with each other. In this context, in this study a general framework was drawn related to the concept of career and career development. In addition, due to its unique structure, career development in the tourism sector is discussed within the frame of situations affecting career development and stakeholders who should take a role in supporting career development.

References

Arthur, M. B. (1994). The boundaryless career: A new perspective for organizational inquiry. *Journal of Organizational Behavior. 15*, 295–306.

Arthur, M. B. & Rousseau, D. M. (1996). A career lexicon for the 21st century. *Academy of Management Executive, 10*(4), 28–39.

Arthur, M. B., Hall, D. T. & Lawrence, B. S. (2010). *Generating new directions in career theory: The case for a transdisciplinary approach.* In M. B. Arthur, D. T. Hall, and B. S. Lawrence (Eds.), Handbook of Career Theory (pp. 7–25). New York: Cambridge University Press.

Ayres, H. (2006). Education and opportunity as influences on career development: Findings from a preliminary study in Eastern Australian tourism. *Journal of Hospitality, Leisure, Sport and Tourism Education, 5*(1), 16–28.

Barron, P. & Ali-Knight, J. (2017). Aspiration and progression of event management graduates: A study of career development. *Journal of Hospitality and Tourism Management, 30,* 29–38.

Baruch, Y. & Rosenstein, E. (1992). Human resource management in Israeli firms. *The International Journal of Human Resource Management, 3*(3), 477–495.

Brennan, J., Williams, R. & Woodley, A. (2005). Alumni studies in the United Kingdom. *New Directions for Institutional Research, 126,* 83–94.

Brunton, C. (2016). Literature review-tourism and hospitality careers international benchmarking. Web: https://www.austrade.gov.au/ArticleDocuments/5499/Tourism-and-hospitality-Careers-International-Benchmarking-Report.pdf.aspx(Accessed: 19.10.2020).

Cabrera, A. F., Weerts, D. J. & Zulick, B. J. (2005). Making an impact with alumni surveys. *New Directions for Institutional Research, 126,* 5–17.

Chang, S. & Tse, E. C. (2015). Understanding the initial career decisions of hospitality graduates in Hong Kong: Quantitative and qualitative evidence. *Journal of Hospitality & Tourism Research, 39*(1), 57–74.

Collin, A. & Young, R. A. (2000). *The future of career.* In A. Collin and R. A. Young (Eds.), The Future of Career. (pp. 276–300). Cambridge: Cambridge University Press.

Dale, C. & Robinson, N. (2001). The theming of tourism education: A three-domain approach. *International Journal of Contemporary Hospitality Management, 13*(1), 30–34.

DeFilippi, R. J. & Arthur, M. B. (1994). The boundaryless career: A competency-based perspective. *Journal of Organizational Behavior, 15*(4), 307–324.

Duncan, T., Scott, D. G. & Baum, T. (2013). The mobilities of hospitality work: An exploration of issues and debates. *Annals of Tourism Research, 41,* 1–19.

Gunz, H. P. & Heslin, P. A. (2005). Reconceptualizing career success. *Journal of Organizational Behavior, 26,* 105–111.

Gutridge, T. G., Leibowitz, Z. B. & Shore, J. E. (1993). A new look at organizational career development. *Human Resource Planning, 16*(2), 71–84.

Gyansah, S. T. & Guantai, H. K. (2018). Career development in organizations: Placing the organization and the employee on the same pedestal to enhance maximum productivity. *European Journal of Business and Management, 10*(14), 40–45.

Hall, D. T. (1996). Protean careers of the 21st century. *Academy of Management Executive, 10*(4), 8–16.

Hall, D. T. (2002). *Careers in and out of Organizations.* California: Sage Publications. Web: https://books.google.com.tr/books?hl=tr&lr=&id=tTQ5DQAAQBAJ&oi=fnd&pg=PP1&dq=Careers+in+and+out+of+organizations&ots=nwqL252N3x&sig=8CJbeJ9sgmVYTH8A0ZkuAtmwyTY&redir_esc=y#v=onepage&q=Careers%20in%20and%20out%20of%20organizations&f=false (Accessed: 09.07.2020).

Hall, D. T. & Moss, J. E. (1998). The new protean career contract: Helping organizations and employees adapt. *Organizational Dynamics, 26*(3), 22–37.

Holman, D. & McClelland, C. (2011). Job quality in growing and declining economic sectors of the EU. Walqing working paper 2011.3.

Iles, P. (1997). Sustainable high-potential career development: A resource-based view. *Career Development International, 2*(7), 347–353.

Janta, H., Ladkin, A., Brown, L. & Lugosi, P. (2011). Employment experiences of Polish migrant workers in the UK hospitality sector. *Tourism Management, 32,* 1006–1019.

Jiang, B. & Tribe, J. (2009). Tourism jobs-short lived professions': Student attitudes towards tourism careers in China. *Journal of Hospitality, Leisure, Sport and Tourism Education, 8*(1), 4–19.

Kerka, S. (1998). Career development and gender, race, and class. ERIC Clearinghouse on Adult, Career, and Vocational Education, Center on Education and Training for Employment, College of Education, the Ohio State University. Digest No: 199.

Kong, H., Cheung, C. & Zhang, H. Q. (2010). Career management systems: What are China's state-owned hotels practising? *International Journal of Contemporary Hospitality Management, 22*(4), 467–482.

Ladkin, A. & Kichuk, A. (2017). *Career progression in hospitality and tourism settings.* In S. Horner (Ed), Talent Management in Hospitality and Tourism (pp. 69–83). Oxford: Goodfellow Publishers.

Ladkin, A. & Weber, K. (2009). *Career profiles of convention industry professionals in Asia: A case study of Hong Kong*. In T. Baum, M. Deery, C. Hanlon, L. Lockstone and K. Smith (Eds.), People & Work in Events & Conventions (pp. 39–50). Oxfordshire: Cab International.

Lewis, S. & Arnold, J. (2012). Organisational career management in the UK retail buying and merchandising community. *International Journal of Retail & Distribution Management, 40*(6), 451–470.

Nabi, G. R. & Bagley, D. (1998). Graduates' perceptions of transferable personal skills and future career preparation in the UK. *Career Development International, 3*(1), 31–39.

O'Leary, S. & Deegan, J. (2005). Career progression of Irish tourism and hospitality management graduates. *International Journal of Contemporary Hospitality Management, 17*(5), 421–432.

O'Neil, D. A. & Bilimoria, D. (2005). Women's career development phases-Idealism, endurance, and reinvention. *Career Development International, 10*(3), 168–189.

OECD Tourism Trends and Policies (2012). Policy Approaches to Skills Development in Tourism. Web: https://www.oecd-ilibrary.org/docserver/tour-2012-5-en.pdf?expires=1594906277&id=id&accname=guest&checksum=FC02F97148DFF6BE531E182DBA26BA11 (Accessed: 03.07.2020).

Orpen, C. (1994). The effects of organizational and individual career management on career success. *International Journal of Manpower, 15*(1), 27–37.

Özdaşlı, K., Kanten, S. & Kanten, P. (2009). Yöneticilerin kariyer ilerleme arzusu ile örgütsel bağlılıklarının, akıl hocalığı eğilimlerine etkisini belirlemeye yönelik bir araştırma. *Süleyman Demirel Üniversitesi İktisadi İdari Bilimler Fakültesi Dergisi, 14*(3), 229–243.

Pazy, A. (1988). Joint responsibility: The relationships between organizational and individual career management and the effectiveness of careers. *Group & Organization Studies, 13*(3), 311–331.

Pieperl, M. & Baruch, Y. (1997). Back to square zero: The post-corporate career. *Organizational Dynamics, 25*(4), 7–22.

Rosenbaum, J. E. (1979). Tournament mobility: Career patterns in a corporation. *Administrative Science Quarterly, 24*(2), 220–241.

Rowley, G. & Purcell, K. (2001). Up to the job? Graduates' perception of the UK higher education careers service. *Higher Education Quarterly, 55*(4), 416–435.

Schein, E. H. (1974). Career anchors and career paths: A panel study of management school graduates. *Technical Report, 1*.

Schein, E. H. (1990). Career anchors and job/role planning: The links between career pathing and career development. Web: https://dspace.mit.edu/bitstream/handle/1721.1/2315/SWP-3192-22603401.pdf (Accessed: 03.07.2020).

Stewart, J. & Knowles, V. (1999). The changing nature of graduate careers. *Career Development International, 4*(7), 370–383.

Tuna, M., Kanten, P., Yeşiltaş, M., Kanten, S. & Alparslan, M. (2014). The effect of academic advising on career adaptabilities: A study on tourism and hotel management' students. *The Macrotheme Review, 3*(8), 139–155.

Weber, K. & Ladkin, A. (2008). Career advancement for tourism and hospitality academics: Publish, network, study, and plan. *Journal of Hospitality & Tourism Research, 32*(4), 448–466.

World Travel & Tourism Council (2018). Travel and Tourism Economic Impact 2018 World. Web: https://www.tourisminvest.co.nz/library/pdf/wttc-world2018.pdf(Accessed: 19.10.2020).

Yarnall, J. (1998). Career anchors: Results of an organisational study in the UK. *Career Development International, 3*(2), 56–61.

Zopiatis, A., Theocharous, A. L. & Constanti, P. (2016). Adult vocational decision, career satisfaction and future intention-Insights from the hospitality industry. *International Journal of Contemporary Hospitality Management, 28*(12), 2696–2720.

Vedat YİĞİTOĞLU & Hüseyin KELEŞ

6 Overbooking and Recovery Practices in Accommodation Enterprises

1 Introduction

Overbooking is one of the oldest and most common practices used in the hospitality industry which has been applied to increase room revenues (Rothstein, 1974; Toh, 1985). This practice, which is defined as accepting more reservations than current or available service capacity of an enterprise, has been the subject of many studies (Ivanov, 2006; Wangenheim and Bayon, 2007; Noone and Lee, 2011). A significant portion of enterprises with limited capacity, especially service businesses such as transportation and hospitality sectors, use revenue management systems to maximize their revenues. Overbooking is considered one of the most important strategies of revenue management (Noone and Lee, 2011, p. 350). Therefore, many service businesses which adopt this approach can systematically use overbooking strategies and thus considerably increase their income (Dekay, Yates, and Toh, 2004; Wangenheim and Bayon, 2007; Todorov and Zhechev, 2010).

Accommodation establishments apply overbooking practices in order to prevent their rooms from being vacant due to various reasons such as cancellation of reservations, no shows and early departures and to maximize the utilization rates of their capacity. Because even one of the rooms, which is regarded as a non-durable product due to the non-storage feature of the service, being vacant for a single day is considered as an unrecoverable loss of income for the business. However, overbooking is not a simple application the results of which can always be positive. On the contrary, it is a very risky, double-headed application that can also cause serious problems. The most important risk of overbooking applications is that the reservations taken may turn into oversold rooms. Oversold hotels become unable to serve some of the guests who come in accordance with their reservations as their capacity is completely full. From the customer's point of view, any service which is not fulfilled by a company or fulfilled deficiently is considered as a service failure (Levesque and McDougall, 2000, p. 33). In this context, the inability to provide a room for hotel guests who have booked due to overbooking is accepted as a serious service failure by the hotel customers (Wangenheim and Bayon, 2007, p. 43).

Accommodation businesses can't avoid service failures but they can avoid the customer losses caused by service failures (Michel, 2001, p. 20; Mount, 2012, p. 160). This can be possible with clear, simple and effective "service recovery" efforts, which can be defined as "replacing wrong with right" (Mount and Mattila, 2009, p. 302). Service Recovery (SR) includes all kinds of activities to compensate service failures, to transform negative attitudes of unsatisfied customers into positive and ultimately retain customers (Lewis and McCann, 2004, p. 7). Typical recovery activities used by accommodation businesses for overbooking-based service failures usually consist of free accommodation at another hotel of similar standards, with supportive services such as transportation, shuttle and telephone (Hwang and Wen, 2009, p. 660; Noone and Lee, 2011, p. 335). This recovery practice is used as a standard application especially by leading hotel chains. However, in recent years, a significant portion of hotel businesses can offer additional compensations such as a free night's accommodation, additional promotions and making payments in order to satisfy their customers (Noone and Lee, 2011, p. 336). With effective service recovery customer satisfaction can be increased and consequently, customer losses can be reduced (DeTienne, Seawright, and Brough, 2005). Another indicator of effective service recovery in terms of business is that recovery efforts are carried out at the lowest cost (Zhu, Sivakumar and Parasuraman, 2004, p. 497). Although the literature supports that effective service recovery is good, it is difficult to say that there is a consensus on the type and amount of the most appropriate practice in the event of service failure (Noone and Lee, 2011, p. 335).

Overbooking and SR practices are basically carried out by the front office department. Front desk employees communicate with customers face to face and directly experience the impact of related practices on customers. On the other hand, being confronted with customers in a situation such as oversold rooms, which is a business based service failure, can also cause negative effects on employees. For this reason, it is very important to take into account the opinions and thoughts of the front office employees in the process of developing effective overbooking and SR practices.

1.1 Literature Review

If a hotel only gets reservations for the number of available or saleable rooms, it may run the risk of rooms staying vacant. Because some guests who have made a reservation can cancel their reservations before or on the day of check-in. On the other hand, it is possible for some of the guests with reservations to no show even without canceling their reservations. Considering these negative aspects, many hotel managers may accept more reservation at certain rates in the periods during peak seasons. This issue, which attracted attention in airline businesses before hotel businesses, has been taken into consideration especially in the scope of revenue management (Ivanov, 2007; Riasi, Schwartz, and Beldona, 2019). Overbooking is one of the most important revenue management strategies that can be used not only in airline and hotel businesses, but also in many service businesses with limited capacity, such as car rental and restaurants. The main purpose of hotels using overbooking practices, like other businesses, is to maximize capacity utilization rates, in other words increasing it to 100 %, and avoid possible revenue losses. In this context, the main reasons for hotels to overbook can be explained as follows:

- ✓ Hotel managers have experienced that all reservations approved for a certain day will not be used completely (Ivanov, 2006, p. 23). For many reasons, some reservations may be canceled, changed at the last minute, or may be no show without any prior notice. On the other hand, some reservations may not be checked in due to unexpected external factors. For example, a guest arriving with the airline may not be able to reach the hotel in accordance with his/her reservation, due to factors such as the flight delay, cancellation, or the airline's overbooking. International hotel reports show that 40 % of hotel overbookings are due to overbooking practices carried out by airlines.
- ✓ The main services offered in accommodation businesses, namely rooms, show the characteristics of non-durable products. Therefore, they must be consumed as they are produced simultaneously. For example, it is not possible to stock up a hotel room which is vacant on any given day and resell it on the following days. Many hotels take into account this feature of the services and apply certain rules to cancel reservations and no show conditions in order to reduce possible revenue losses. They demand certain fees for the guaranteed reservations that are not cancelled or used depending on the specified rules. In a study conducted by Chen, Schwartz and Vargas, (2011, p. 129), it is stated that approximately 80 % of the hotels have got a certain amount of penalty fees for late cancellations in recent years. Thanks to such practices, no show rates have dropped to 5 %, which were up to 15 % in the past (Dekay, Yates, Toh, 2004, p. 282).
- ✓ In addition to situations that cause revenue losses in accommodation businesses, there are also some possible situations that increase revenue such as extensions, last-minute reservations and walk-ins. However, research has revealed that last minute reservations made on the day of arrival and income

contributions from non-reserved arrivals are less than the losses caused by cancellation and no show reservations (Toh and Dekay, 2002, p. 81). Similarly, fines taken from rooms in case of early departure cannot compensate for the loss of rooms that have been vacant.

- ✓ Even if reservations are guaranteed, these guarantees are generally limited to a few days. In case of inappropriate use of guaranteed reservations (such as cancellation after the deadline, no show or early departure), hotels may charge some fines. In this case, the most common practice is demanding a daily room rate as penalty (Dekay, Yates, and Toh, 2004, p. 277). According to the national tourism legislation in Turkey, a fee can be demanded for 1 day in low season and 3 days in high season by accommodation enterprises (T.C. Cumhurbaşkanlığı Mevzuat Bilgi Sistemi, 2020). However, if reservations made for a week or more are not used properly, the possible loss of revenue may be much higher than these amounts.
- ✓ The revenues of restaurants and bars in hotels largely depend on occupancy rates. Therefore, vacant rooms causes a decrease in such revenues (Ivanov, 2006, p. 23). This effect can be felt much more especially in businesses where additional income items can be high. For example, a hotel with a reputation for spa services, restaurants, bars or sports facilities may lose revenue much more than the potential room revenues.

Overbooking levels may vary depending on different factors such as the characteristics of companies, location, guest types, booking patterns and legal regulations. After all, it is stated that the overbooking rates observed in hotel enterprises on an international scale is between 5 % and 13 % on average (Gökdeniz and Dinç, 2011, p. 243). The most unfavorable consequence of overbooking is oversold rooms. Oversold hotels in this situation have to perform some service recovery efforts. In this direction, studies on overbooking generally focus on two points. The first of these is the determination of the most appropriate overbooking levels (Toh and Dekay, 2002; Ivanov, 2007; Phumchusri and Maneesophon, 2014; Ivanov, 2015; Yiğit and Esnaf, 2018), and the second is the determination and implementation of the best service recovery which can be applied in oversold rooms (Wangenheim and Bayon, 2007; Hwang and Wen, 2009; Noone and Lee, 2011).

2 Method

This study, which is based on overbooking and the service recovery practices used in oversold rooms, has primarily been conducted to reveal the answers to the following research questions:

Question 1. What are the participants' opinions about overbookings?
Question 2. Which recovery practices are applied in oversold rooms?
Question 3. How are the recovery practices used in oversold rooms ranked in terms of applicability and cost?

There is no clear research data on the level of overbooking carried out by which type of enterprises. However, it is known that the related application in accommodation management is widely used especially by large-scale enterprises (Notarius and Brewer, 2001; Ivanov, 2015). For this reason, the population of the research consists of 4 and 5 star accommodation enterprises. By 2017 there were 595 5-star hotels, 781 4-star hotels, 67 first class in other words 5-star resorts and 11 second class 4-star resorts in Turkey (T.C. Kültür ve Turizm Bakanlığı, 2018). In accommodation businesses, overbooking and recovery practices are generally carried out under the authority and responsibility of the front office department. Therefore, front office managers are the most competent business employees in related applications. For this reason, research data were collected from front office managers representing accommodation businesses.

Research data were collected through a questionnaire generated as a result of literature review and interviews with front office managers. The questionnaire form consists of three parts. The first part includes questions about accommodation businesses, questions in the second part are about overbooking practices, and the third and last part includes questions about service recovery practices used in overbooking. In order to determine the recovery practices in the third part of the survey, initially, phone calls were made with the front office officials of 5 different hotels in Manavgat, Fethiye and Bodrum regions on May 10–12, 2018. As a result of these interviews, seven different practices were determined as shown in Tab. 1.

In the questionnaire form, seven different recovery practices described above are listed first. Later, they were asked to indicate if there are other practices used by enterprises other than these practices. The questionnaire was sent to the front office managers of 4 and 5 star resort hotels, where they can be contacted via e-mail, using a combination of convenience and snowball sampling methods in August-September 2018. A face-to-face questionnaire was applied to the managers of 21 accommodation establishments in the Manavgat region and a total of 47 valid samples were reached.

Tab. 1: Possible Service Recovery Practices Used by Hotels

Recovery Code	Recovery Content
R1	Transfer to another hotel of similar standard for the entire vacation.
R2	Transfer to a hotel of similar standard for a temporary period (such as one or two days) and then providing free services such as dinner, fruit basket and wine.
R3	In addition to the second practice, providing free additional services such as boat tours, safari, and city tours outside the hotel.
R4	After hosting temporarily in a hotel of similar standard bringing the guest back to the hotel and providing a better quality room (like a suite) without extra charge.
R5	After hosting temporarily in a hotel of similar standard bring the guest back to the hotel and provide discounts at certain rates for some paid services (such as spa service, Turkish bath).
R6	After hosting temporarily in a hotel of similar standard bring the guest back to the hotel and giving a free night stay for the guest (in the form of a one-day stay coupon or ticket that can be used whenever and under any circumstances).
R7	After hosting temporarily in a hotel of similar standard bringing the guest back to the hotel and giving an extra free night stay on the next vacation.

3 Results

General characteristics of accommodation enterprises are included in Tab. 2. As seen in the table, most of the participating businesses consist of 5-star hotels (75 %) and show seasonal activities (72 %). However, the number of rooms of approximately half of the enterprises (49 %) is between 201–400 rooms. According to the table, nearly 70 % of businesses are overbooked.

On the other hand, the rate of enterprises faced with oversold rooms has emerged as approximately 87 % at a much higher level. This data shows that even hotels that do not apply overbooking are at risk of oversold rooms in some way (such as technical problems in rooms, customer oriented situations like illnesses, accidents, and duplicate or incorrect reservation records). Therefore, even any hotel which doesn't use overbooking policy is faced with oversold rooms. Therefore, the results of the relevant service recovery efforts are important not only for businesses implementing overbooking but also for businesses which do not implement. On the other hand, looking at the number of oversold rooms, it is seen that 67 % of the accommodation establishments sell 20 rooms or less.

Tab. 2: General Characteristics of Accommodation Enterprises

Characteristic	Options	frequency	%
	5* Hotel	35	74,5
Business Type	4* Hotel	10	21,3
	5* Resort	2	4,3
Activity Period	Seasonal	34	72,3
	Permanent	13	27,7
	200 and less	7	14,9
Number of Rooms	201–400	23	48,9
	401–600	15	31,9
	601 and more	2	4,3
Overbooking	Yes	33	70,2
	No	14	29,8
Oversold Rooms	Yes	41	87,2
	No	6	12,8
	10 Rooms and Less	15	36,6
	11–20 Rooms	12	29,3
Number of Oversold	21–30 Rooms	8	19,5
Rooms	31–40 Rooms	2	4,9
	41 Rooms and More	4	9,8

In a study conducted by Ivanov and Ayas (2017, p. 143), it was revealed that a large portion of accommodation businesses, such as 76 %, do not prefer overbooking. This rate was determined as 36 % in another study conducted by Riasi et al. (2019, p. 2), with the sample of upper class accommodation enterprises. However, in the same research, it was concluded that most of the hotels (68 %) which had overbooking were cautious and did not get any reservations above 5 %. When these two studies are taken together, it is possible to say that overbooking policies are beginning to lose their former popularity.

Tab. 3 below contains data regarding the opinions of the participants in terms of over-booking practices. When the mean of practices is considered, it can be seen that the participants do not fully support the over-booking practices and partially participate in the related applications. However, participants find overbooking applications more normal, especially in resort hotels. Front office officials' views on falling short are relatively stricter.

The recovery practices used in service failures originating from overbooking by accommodation enterprises are listed in Tab. 4. As indicated in Tab. 2, the

Tab. 3: Opinions of Participants about Overbooking (n=47)

	Mean	Std. Deviation
Overbooking can be applied in every hotel.	4,7660	1,86749
Overbooking can be applied in every resort.	5,5319	1,71757
Oversold rooms is natural consequence of overbooking.	4,5745	2,02961

Response categories: 1.Strongly disagree ----- 7.Absolutely agree

Tab. 4: Service Recovery Practices Used by Accommodation Enterprises

Recovery Practices	frequency	%
1. Recovery Practice	34	83
2. Recovery Practice	41	100
3. Recovery Practice	24	59
4. Recovery Practice	40	98
5. Recovery Practice	35	85
6. Recovery Practice	8	19
7. Recovery Practice	12	29

Note: The total exceeds 100 % since more than one option was selected.

number of enterprises that have experienced oversold problems in the past and therefore offered a recovery activity is 41 in total. Some of these participants also stated that they apply room share practice as a solution. Room share is generally possible for reservations that come with at least two or more rooms, or for customers who come with acquaintances (friends and relatives). Since individual reservations are taken into account in the study, this room share practice has been ignored. As can be seen from the relevant table, almost all hotels that use overbooking have applied their 2nd and 4th recovery practice at least once in the past. The least preferred service practice were 6th and 7th ones, respectively. The minimal use of these scenarios can be explained by the fact that customers come with a package tour (or under inflexible conditions in terms of vacation time) and most accommodation businesses operate seasonally. Because it is very difficult for the guests extending their holiday for an extra day or coming to stay again in the same season.

In the process of evaluating the related service recovery efforts from the point of view of the enterprise, it was asked to rank the related recovery efforts in terms of both implementation and cost. In this context, first of all, each recovery was

Tab. 5: The Ranking of Recovery Practices in Terms of Applicability and Cost

Recovery Practices	**Applicability**		**Cost**	
	f	x̄	f	x̄
1. Recovery Practice	43	3,16	44	3,64
2. Recovery Practice	46	***1,80***	46	***2,57***
3. Recovery Practice	43	4,53	45	5,29
4. Recovery Practice	46	*2,83*	46	4,00
5. Recovery Practice	46	3,58	46	*3,48*
6. Recovery Practice	37	5,57	41	5,05
7. Recovery Practice	36	***6,08***	38	4,70

listed from the most applicable to the least applicable. In the ranking made in terms of cost, a ranking was followed from the least costly to the most costly. In cost ranking, indirect costs such as negative customer attitudes and behaviors and possible legal situations such as reclamation or compensation were ignored, only short-term expenses directly reflected in operating expenses were taken into account. The results of the application and cost ranking related to the service recovery options specified by all participants who received and did not get overbooking are shown in Tab. 5 below.

When the Tab. 5 has been analyzed, it is seen that the most applicable recovery is the 2nd practice (x̄=1,80), and the least applicable one is the 7th one (x̄=6,08). In the cost ranking, the least costly recovery is again the 2nd recovery (x̄= 2,57), and the most costly one is the 5th recovery practice (x̄=5,29). The 2nd recovery, which is expressed as bringing back the guests after transferring to a hotel of similar standards for a temporary period and providing free services such as dinner, fruit basket and wine, has emerged as the best option both in terms of application and cost. And it is possible to associate this situation with the form of service concepts applied by businesses. Because, most of the resort hotels participating in the research provide all-inclusive concept. In other words, complimentary food and drink offered to customers do not directly bring extra costs to the hotel. On the other hand, most of the hotels provide similar services for many customers such as honeymoon and repeat customers. In this context, the related services might be considered as catering, but not cost.

Each recovery effort used in oversold rooms are summarized comparatively in terms of applicability and cost in Fig. 1 below. As can be seen in the figure, the 2nd recovery is the most advantageous option in terms of both applicability and cost. The most disadvantageous options were the 7th and 6th recoveries.

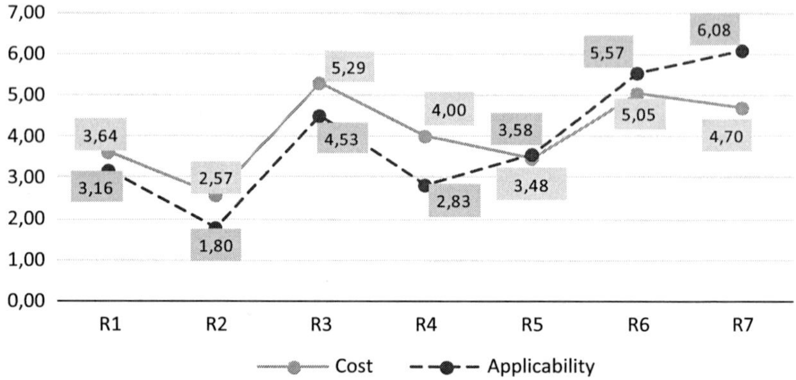

Fig. 1: Comparison of Service Recovery Practices

When evaluated from a business point of view, employees should start from low-cost recovery strategies. On the other hand, the effects of the recoveries used on customers may vary (Wangenheim and Bayon, 2007). The results of overbooking can become much more costly if an improper recovery is presented to customers or if a customer is not eligible for the recovery used in the hotel selected. For this reason, determining customers who will give more positive reactions to low-cost alternatives should be seen as a part of successful overbooking policies.

As it can be seen in the figure above, recovery practices have been listed differently in terms of cost and applicability. In other words, a recovery practice that ranks high in terms of applicability can be ranked lower in terms of cost. Similarly, the reverse situation may also be possible. For example, the 4th recovery practice, which ranked 2nd in terms of applicability, ranked 4th in terms of cost. In this situation, an accommodation enterprise should act either implementation or cost-oriented. However, it is possible to evaluate both rankings on a single score. For that, the average of the relevant recoveries can be taken into account. In this way, the score of the 1st recovery score was calculated as $\bar{x}$= 3,40 ((3,16+3,64)/2). According to the calculations made similarly among other recoveries, the ones with the best score were R2 ($\bar{x}$=2,19), R1 ($\bar{x}$=3,40), R3 ($\bar{x}$=3,42), R4 ($\bar{x}$=3,53), R5 ($\bar{x}$=4,91), R6 ($\bar{x}$=5,31) and R7 ($\bar{x}$=5,39).

4 Conclusion and Recommendations

Overbooking is a very common revenue management practice in accommodation businesses. The main objective of the application is preventing possible revenue losses that may occur due to reservation cancellations, no shows and early departures during periods of peak season. According to the results of this research, more than two-thirds of the accommodation enterprises (about 70 %) continue to use the overbooking policy. Overbooking may vary according to the characteristics of the businesses, market structure and activity levels. However, overbooking has become a riskier application compared to the past and has been less preferred. It is thought that the automation systems used in reservation management and the reservation cancellation or change policies' being bound to certain conditions have been causing this application to be less popular. On the other hand, customer based legal regulations also increase the risk of the relevant practices. Because, in recent years, service enterprises have to pay compensation to their customers who are walked because of overbooking-oriented service failures. For example, with a legal regulation recently made in Turkey, a fine of 8,500 TRY is imposed for each room of which cannot have accommodation service due to overbookings or duplicate bookings.

Accommodation enterprises that do not use overbooking may also have to transfer their customers with guaranteed reservation to other hotels because of some factors such as a decrease in the number of saleable rooms or the customers who are unable to vacate the room due to serious health problems. In the research, it has been understood that this situation is an important issue which needs to be taken into account. Accordingly, it has been revealed that more than half of the hotels that do not use overbooking policies face a similar situation. For this reason, every business should be ready for the oversold room situation. The success of overbooking applications should be considered from two different perspectives. The first of these is the data on oversold rooms occur at a minimum level. The other is the effect of recoveries on customers used in oversold rooms. From the point of view of hotel businesses, the most important cost of overbooking applications is the negative guest attitudes and behaviors that cause customer losses. For this reason, starting from the service recovery efforts used in overbooking, it is important to reveal the effect of each recovery that is considered to be used on guests.

Providing effective service recovery is a very difficult and risky issue for any business. Because the effectiveness of service recovery efforts may vary depending on the characteristics of service failures. On the other hand, it may be more difficult to please the guests who have reexperienced service failure. In addition, if the applied recovery efforts fail to show the expected effect, this opportunity, which is described as a second chance, can turn into a major failure (double deviation) which has deeper negative consequences. On the other hand, in some cases, service recovery practices used by businesses can surpass the feature of compensating a failure and gain the quality of a positive marketing action. Some guests who received service recoveries after a failure show more positive attitudes and acts than those who did not experience any failures. It is claimed that this situation, which is described as a service recovery paradox in the literature, is not valid for service failures originating from overbooking (Noone, 2011, p. 334). For this reason, it is recommended for hotels to focus more on overbooking than service recoveries.

References

Chen, C. C., Schwartz, Z. & Vargas, P. (2011). The search for the best deal: how hotel cancellation policies affect the search and booking decisions of deal-seeking customers, *International Journal of Hospitality Management*, 30(1), 129–135.

Dekay, F., Yates, B. & Toh, R. S. (2004). Non-performance penalties in the hotel industry, *International Journal of Hospitality Management*, 23(3), 273–286.

DeTienne, K. B., Seawright, K. K. & Brough, A. R. (2005). Cost-effective service recovery knowing which customers to keep, *International Journal of Applied Marketing*, 1(3), 1–22.

Gökdeniz, A. & Dinç, Y. (2011). *Konaklama işletmelerinde önbüro operasyonları ve yönetimi* (7. Baskı). Ankara: Detay Yayıncılık.

Hwang, J. & Wen, L. (2009). The effect of perceived fairness toward hotel overbooking and compensation practices on customer loyalty, *International Journal of Contemporary Hospitality Management*, 21(6), 659–675.

Ivanov, S. (2006). Management of overbookings in the hotel industry–basic concepts and practical challenges, *Tourism Today*, 6, 19–32.

Ivanov, S. (2007). Dynamic overbooking limits for guaranteed and nonguaranteed hotel reservations, *Tourism Today*, 7, 100–108.

Ivanov, S. H. (2015). Optimal overbooking limits for a hotel with three room types and with upgrade and downgrade constraints. *Tourism Economics*, 21(1), 223–240.

Ivanov, S. & Ayas, Ç. (2017). Investigation of the revenue management practices of accommodation establishments in Turkey: An exploratory study. *Tourism Management Perspectives*, 22, 137–149.

Levesque, T. J. & McDougall, G. H. (2000). Service problems and recovery strategies: An experiment, *Canadian Journal of Administrative Sciences*, 17(1), 20–37.

Lewis, B. R. & McCann, P. (2004). Service failure and recovery: Evidence from the hotel industry, *International Journal of Contemporary Hospitality Management*, 16(1), 6–17.

Michel, S. (2001). Analyzing service failures and recoveries: A process approach, *International Journal of Service Industry Management*, 12(1), 20–33.

Mount, D. J. (2012). Putting the dollar signs on quality: the benefits of service recovery in the hotel industry, *Journal of Quality Assurance in Hospitality & Tourism*, 13(2), 152–163.

Mount, D. J. & Mattila, A (2009). The relationship of reliability and recovery to satisfaction and return intent at the hotel unit level, *Journal of Quality Assurance in Hospitality & Tourism*, 10(4), 301–310.

Noone, B. M. & Lee, C. H. (2011). Hotel overbooking: The effect of overcompensation on customers' reactions to denied services, *Journal of Hospitality & Tourism Research*, 35(3), 334–357.

Notarius, B. & Brewer, G. S. (2001). *Open your own bed & breakfast* (4 Edition). New York: John Wiley & Sons.

Phumchusri, N. & Maneesophon, P. (2014). Optimal overbooking decision for hotel rooms revenue management, *Journal of Hospitality and Tourism Technology*, 5(3), 261–277.

Riasi, A., Schwartz, Z. & Beldona, S. (2019). Hotel overbooking taxonomy: Who and how? *International Journal of Hospitality Management*, 82, 1–4.

Rothstein, M. (1974). Hotel overbooking as a Markovian sequential decision process, *Decision Sciences*, 5, 389–404.

Toh, R. (1985). An inventory depletion overbooking model for the hotel industry, *Journal of Travel Research*, 23(4), 24–30.

T.C. Cumhurbaşkanlığı Mevzuat Bilgi Sistemi (2020). Turizm İşletmelerinin Bakanlıkla, Birbirleriyle Ve Müşterileriyle İlişkileri Hakkında Yönetmelik, https://www.mevzuat.gov.tr/mevzuat?MevzuatNo=16848&MevzuatTur=7&MevzuatTertip=5, Accessed: 22.02.2020.

T.C. Kültür ve Turizm Bakanlığı (2018). https://yigm.ktb.gov.tr/TR-201140/yillik-bultenler.html, Accessed: 02.01.2020.

Toh, R. S. & Dekay, F. (2002). Hotel room-inventory management: an overbooking model. *Cornell Hotel and Rest. Administration Quarterly*, 43(4), 79–90.

Wangenheim, F. V. & Bayon, T. (2007). Behavioral consequences of overbooking service capacity, *American Marketing Association*, 71(4), 36–47.

Yiğit, F. & Esnaf, Ş. (2018). Gelir yönetiminde kapasite üzeri rezervasyon için bir karma dinamik model önerisi. *İstanbul Üniversitesi İşletme Fakültesi İşletme İktisadı Enstitüsü Yönetim Dergisi*, 24(75), 105–120.

Zhechev, V. S. & Todorov, A. (2010). The impact of overbooking on hotels' operation management. Available at SSRN 1698103.

Zhu, Z., Sivakumart, K. & Parasuraman, A. (2004). A mathematical model of service failure and recovery strategies, *Decision Sciences*, 35(3), 493–525.

Özgür YAYLA

7 The Evaluation of Disabled Tourism Market in Turkey

1 Introduction

Tourism has some crucial effects such as the balance of payments, employment creation, development of infrastructure and superstructures, and elimination of inequalities between regions (Tunçsiper and Bezirgan, 2010). Besides these afore-mentioned effects, travelling has become a social need and its positive effects on the people benefiting from tourism have been supported by the studies carried out (Daniels, Drogin Rodgers and Wiggins, 2005; Minnaert, Maitland and Miller, 2009). With reference to this need, the World Tourism Organization (UNWTO) has set a number of ethical rules. In the 7th article of the global ethical rules, there is a statement that participation in tourism activities is an equal right for all humans, and moreover; families, youngsters, students, elderly and disabled people should also be encouraged to participate in tourism activities (TURSAB, 2020). These principles demonstrate that tourism appeals to all humanity and that the obstacles to participate in tourism activities should definitely be eliminated.

According to the data taken from the Economic Cooperation and Development Organization (OECD-EU), 15 % of the world population consists of disabled individuals. When the data are evaluated, it can be stated that approximately one of every seven people in the world is disabled. In terms of the statistics of the European Statistical Institute (EUROSTAT), the proportion of the disabled population in the European Union countries compared to the total population was determined as 16.2 % in 2010 (EUROSTAT, 2010). This proportion is 12.1 % in the United States (Cornell University, 2012). Despite the decline in child and youth mortality, the number of disabled people is expected to increase more in the future. For example, there is information that in 2030, about 100 million people will be disabled in the USA (Burnett and Bender, 2001). When the situation is evaluated concerning Turkey, today it is known that about 1.5 million people are disabled in Turkey (ERT's, 2020).

In Europe more than 8 million people travel abroad for tourism purposes at least once a year (Zengin and Eryilmaz, 2013). The participation of such a large group in tourism activities has made the disabled tourism market as one of the important markets throughout the world. It is a universal right of disabled people to travel and have a good time in a touristic destination like the people without any disabilities. It can be uttered that having sufficient facilities and qualified human resources is a prerequisite in order to be successful in this market (Ozturk, Yayli and Yesiltas, 2008). Because, the behaviors of the employees working in tourism business that include sincerity, superficiality or professionalism may affect the satisfaction levels of disabled guests positively or negatively. On the other hand, despite the size of the disabled tourism market on such a great scale in the world, when the structure of the tourism enterprises is analyzed, it has been determined that they are generally designed for people without any disabilities and they do not think of doing more than meeting the minimum legal requirements such as parking lots, rooms and toilets (Darcy, Cameron and Pegg, 2010).

Tourism for disabled individuals is an important area for both academic studies and tourism sector (Richter and Richter, 1999). When the literature and tourism for the disabled are examined, it can be stated that it is frequently studied in America (Huh and Singh, 2007) and in the countries affiliated to the European Union (Roth, 1993; Deakin, 1996). In the study conducted by Yau, McKercher and Packer (2004), it was found out that the people with disabilities had the same wishes and expectations just like other individuals. However, the individuals with disabilities may face a number of different difficulties in travel and accommodation activities compared to individuals without disabilities. Within this context, tourism enterprises have put different practices on the agenda by virtue of the increasing number of disabled individuals worldwide (Kim, Wade Stonesifer and Han, 2012). In addition to these, policy makers have been making some decisions on the development of disabled tourism in their countries. For example, in Canada and France, the legislation enabling all the people to access tourism has been developed (Freeman and Selmi, 2010).

Disabled tourism is a type of tourism that can be realized only by making necessary arrangements (Tozlu, Mercan and Atay, 2012). This research has been prepared to evaluate the position of Turkey in terms of disabled tourism market. In the study, first of all the concept of disability has been presented and then the required information about disabled tourism and current applications in Turkey has been given.

1.1 The Concept of Disability

The World Health Organization (WHO) defines the disability concept as "the situation of the individual not being able to comply with the normal requirements of life as a result of the absence or disorder of an organ, which causes permanent loss of function and image to a certain extent in terms of physical, mental or spiritual characteristics" (WHO, 2020). It is understood in the definition that disability is not only related to the absence or dysfunction of the bodily organs, but also to the mental and spiritual health of the individual. Disability is a multi-faceted structure and each disabled person has different needs from the others (Darcy, 2010).

Today, when the statistics related to disability are evaluated, the most common obstacles in humans can be grouped as follows (Darcy, et al., 2010):

- Physical
- Hearing
- Visual
- Cognitive
- Spiritual
- Long-term health threatening diseases

When the literature is examined, the concepts of impairment or handicapped can be used instead of disability. Besides, different definitions have been made for the concept of disability with the introduction of new regulations. For example, according to the laws in the USA, disability is defined as all of the mental and physical problems that will limit the activities which people must normally do (Çakmak, 2008). Bilir (1986), using the concepts of disabled and impaired together, defines the concept of disability as the health problems that fully disrupt our participation in and adaptation to life. Disability can happen later in life as well as being congenitally (Eryılmaz and Zengin, 2013). In other words, every healthy individual is a candidate for being disabled.

The concept of the disability has taken more part in the literature in the periods following the Industrial Revolution. Because, apart from congenital disability, both during the Industrial Revolution and afterwards a serious increase was observed in the number and proportion of disabled individuals resulting from different work accidents (Tellioğlu, 2018).

The disabled individuals face many obstacles that healthy people do not encounter in accessing the services such as primarily health, education, employment, travel and tourism. Besides, compared to other people, their health conditions should be kept under constant control. What's more, it can be said that disabled people are generally individuals with low education and welfare levels in underdeveloped and developing countries (WHO, 2011). The disabled individuals' having difficulties in performing some social roles such as marrying, starting a family, being adapted to social life and having children, etc. is an indicator of the obstacle that the disability is felt not only physically or mentally but also sociologically (Karataş and Çiftçi, 2010).

1.2 Disabled Tourism

Not all people can benefit from tourism in the same way. For instance, while economically low-budget people can travel for shorter periods and to shorter distant places, economically strong people can participate in longer-term tourism activities in overseas countries. Other than these, the opportunities to benefit from tourism may pose some problems for the disabled individuals, as well.

The tourists with disabilities may face a number of obstacles in their decision to travel in terms of environmental or economical aspects. Disabled individuals carry their limitations directly or indirectly to other members of their families, and as a result, both the disabled individuals and their families can be affected from the situation (Nalçacı İkiz and Solunoglu, 2017). In particular, the disadvantaged individuals should get away from the problems that they encounter both in the community and the family by getting integrated into the society with their family members, renewing themselves in social and spiritual terms, having a more enjoyable time through different activities (Kim and Lehto, 2013). McKercher, Packer, Yau and Lam (2003) argue that individuals with disabilities face certain problems in participating in tourism activities, and these problems can be grouped into two different groups as internal and external factors.

In Tab. 1, the obstacles that individuals with disabilities encounter in the process of participating in tourism activities are presented. While the problems that are related to personality and economic factors are under the category of internal problems environmental and communication obstacles are classified under external problems. The lack of information for the tourism destination or activities organized at the destination or problems regarding their social skills constitute personality-related obstacles. Besides, the economic difficulties that arise due to the problems experienced at the point of employment of disabled individuals or the fact that traveling alone is not possible is stated as an internal problem in the category of economic obstacles. On the other hand, the problems that are not directly linked to the disabled individuals are gathered under external problems. As a matter of fact, the physical and architectural structures of tourism facilities or the construction of public areas in the destination by ignoring the disabled people can be stated as an external problem that makes it difficult for the disabled individuals to participate in tourism. Furthermore, the attitudes of tourism sector employees toward the disabled individuals or communicative problems show the external problems faced by the disabled individuals.

Tab. 1: The Problems Encountered by Disabled Individuals while Participating in Tourism Activities

INTERNAL PROBLEMS		**EXTERNAL PROBLEMS**	
Personality-related obstacles	**Economic Obstacles**	**Environmental Obstacles**	**Communication Obstacles**
Lack of Information	Economic difficulties	Physical structure of tourism facilities	Disputes
Lack of Social Talent	Injustice in income distribution	Environmental obstacles such as paths and hills	Communication Difficulties
Physical or psychological problems	The need for someone while travelling	Transportation obstacles	Helplessness in activities
Considering travel as a right only to healthy people		Legal regulations	The attitudes of employees working in tourism facilities
		Safety Obstacles	The inconsistency of information

Source: Mckercher, et al., (2003)

It's an undeniable fact that, all the areas of social life should be designed in a way that every individual can reach. The approach of design for everyone includes the disabled individuals, as well. In order to meet these needs, the concept of accessibility has emerged (Sirel, Yücel Caymaz and Sirel, 2018). Due to the fact that tourism should be accessible for every individual, the situation of making facilities, places and services accessible, in other words, the concept of accessible tourism has emerged.

The types of disability types and even if they are the same type, the levels of disability are differentiating the tourist experiences. For this reason, travel or service situations should be shaped according to disability levels and disability types (Small, et al., 2012; Erkılıç and Eren, 2020).The fact that people with disabilities, who are called the most crowded minority of the world by the International United Nations Organization and the reality that everyone is a candidate for being disabled, also expresses a population that cannot be underestimated for tourism businesses both in terms of social responsibility and economical aspects (Artar and Karabacakoğlu, 2003). The former studies have shown that 30 % of the population will become disabled individuals in some part of their lives and will need a certain level of accessibility. Accessible tourism should be globally designed for individuals who need ease of access in terms of equality, independence and honorific aspects, by covering suitable mobility and the dimensions of visual, hearing and mental access. It will be possible to talk about accessible tourism only if the improvements are made to facilitate the use of tourism products, services and the environment (Darcy and Dickson, 2009).

The disabled individuals, who have a considerable rate in societies, are in a constant interaction with the society in terms of both accessibility and social life. It has been supported by the studies carried out that tourism has a role on all the people in increasing the quality of life (Kim, Uysal and Sirgy, 2013; Kim, Woo and Uysal, 2015). Similarly, tourism can help the development of physical, psychological and mental health of disabled people (McCabe, Joldersma and Li, 2010; Small, Darcy and Parcker, 2012; McCabe and Johnson, 2013).

The academic studies concerning the experiences of disabled individuals related to tourism activities have started to take part in the literature since the late 1970s (Allon, 2013). In 1980 and afterwards, especially with the realization of legal arrangements, the activities for the disabled tourism gained speed (Yaylı and Öztürk, 2006). In addition to these, in the current century different studies have taken place in the literature on disabled tourism. For example, the travels of disabled individuals with their families mostly consist of the visits made to spouses, friends and relatives (Nalçacı İkiz and Solunoglu, 2017). Pagan (2012) examined the tourism behaviors of the disabled individuals comparing them to those of the healthy individuals and found that the individuals with disabilities are more reluctant to participate in tourism activities. According to Darcy (2010), the size of the handicapped guests and the support they need in hotel choices are of great importance. Zengin and Eryilmaz (2013), on the other hand, examined the awareness of disabled tourism from the perspective of hotel managers and found that hotel managers are not sufficiently aware of the relevant market and that the disabled tourism legislation is insufficient compared to European countries. Zbikowski Kuzmicki, Dabrowski, and Soroka (2011) found that occupational conditions have a significant impact on the participation of disabled people in tourism. Zajadacz (2014) investigated the sources of information in the choice of travel destinations for the people with hearing impairments and found that the people in the immediate environment and the internet are the most preferred sources of information.

1.3 The Evaluation of Disabled Tourism Market in Turkey

In the current era, as in other European countries, in Turkey, the travel conditions of the disabled individuals and the academic studies conducted concerning the legislation have increased (Şahin, 2012). A number of legal regulations have been set in Turkey with the purpose of enabling the disabled people to have comfortable and relaxed vacations.

In 1982, the Regulation on Certification and Qualifications of Tourism Facilities prepared under the Tourism Incentive Law with the number of 2634 became valid by being published in the Official Gazette on 21 June 2005. Along with this arrangement, in the regulation of the Ministry of Culture and Tourism dated March 5, 2011 with the number of 27865, there are statements regarding the necessity of arranging at least one room in accordance with the use of physically disabled people in the hotels and holiday resorts that have a minimum of 80 rooms and more (Culture and Tourism Ministry, 2005). The Ministry of Culture and Tourism also issued a circular in 2011 that contributes to the freedom of travelling and tourism activities of the people with disabilities on the subject of 'Barrier-Free Tourism'.

One of the other important institutions is the Association of Turkish Travel Agencies (TÜRSAB) that has an important mission for the development of disabled tourism. In accordance with this development, in 2006 within the body of TÜRSAB the committee of 'Accessible Tourism for Everyone' was established in Turkey. TÜRSAB has been conducting scientific studies to support disabled tourism for the determination of the problems experienced by disabled individuals and possible solutions of these problems, as well (Akdu and Akdu, 2018). In addition to this, since Turkey is a candidate country to the European Union, it has been implementing the encouraging recommendation which was put forth in 2013 by the Council of European Ministers Committee in terms of encouraging the disabled people to participate in tourism activities (Olcay, Giritlioğlu and Parlak, 2014).

Disabled tourism sector is an emerging tourism type in Turkey, similar to many developing countries in the tourism sector. This situation causes problems arising from the lack of infrastructure and service understanding in tourism facilities for the individuals with disabilities. (Zengin and Eryılmaz, 2013). As a result, in spite of the laws and regulations, it is possible to say that disabled individuals experience serious problems in accessing tourism services. As a matter of fact, Pehlivanoğlu (2012) found in his study in the Antalya region that minimum number of disabled rooms were arranged in accommodation establishments because of the obligations of the legal requirements. As parallel to this fact, Tozlu, Mercan and Atay (2012) determined in their study in Çanakkale destination that the disabled hotel rooms hadn't been designed suitable for the use of disabled people, but for the use of physically handicapped guests only. In another study, Yaylı and Öztürk (2006) found a difference between the number of stars of accommodation establishments and the perspective of tourism for disabled individuals. According to the results of the study, as the number of stars increases, the perspective toward the disabled tourism market becomes more positive. In their study, Öztürk, Yaylı and Yeşiltaş (2008) put forth that the business managers who serve in the tourism sector claim that the disabled tourism market has a great potential, on the other hand, due to insufficiency in environmental conditions, personnel training and physical areas, they do not adequately dominate this market.

Turkey is a powerful destination in the tourism sector, in terms of not only its historical features but environmental features, as well. On the other hand, as a result of the findings obtained from the studies, it can be stated that the disabled tourism market hasn't been sufficiently focused. The areas for the disabled individuals, which are organized solely to fulfill legal obligations, both limit the right to travel of potential tourists with disabilities and pose a problem for the destination to take the value it deserves from the disabled tourism market.

2 Conclusion and Suggestions

The international studies conducted in the literature, indicate that tourists or local people living in a destination cannot benefit from tourism services due to the presence of physical, economic, social or some other obstacles (Eichhorn, Miller, Michopoulou and Buhalis, 2008; Small and Harris, 2011).

The legal and physical regulations that will be made for the disabled individuals to participate in tourism activities in Turkey, must be organized and implemented within the scope of a plan. Because every disabled has his/her own personal needs (Daniels, et al., 2005). The studies carried out have shown that when the existing services that are offered in hotels or destinations are adapted to a particular disability, the people with that disability increase their tendency to purchase those services (Figueiredo, Eusebio and Kastenholz, 2012). At this point, the cooperation of enterprises, organizations of destination management, local administrations and local people is of great importance in order to gain success. It can be stated that conducting the development process of tourism destinations in a planned manner will facilitate the positioning and marketing of the product or service desired to be presented in the region.

Improving the qualifications of the accommodation businesses will not be enough on its own in order to increase the number of disabled tourists coming to these businesses in Turkey. Besides, it is necessary to make all communal areas, walkways and other centers in the destination suitable for the use of disabled individuals. What's more; if the travel agencies and tour operators that direct tourists to Turkey make necessary arrangements in the transportation vehicles in accordance with the needs of the disabled people, this will accelerate the decision making and purchase processes of the disabled tourists.

Turkey has a considerable potential to be able to realize alternative tourism types such as culture (Arslan, Kendir & Özkoç, 2019), sports and activity tourism facilities (Argan, 2004), online tourism (Kendir, Arslan and Özkoç, 2019) along with having the sea, sand and sun trio. However, in all these mentioned touristic activities, Turkey is in a severe competition with strong countries in the field of tourism. As for disabled tourism, it can be said that it is a more niche market compared to other alternative tourism types. Although disabled tourism is an important market in the world, it is an area that has been overlooked by many countries including Turkey. Within this context, it is important to analyze the disabled tourism market in detail and make the businesses and destinations suitable for the use of disabled individuals, as well as carrying out the positioning activities of the destination in accordance with the disabled tourism in order to get the value deserved from the market.

The people with disabilities living in Turkey should particularly be encouraged to participate in domestic tourism activities. In this sense, necessary plans can be made for disabled people, who have economic problems, such as giving right to vacation or increasing their employment rates. On the other hand, in Turkey the accommodation businesses that have a total of 80 rooms or more have been making the required arrangements, as 1 % of the total room number, for disabled individuals in accordance with the legal regulations. But, this situation is insufficient considering the number of disabled people worldwide. The mentioned rate can be revised as 4 rooms in every 100 rooms as in the similar applications implemented in developed countries in the field of tourism.

Employing staff who know the use of sign language in all tourism establishments serving in touristic destinations, especially in accommodation businesses, will play an important role in the decision-making process of the individuals with hearing impairment. Moreover, the facilities that are going to be built for all the disabled individuals can be supported by the state. In particular, some encouraging incentive policies should be made up for accommodation businesses that will be built only for the individuals with disabilities.

This research, which was carried out in order to determine the disabled tourism potential of Turkey, was completed by benefiting from secondary. In the future, the studies that will be done to determine the regional potential disabled tourism will be beneficial. Because each region has its own unique transportation and infrastructure conditions. That's why, it may be difficult to make all the destinations a suitable tourism destination for the disabled people. Instead, determining the destinations suitable for disabled tourism by considering the variables such as transportation, environmental factors, geographical attractiveness, and the level of awareness of the local people through studies in different destinations will increase the share that is targeted to be taken from the disabled tourism market.

References

Akdu, U., and Akdu, S. (2018). Engelli turizmi: Yasal düzenlemeler ve uygulamalar, *Gümüşhane Üniversitesi Sosyal Bilimler Enstitüsü Dergisi*, 9 (23), 99–123.

Akıncı, Z., and Sönmez, N. (2015). Engelli bireylerin erişilebilir turizm beklentilerinin değerlendirilmesine yönelik nitel bir araştırma. *Anatolia: Turizm Araştırmaları Dergisi*, 26 (1), 97–113.

Allon, M. (2013). Disability tourism: Why do disabled people engaging in tourism activities? *European Journal of Social Sciences*, 39 (3), 480–486.

Argan, M. (2004). Spor ve turizm pazarlamasının kesişim noktası olarak spor turizmine kuramsal bir bakış, Anatolia: Turizm Araştırmaları Dergisi, 15 (2), 158–168.

Arslan, E., Kendir, H., and Özkoç, G. (2019). *Niksar'ın kültürel miras turizmi potansiyelinin belirlenmesi*, VIII. National VI. International Eastern Mediterranean Tourism Symposium, 262–272.

Artar, Y., and Karabacakoğlu, Ç. (2013). *Türkiye'de özürlüler turizminin geliştirilmesine yönelik olarak konaklama tesislerindeki altyapı olanaklarının araştırılması*. Özürlülerin Toplumsal Gelişimine Yönelik Proje. Milli Prodüktivite Merkezi, Ankara.

Bilir, Ş. (1986). *Özürlü çocuklar ve eğitimleri*. Ankara: Ayyıldız Matbaası.

Burnett, J., and Bender, B. H. (2001). Assessing the travel-related behaviors of themobility-disabled consumer. *Journal of Travel Research*, 40, 4–11.

Cornell University (2012). Disability Report United States, http://www.disabilitystatistics.org/ (Accessed: 22.06.2020).

Culture and Tourism Ministry (2020). *Turizm tesislerinin belgelendirilmesine ve niteliklerine ilişkin yönetmelik.2005*. http://teftis.kulturturizm.gov.tr/ (Accessed: 18.07.2020).

Çakmak, M. (2008). Amerika Birleşik Devletleri'nde engelli tanımı hakkında bir inceleme. *Ankara Üniversitesi Hukuk Fakültesi Dergisi*, 57 (2), 51–62.

Daniels, M. J., Drogin Rodgers, E. B., and Wiggins, B. P. (2005). "Travel Tales": An interpretive analysis of constraints and negotiations to pleasure travel as experienced by persons with physical disabilities. *Tourism Management,* 26, 919–930.

Darcy, S. (2010). Inherent complexity: Disability, accessible tourism and accommodation information preferences. *Tourism Management*, 31 (6), 816–826.

Darcy, S., and Dickson, T. (2009). A Whole-of-Life approach to tourism: the case for accessible tourism experiences. *Journal of Hospitality and Tourism Management,* 16 (1), 32–44.

Darcy, S., Cameron, B., and Pegg S. (2010). Accessible tourism and sustainability: A discussion and case study, *Journal of Sustainable Tourism,* 18 (4), 515–537.

Deakin, A. (1996). Body language. *Marketing Week*, 19 (26), 37.

Eichhorn, V., Miller, G., Michopoulou, E., and Buhalis, D. (2008). Enabling access to tourism through information schemes? *Annals of Tourism Research*, 35(1), 189–210.

Erkılıç, C. E., and Eren, A. (2020). Seyahat acentalarında yaşlı ve engelli yerli turistlere sunulan hizmetlerin değerlendirilmesi, *Üçüncü Sektör Sosyal Ekonomi Dergisi*, 55 (1), 564–585.

Eryılmaz, B., and Zengin, B. (2013). Bodrum destinasyonunda engelli turizm pazarının değerlendirilmesi, *Uluslararası İktisadi ve İdari İncelemeler Dergisi*, 11, 51–74.

EUROSTAT (2010). The European Social Fund and Disability. White Chlorine-free Paper. Belgium, https://ec.europa.eu/employment_social/esf/docs/br_disability_en.pdf (Accessed: 23.06.2020).

EYDER (2020). Sayılarla Türkiye'de ve Dünyada Engellilik. https://ey-der.com/ana-sayfa/turkiye-ve-dunyada-engelliler/ (Accessed: 25.06.2020).

Figueiredo, E., Eusebio, C., and Kastenholoz, E. (2012). How diverse are tourists with disabilities? A pilot study on accessible leisure tourism experiences in Portugal. *International Journal of Tourism Research*, 14, 531–550.

Freeman, I., and Selmi, N. (2010). French versus Canadian tourism: Response to the disabled. *Journal of Travel Research*, 49 (4), 471–485.

Huh, C., and Singh, A. J. (2007). Families traveling with a disabled member: analysing the potential of an emerging niche market segment. *Tourism and Hospitality Research*, 7 (3/4), 212–229.

Karataş, K., and Çifci, E. (2010). Türkiye'de engelli kadın olmak: deneyimler ve çözüm önerileri. *Uluslararası Sosyal Araştırmalar Dergisi*, Kadın Çalışmaları (Özel Sayı), 3(13), 147–153.

Kendir, H., Arslan, E., and Özkoç, G. (2019). *Kaz Gölü'nün ornitoturizm açısından değerlendirilmesi*, VIII. National VI. International Eastern Mediterranean Tourism Symposium, 334–345.

Kim, S., and Lehto, X. Y. (2013). Travel by families with children possessing disabilities: Motives and activities. *Tourism Management*. 13–24.

Kim, H., Woo, E., and Uysal, M. (2015). Tourism experience and quality of life among elderly tourists. *Tourism Management*, 46, 465–476.

Kim, K., Uysal, M., and Sirgy, M. J. (2013). How does tourism in a community impact the quality of life of community residents?. *Tourism Management*, 36, 527–540.

Kim, W. G., Stonesifer, H. W., and Han, J. S. (2012). Accommodating the needs of disabled hotel guests: Implications for guests and management, *International Journal of Hospitality Management*, 31, 1311–1317.

McCabe, S., and Johnson, S. (2013). The happiness factor in tourism: Subjective well-being and social tourism. *Annals of Tourism Research*, 41, 42–65.

McCabe, S., Joldersma, T., and Li, C. (2010). Understanding the benefits of social tourism: Linking participation to subjective well-being and quality of life. *International Journal of Tourism Research*, 12, 761–773.

Mckercher, B., Packer, T., Yau, M. K., and Lam, P. (2003). Travel agents as facilitators or inhibitors of travel: Perceptions of people with disabilities, *Tourism Management*, 24 (4), 465–474.

Minnaert, L., Maitland, R., and Miller, G. (2009). Tourism and social policy. The value of social tourism, *Annals of Tourism Research*, 36 (2), 316–334.

Nalçacı İkiz, A., and Solunoğlu, A. (2017). Engelli çocuğa sahip ebeveynlerin turizm davranışları, *Journal of Recreation and Tourism Research*, 4 (2), 92–104.

Olcay, A., Giritlioğlu İ., and Parlak Ö. (2014). ENAT (European Network For Accessible Tourism) ile Türkiye'nin erişilebilir turizme yönelik otel işletmelerini kapsayan düzenlemeleri ve bu düzenlemelerin karşılaştırılması, *Gazi Üniversitesi Turizm Fakültesi Dergisi*, 2, 127–144.

Ozturk, Y., Yayli, A., and Yesiltas, M. (2008). Is the Turkish tourism industry ready for a disabled customer's market? The views of hotel and travel agency managers, *Tourism Management*, 29, 382–389.

Pagán, R. (2012). Time allocation in tourism for people with disabilities. *Annals of Tourism Research*, 39 (3), 1514–1537.

Pehlivanoğlu, B. (2012). Konaklama yapılarının engellilere yönelik oda düzenlemelerinin irdelenmesi, *İnönü Üniversitesi Sanat ve Tasarım Dergisi*, 2 (4), 27–35.

Richter, L. K., and Richter, W. L. (1999). Ethics challenges: Health, safety and accessibility in international travel and tourism. *Public Personnel Management*, 28, 595–615.

Roth, S. (1993). EC making progress on access, *Hotel and Motel Management*, 208 (6), 29.

Sirel, A., Yücel Caymaz, G. F., and Sirel, O. Ü. (2018). *The universal design approach in barrier-free shopping centers: evaluation on selected examples*, 14th International Conference in Standardization, Protypes and Quality: A Means of Balkan Countries' Collaboration, Tirana, Albania.

Small, J., and Harris, C. (2011). Obesity and tourism: Rights and responsibility. *Annals of Tourism Research*, 39 (2), 686–707.

Small, J., Darcy, S., and Packer, T. (2012). The embodied tourist experiences of people with vision impairment: Management implications beyond the visual gaze. *Tourism Management*, 33, 941–950.

Şahin, H. (2012). *Engelli bireylern konaklama tesislerinden memnuniyet durumlarının incelenmesi*, Doktora tezi, Hacettepe Üniversitesi, Ankara.

Tellioğlu, S. (2018). Engelli turizm pazarlaması: Alanya araştırması. *Doktora Tezi*, Selçuk Üniversitesi, Sosyal Bilimler Enstitüsü, Konya.

Tozlu, E., Mercan, O. Ş., and Atay, L. (2012). Çanakkale'nin engelli turizmine ilişkin durumunun belirlenmesine ve planlanmasına yönelik bir çalışma, *Aksaray Üniversitesi İktisadi ve İdari Bilimler Fakültesi Dergisi*, 4 (1), 1–16.

Tunçsiper B., and Bezirgan, M. (2010). Termal turizmin ekonomik etkilerinin algılanması, emet ilçesinde bir araştırma. *Gümüşhane Üniversitesi Sosyal Bilimler Elektronik Dergisi*, 2, 1–20.

TURSAB (2020). UNWTO Global Code of Ethics for Tourism. https://www.tursab.org.tr/apps/OldFiles//dosya/14733/bmdto-turizm-icin-kuresel-etik-kurallari-unwto-global-code-of-ethics-f_14733_4881829.pdf (Accessed 23.06.2020).

WHO (2020). World Report on Disability. https://www.who.int/disabilities/world_report/2011/report.pdf?ua=1 (Accessed 25.06.2020).

Yau, M. K, McKercher, B., and Packer, T. L. (2004). Traveling with a disability more than an access issue, *Annals of Tourism Research*, 31 (4), 946–960.

Yaylı, A., and Öztürk, Y. (2006). Konaklama işletmeleri ymneticilerinin bedensel engelliler pazarına bakış açıları üzerine bir araştırma. *Anatolia: Turizm Araştırmaları Dergisi*, 17 (1), 87–97.

Zajadacz, A. (2014). Sources of tourist information used by Deaf people. Case study: the Polish Deaf community. *Current Issues in Tourism*, 17 (5), 434–454.

Zbikowski, J., Kuzmicki, M., Dabrowski, D., and Soroka, A. (2011). Vocational activity as a determinant of participation in tourism of disabled people from the area of Eastern Poland. *Oeconomia*, 10 (2), 121–129.

Zengin, B., and Eryılmaz, B. (2013). Bodrum destinasyonunda engelli turizm pazarının değerlendirilmesi, *International Journal of Economic and Administrative Studies*, 6 (11), 51–74.

Ali İSKENDER

8 Wedding Tourism in Turkey

1 Introduction

Millions of tourists have been traveling to different regions and countries in current times. The trends in tourism sector are in dynamic and are tend to be change year by year. So It's becoming harder and harder to sort the suitable market through different profiles of tourists. Therefore, it is necessary to develop new strategies to make a profit within the highly competitive tourism industry.

The countries that want to get more shares from tourism revenues need to implement policies in order to create new demands. For this purpose, these countries have been competing with the aim of developing new touristic products. Turkey also wants to increase the tourism demand through diversifying touristic products. In this context, in Turkey, an awareness has aroused for wedding tourism which is an alternative tourism type.

Since it is an innovative niche market and has a high economic return, more investments for wedding tourism have started to be included. Turkey has been attaching more importance to the promotional activities for wedding tourism aiming at the countries that have been chosen as a target market. Introductory activities have been carried out at the wedding fairs organized in these countries. At the same time, the wedding organizers in target market countries are invited to Turkey and necessary organizations have been carried out to demonstrate the potential of Turkey.

In Turkey, wedding tourism is perceived as an alternative tourism type which is comprised of foreign tourists and their relatives that prefer Turkey for their wedding organizations. Therefore, in this study the international potential, targets and expectations of Turkey concerning the wedding tourism have been focused on.

2 Wedding Tourism

The term "wedding tourism" generally refers to the weddings that are organized where neither the bride nor the groom lives, and direct the influx of tourists to the region in which the weddings are organized. Wedding tourism can also be associated with the tourism activities concerning visiting friends and relatives (Bertella, 2015: 398). According to the Oxford dictionary, wedding is defined as a marriage ceremony that includes eating, drinking and entertainment along with the wedding. As for marrying, it is explained as the legal coupledom of two people (www.oxfordlearnersdictionaries.com). Major et al. expressed wedding tourism as international travels in order to marry or attend a marriage ceremony (Major, McLeay and Waine, 2010: 249). Wedding tourism involves the activities carried out with the participation of friends and relatives for marriage purposes both in home country and abroad. Wedding tourism classification can be made in two ways. These are as follows (Suklabaidya, 2008: 27–28).

- **Destination Wedding:** The wedding ceremonies organized by the couples in impressive destinations that everyone admires such as Hawaii or New Zealand. Here, the scenery and atmosphere offered by the destination play a major role. Besides, not only legal regulations and rules related to marriage in the destination, but marriage traditions and local customs in the region are also influential in the preference.
- **Unique or Exotic Wedding Ceremony:** The couples who are in search of romance and adventure dream of a wedding ceremony apart from the traditional wedding concept. They may have unlimited preferences and wishes such as marrying underwater or in a hot air balloon, or perhaps marrying by doing bungee jumping. Needless to say that, these couples may not find the opportunities to meet their desires in every city. That's why; they want to perform their marriage ceremonies in the cities that have the infrastructure and facilities to meet their desires.

People may want to get married in a different country because they want to marry abroad at a lower cost since the traditional weddings are much more expensive. Another advantage of the abroad marriages is that the couples have the opportunity to make honeymoon at the same destination after the wedding. It is quite common for abroad weddings to be in the form of both wedding and honeymoon. The opportunities such as getting affordable prices, having honeymoon, the comfort and frequency of transportation and flight increase the attractiveness of weddings abroad day by day (Major, McLeay and Waine, 2010: 253).

Another reason that encourages people to marry in different countries is that, by gender, limited countries allow individuals of "the same sex" to get married. These countries that legally allow gay, lesbian, etc. marriages are the main reason of preference for these couples. This tourism type, which can also be named as gay marriage tourism, can positively affect the image of the destination in the international community. The destinations in which gay marriages are allowed can be described as more humane, hospitable and libertarian (Suklabaidya, 2008: 30).

There are actually four general characteristics in the definitions of wedding tourism. The first of these is an international tourism movement that arises from the friends and relatives of the bride and groom traveling to the country with the organization of the wedding outside the country where the bride and groom live. The second one is generally the domestic tourism movements that arise from the travels made by friends and relatives of the couples through the weddings organized in a different city within the borders of the country where the bride and groom live. The third one is the domestic and foreign tourism movements, which are the results of the participation of friends and relatives from different cities in the same country or from a different country in the wedding ceremony held within the borders of the province where the bride and groom live. And the fourth one is, since the bride and the groom are from different cities, the tourism movement that arises from the travel of friends and relatives of either the bride or the groom to the hometown of the bride or the groom where the wedding is organized (Daniels and Loveless, 2007; Schumann and Amado, 2010: 176; Del Chiappa and Fortezza, 2016: 57).

Wedding tourist is defined as the person who travels to get married or to attend the wedding of a relative, friend or an acquaintance (Bertalla et al., 2018: 55). Wedding tourism is considered as a relatively new phenomenon. Academic studies carried out in this area are rather focused on wedding planning and management (Blakely, 2008; McLeay, and Waine, 2010; Schuman and Amado, 2010).

Nowadays wedding tourism is considered as a niche market. Wedding tourism offers a number of opportunities to the wedding tourism destination. The invitees attending a wedding ceremony organized abroad may spend money on different things such as accommodation, car rental, souvenir shopping, transportation, etc. Apart from this, it is mostly observed that tourists traveling for wedding tourism spend much more than the typical tourists because they want to have a unique and unforgettable experience as much as possible (Kim and Agrusa, 2005: 890). Besides, new lines of business and employment areas are formed in wedding tourism destinations. In order to meet this demand, new businesses will be opened and especially local businesses and local people will get more economic gains from this demand.

The culture, history, gastronomy, architectural works, monuments, romantic sceneries in wedding tourism destinations affect the demands of people. But there are a great many destinations in the world that offer such features. At this point, the main characteristics that will distinguish the wedding tourism destination from the others are the differences in accommodation, transportation, entertainment, eating and drinking facilities and the adequacy of infrastructure and superstructure, as well. Apart from this, advertising and promotional activities are among the most important elements that distinguish these centers from others. At the same time, the laws of the wedding tourism destination country should be clear and legally permit foreign couples for such ceremonies within the country (Khodzhaeva, 2016: 7–8).

In recent years, according to the new wedding tourism trend, it has been determined that the couples have been preferring the destinations such as Hawaii, the Caribbean Islands, Mexico, Fiji, Seychelles, Maldives, Mauritius, Jamaica, Bali, New Zealand, in the European Region, especially Italy, Greek Islands and Cyprus, and in the United States, Las Vegas and New York in order to celebrate their wedding ceremonies (Vidauskaite, 2015: 21).

The main driving factors of the growth in wedding tourism; can be listed, with the rise of new markets, as the desire to get extraordinary, unforgettable experiences, the desire to avoid the stress created by traditional weddings, wow renewals of the couples and the same sex marriages (Etemaddar, 2018: 423). There are a number of wedding tourism. Attending a wedding organized abroad may give some messages that include family solidarity and family loyalty, and it is also an indicator of sincerity among friend groups. In this context, weddings are indicators of social sincerity (Bertella, 2015).

Abroad weddings have been becoming more and more attractive for the couples who do not want to spend much more money on traditional weddings. The couples may want to spend the extra money they will pay for traditional weddings on their honeymoon after the weddings. Abroad weddings are organized in two concepts. This dual concept, which includes both wedding and honeymoon, is specified as a single price. Therefore, wedding tourism has become an important part of the tourism industry for some socio-economic reasons. Exotic package weddings are becoming increasingly popular (Moira, Mylonopoulos and Parthenis, 2011; Seebaluck et al., 2014: 202).

In Fig. 1, the components of destination weddings are shown. Within this context, destination weddings are classified as first-time marriages, re-marriages, wow renewals and same sex marriages.

2.1 The Features and Components of Wedding Tourism

- ***Abstractness:*** Wedding services are abstract. Any kind of wedding service cannot be tried, touched, smelled or tasted before purchasing.
- ***They are inseparable:*** The employees and customers at the weddings are in constant contact.

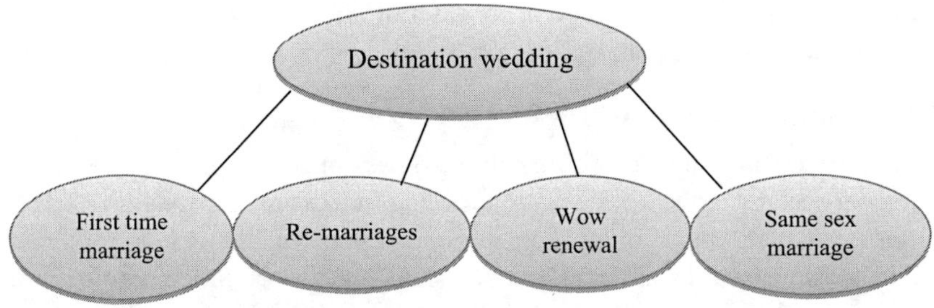

Fig. 1: Destination Wedding Composition.
Source: Major et al., 2010: 252

- *Perishability:* Wedding services cannot be stored to be sold at another time or returned.
- ***Atmosphere and service:*** Ambiance plays a crucial role for the result of a wedding. Good foods, expensive and beautiful decor can help to create the right atmosphere, but they cannot guarantee the success of an event.
- ***Labor intensive:*** A large number of staff is employed in weddings that are organized within the context of wedding tourism.
- ***Fixed time scale:*** Although the duration of the weddings varies according to the wedding type, they are approximately similar to each other and has some common characteristics.
- ***The Sense of belonging:*** One of the main reasons for the people attending the weddings is to interact with families, friends and other guests and gain a sense of belonging by attending a wedding (De Witt, 2006: 66–67; Kanca, 2018: 21).

To be able to coordinate all the components of a wedding, it is necessary to know what these are. Weddings consist of the following components (Wolf, Wolf and Levine, 2005: 192; De Witt, 2006: 69; Kanca, 2018: 22).

- ***Guests:*** The number of guests, guest lists, invitation cards
- ***Transportation:*** The guests flying for the wedding, transportation of the bride and groom for the wedding
- ***Wedding Clothes:*** Wedding dress, groom's suit, bridesmaid clothes, groomsman clothes, rings, jewelry, hairdressing
- ***Ceremony:*** Ceremonial program, venue, permissions, music, confetti, flowers, registrar
- ***Reception:*** Menu selection, the number of service plates, wedding cake, decor, flowers, layout, technical needs, candies, speeches

- ***Entertainment:*** Musicians and other entertainment decisions
- ***Accommodation:*** Bride and groom, people other cities, reservation confirmations

2.2 Wedding Tourism in Turkey

Turkey, under the influence of Turkish TV series in recent years, has been preferred particularly in the Middle East, including Indian and European markets (by Turkish-origin citizens). Turkey is an appropriate wedding destination for tourists thanks to its historical and visual beauties, shopping opportunities, product diversity and affordable prices. The main target market of Turkey in wedding tourism is India. While about 30 million of Indian tourists travel abroad annually only 100 thousand of them prefer to spend their holidays in Turkey. The budget allocated for the wedding by the Indians ranges from 1 million to 20 million dollars. In 2018, 15 major Indian major weddings have been organized in Turkey. Especially when the problems concerning the flights are corrected, about 1 million Indian tourists are expected to visit Turkey (www.timeturk.com). In India, around 10 million weddings are held on average every year. Some of the rich families tend to have their weddings in different countries. Approximately 10 thousand of these weddings are held in various countries around the world (www.izmirdergisi.com).

Antalya is a particularly preferred wedding tourism destination for Indian weddings. Luxury concept hotels in Antalya specifically attract the Indian wedding tourists. What's more; hospitality, friendship and food quality make Antalya rather attractive for Indian tourists. While the wedding season in India is restricted between 4–5 months, Antalya offers adequate infrastructure and superstructure facilities and natural opportunities (climate, etc.) that give the Indian wedding tourist the chance to have weddings for 12 months. The low season for Antalya actually coincides with the period when weddings start to be organized in India. This coincidence also reduces the costs for the Indian wedding tourists. Therefore; Turkey has the potential to get more shares from the Indian wedding market that has a volume of over 40 billion dollars annually (www. baka.gov.tr).

The Indian weddings present valuable contributions to the country and the region since they are held in the periods when the tourism season and the number of tourists start to decrease in Turkey. These weddings, which last for days with the participation of thousands of people, are mainly concentrated in Antalya, Bodrum and Cappadocia. Throughout 2019 about 20 Indian weddings were organized in Turkey and about 50 million dollars were gained from these weddings (www.haber3.com). When deciding a destination in wedding tourism, visa applications could be the reason of preference. The countries that are preferred most by the Indian wedding tourists, who spend the most money in international wedding tourism in the world, are Thailand, Dubai and Singapore and these countries do not require a visa from the Indians. Having taken a place in the Indian market through the promotional activities carried out, Turkey has begun to focus on other markets such as China, Japan, South Korea, Indonesia, and South American countries. Especially Turkish TV series that are quite influential in the promotion of Turkey are directed to the target market. In these series the historical, natural and cultural beauties are brought to the forefront and the interests of the wedding tourists have been tried to be attracted.

There is an increase in the number of the tourists coming to Turkey for wedding tourism. Indian, Azerbaijani or Arab weddings, which could be considered in the upper segment, are the most luxurious organizations. In Turkey, when it's considered that the shopping for a wedding organization start from 50 thousand Turkish Liras, the expected financial burden of weddings, events and similar organizations will be about 300 million dollars. The wedding companies have organized "concept weddings" not only in Istanbul but also in Bodrum and Çeşme. The number of Indian tourists who come to Antalya and Alanya to get married is increasing day by day. The couples who want to get married in Istanbul can have a chance to have their weddings in fascinating places. Each of the historical sites immortalizes this special day for the couples (www.barsantourism.com).

Wedding tourism sector has been growing in Turkey where about 650 thousand couples get married each year. The wedding organizations that feed about 20 different business branches, from bakery to cooking, have actually become an industry. Istanbul is the city in the world where most wedding organizations are held after Las Vegas. At the same time, each year, more than 350 international weddings are organized in Turkey.

Turkey is a wedding destination preferred by the Iranian Diaspora, as well. Turkey generally becomes a common meeting destination between the Iranian families living in the West and the Iranians living in Iran. Another aspect is that Turkey doesn't require visa from the Iranians. So, this situation encourages the Iranian wedding to be held in Turkey.

The SWOT Analysis concerning Turkey's Wedding Tourism has been thoroughly examined in Fig. 2.

2.3 Wedding Tourism Politics and Planning in Turkey

Within the scope of 3rd Tourism Council Meeting organized by the Ministry of Culture and Tourism, with the aim of diversifying the tourism types and

S

- Adequate infrastructure and superstructure possibilities
- Fascinating natural beauties
- Rich historical and cultural heritage
- Service quality
- Being a relatively cheap country
- Favorable climatic conditions and the chance of having four seasons
- Being experienced in terms of wedding services
- Turkish hospitality

W

- The deficiencies in promotional activities
- The fact that the strong competitors' in the market starting the activities related to wedding tourism much earlier than Turkey
- No direct flights of Turkish Airlines to major wedding tourism destinations such as Antalya, Muğla, Cappadocia
- Visa barriers

O

- Istanbul is the second city where the most weddings are organized in the world. The experience of Turkish wedding organizers
- The fact that the months of March and November, which are considered to be the low season for Antalya, are the most preferred periods for Indian wedding tourists.
- Facilities with relatively modern, new and technological equipment
- Turkey's being a new and unspoilt geography in terms of wedding tourism
- Popular Turkish TV series

T

- Being geographically away from major target markets such as India and China
- Lack of legal legislation in terms of the same sex marriages, which is a niche market
- The preference of exotic countries such as Maldives and Sri Lanka
- Intense rivalry
- The instability and increased uncertainty in the region where Turkey is located

Fig. 2: The SWOT Analysis Concerning Turkey's Wedding Tourism.

spreading the tourism activities throughout the year, wedding tourism was considered one of the alternative tourism types that needed to be envisaged and supported in Turkey. In this context, the following decisions and suggestions were taken under the relevant headings (3th Tourism Council, 2017: 9–10).

2.4 Suggestions Related to Promotional and Marketing Activities

- Hosting the leading wedding organizers in Turkey and making on-site presentation,
- The creation of a comprehensive promotional program in terms of wedding tourism in Turkey and putting greater emphasis on domestic and overseas promotional activities,

- Along with India, the evaluation of Pakistan, Iran, Lebanon, Russia and CIS countries as target markets of wedding tourism,
- The participation in SATTE and IITTE Tourism Fairs held in Indian cities such as Mumbai and New Delhi and the organizations of promotional events in different cities with the participation of related wedding tourism agencies and hotels in Turkey,
- The appointment of a personnel, who has a full understanding of Antalya and wedding tourism, as the Culture and Promotion Counselor in India,
- The consolidation of Turkey's secure destination image through professional promotional events and PR activities; specifically, to emphasize the promotion of destinations rather than the promotion of the country,
- Emphasizing service, quality and price stability, ultra-luxurious resorts suitable for wedding tourism, the common cultural elements with India and other wedding tourism countries, eating habits and the other competitive advantageous points of our country and particularly Antalya, being in the first place.

Suggestion on Facilities and Standards

- Making agreements with catering companies from India
- Within the scope of wedding requests for palaces and similar areas providing reasonable prices to be requested by wedding venues

Legislative Suggestions

- In visa applications not asking for the documents (Approved Company Certificate, Company Assignment Certificate, Police Clearance Certificate, etc.) that are difficult to get or providing flexibility for such documents, implementing a flexible application in terms of visas of those who will attend the wedding, facilitating visa procedures for wedding participants with the official document attachments written by the Governorships or Development Agency for visa applications,

3 Conclusion

In Turkey tourism has been developing depending on mainly sea, sand, sun tourism in the summer months. The Turkish administrators and representatives of the tourism sector who want to change this perception have focused on alternative types of tourism for a long time. Within this context, wedding tourism is considered as one of the tourism types that should be supported. There is a general perception of Turkey as a cheap country concerning tourism. Tourists visiting Turkey spend 650 dollars per person on average. While Turkey is among the first ten countries that host most tourists in the world, it's not among the first ten in terms of tourism revenues.

The administrators who are aware of this contradiction have started to emphasize that not the number of tourists hosted but the revenue gained from tourism is much more important. In this sense, new plans are being made in order to attract the tourists, who spend more money, to the country. In Turkey, the average expenditure of tourists attending the alternative tourism types such as golf tourism, congress tourism, health tourism is much more than that of the typical tourists. Therefore; attracting the wedding tourists, who like luxury and spending, to Turkey will create a new alternative with regards to tourism revenues.

The couples in the target market countries of Turkey such as India, Iran, Qatar, Dubai, Azerbaijan, Saudi Arabia, Lebanon and Russia prefer luxurious wedding ceremonies. At this point, Turkey has been using the Turkish TV series as a promotional tool with the aim of attracting particularly the Indian market. Indian weddings are generally comprehensive and lucrative ceremonies. Approximately, ten thousand of ten million weddings held in India are held abroad each year. The cost of each wedding ranges from 1 to 20 million dollars and annually there is a wedding market with a total of 40 billion dollars. This situation has dragged many countries into a promotional rush and wedding tourism rivalry. In this respect, this market has become highly attractive for Turkey who wants to get more profit from tourism. Therefore, Turkey has been planning a number of organizations to take place in the Indian market.

The presentation of Turkey's wedding tourism has been planned with "Bollywood" film stars. At the same time, with the support of some Indian films, which are shown in India, to be shot in Turkey the promotion of Turkey has been aimed, as well. One of the most fundamental determinants in the preferences of the Indians has been luxury, vanity and splendor. Here, the presence of quality products and the variety of services provided by the facility become prominent. Another important expectation by Indian wedding tourist is the accessibility and ease of transportation of the facility where the wedding is planned to be held.

Turkey has the required potential to host wedding tourists with its historic and natural wealth, a wide range of product variety, hotel capacity and quality of service. Nevertheless, in order to increase the share gained from Wedding Market, it is essential to pay attention to the advertisement and promotional activities. Another criterion that the participants pay attention to in potential hotels that can host wedding organizations has been the room and bed capacity of the hotels. Since international weddings are mostly organized with crowded guest groups, the room and bed capacities of the regional hotels may remain low. That's why; it will be good idea for the hotels located physically close to each other to open themselves to partner organizations within the framework of cooperation, in order to increase the potential of the region.

References

Bertella,G. (2015). Celebrating the family abroad: the wedding tourism experience, *Annals of Leisure Research, 18*(3), 397–413.

Bertella, G., Cavicchi, A. and Bentini, T. (2018). The reciprocal aspect of the experience value: tourists and residents celebrating weddings in the rural village of Petritoli (Italy). *Anatolia, 29*(1), 52–62.

Blakely, K. (2008). Busy brides and the business of family life: The wedding-planning industry and the commodity frontier. *Journal of Family Issues, 29*, 639–662.

Daniels, M. and Loveless, C. (2007). Wedding Planning and Management. Elsevier, Oxford, UK.

De Witt, L. (2006). Key success factors for managing special events: The case of wedding tourism. Unpublished doctoral dissertation, North-West University, Potchefstroom.

Del Chiappa, G. and Fortezza, F. (2016). Motivations for Wedding Tourism: A Demand-side Perspective. Metin Kozak and Nazmi Kozak (eds.) Tourist Behaviour: An International Perspective (pp. 56–63). CAB, USA.

Etemaddar, M., Thyne, T. and Insch, A., (2018). A taste of home – choosing a destination wedding, *Anatolia, 29*(3), 422–432.

Kanca, B. (2018). Düğün Turizmi Kapsamında Antalya ve İstanbul Destinasyonlarının Değerlendirilmesi. Unpublished doctoral dissertation, Mersin Üniversitesi Sosyal Bilimleri Enstitüsü, Mersin

Khodzhaeva, V. (2016). Internshıp Report: Weddıng Tourısm In Portugal. Unpublished Master's thesis. Estoril Higher Institute for Tourism and Hotel Studies. Portugal.

Kim, S., and Agrusa, J. (2005). The positioning of overseas honeymoon destinations. *Annals of Tourism Research, 32*(4), 887–904.

Major, B., McLeay, F. ve Waine, D. (2010). Perfect Weddings Abroad. *Journal of Vacation Marketing,* 16(3), 249–262.

Moira, P., Mylonopoulos, D., and Parthenis, S. (2011). A sociological approach to wedding travel. A Case Study: Honeymooners in Ioannina, Greece. In Hummel, F., ed. Annual Conference Proceedings of Research and Academic Papers. Miami, Florida, United States.

Schumann, F. R. and Amado, C. (2010). Japanese Overseas Weddings in Guam: A Case Study of Guam's First Hotel Wedding Chapel. *South Asian Journal of Tourism and Heritage, 3*(1), 173–181.

Seebaluck, N.V., Munhurrun, P.R., Naidoo, P. and Rughoonauth, P. (2015). An analysis of the push and pull motives for choosing Mauritius as 'the' wedding destination, *Procedia Social and Behavioral Sciences, 175*, 201–209.

Suklabaidya, P. (2008). Wedding Tourism and India, *Atna Journal of Tourism Studies, 3*(2), 25–35.

Vidauskaite, R. (2015). Destination Branding Through Wedding Tourism; The Case of the Caribbean, Master Thesis, University of Ljubljana Slovenia.

Wolf, P., Wolf, J. and Levine, D. (2005). Event planning made easy. New York: McGraw-Hill. Liite.

3th Tourism Council(2017). https://turizmsurasi.ktb.gov.tr/Eklenti/57385,turizmdeuruncesitliligisurdurulebilirlikkomisyonraporup-.pdf?0 (Accessed: 15.06.2020).

https://www.timeturk.com/hindastan-dan-turkiye-ye-dugun-turizmi/haber-966528 (Accessed: 10.06.2020).

https://www.izmirdergisi.com/tr/dergi-arsivi/338-47-nci-sayi/2781-yaz-geldi-dugunun-sezonu-da-acildi-hint-dugunlerine-kilitlendik (Accessed: 07.06.2020).

https://baka.gov.tr/haber/hint-dugun-turizmi-hedefinde-basarili-antalya-tanitimi (Accessed: 07.06.2020).

https://www.haber3.com/guncel/turkiye-simdi-de-dugun-turizmi-donemi-50-milyon-dolarlik-dugun-haberi-5036412 (Accessed: 04.06.2020).

https://www.barsantourism.com/blog/46/wedding-tourism (Accessed: 10.06.2020).

https://www.oxfordlearnersdictionaries.com/definition/english/wedding?q=wedding (Accessed: 17.06.2020).

Emin ARSLAN & Hakan KENDİR

9 Determination of Trekking Routes Inventory within the Scope of Alternative Tourism in Tokat

1 Introduction

With the occurrence of postmodernism, the changes on the wishes and expectations of individuals began to be felt in the tourism sector as well. As a result of this change, it is known that the intense interest in mass tourism has turned toward alternative tourism types. It is seen that trekking, which is respectful to nature, supports sustainability and can be done individually and it is one of the alternative tourism types, examined in international (Nepal, 2008; Nyaupane et al., 2014; Upadhayaya, 2018; Apollo & Rettinger, 2019) and national (Tekin, 2017; Selim, 2019; Kendir et al., 2019) literature.

In Turkey, on the purpose of the development, diversifying and accordingly increasing the income derived from tourism in recent years, alternative tourism investments have gained momentum. In accordance with this, new touristic supply sources started to emerge with the investments made in the province of Tokat. Touristic investments such as Ballica Cave, Kanal (Canal) Tokat, Zinav Lake and Canyon, Kaz Lake, Gumenek National Garden make Tokat a center of attraction. However, it is thought that the touristic attraction of Tokat will increase by introducing different recreational and sportive activities for alternative tourism within the scope of diversification of tourism.

Spread on a mountainous and forested land, Tokat province is suitable for nature walks with its natural beauties and flora-fauna. The basis of this research is the evaluation of these characteristics of the province within the scope of alternative tourism and the determination of nature trekking routes that may be a different activity for tourists in terms of recreation.

2 Alternative Tourism and Trekking

Individuals participating in tourism activities around the world act massively and prefer the sea-sand-sun trilogy, which is mostly described as mass tourism. However, with the changes in the preferences of the tourists and the emerging trends, it is observed that there is a trend from mass tourism toward alternative tourism products in tourism movements (Gursoy et al., 2010). In order to spread tourism to 12 months and decrease migration from the country to town, alternative tourism types that have occurred against mass tourism are needed. Among the species included in the scope of alternative tourism, there are natural and rural tourism activities such as; Adventure tourism (trekking, mountaineering, rafting, cave), orienteering (target production), windsurfing, hunting, touristic air sports (paragliding, glider, balloon), diving, nature watching (botanical), highland tourism, bird watching (Kucukaslan, 2007). Especially after the COVID-19 pandemic that shook the world deeply, it is thought that touristic activities, such as trekking which can be done individually and evaluated within the concept of healthy tourism and where social distance is at the forefront will gain importance (Seraphin & Dosquet, 2020).

Walking, which is one of the most basic sports activities, is classified separately in the literature as hiking, trekking and expedition. However, they can often be confused in terms of meaning. The explanation of the terms of the walking types mentioned in the literature is given below (Turkish Mountaineering Federation, 2017).

Hiking: It means walking in nature for pleasure and sports. It is a term used for daily walking activity which includes natural beauties such as forest, path and valley. Hiking routes can be easy depending on the geographical structure or have different degrees of difficulty. The walks can vary between 2 and 8 hours.

Trekking: It is used for the walks at long and hilly lands and high altitudes made to walk on more than one day. Trekking is a more difficult activity than hiking because it is required to made with food and beverage materials and camping equipment.

Expedition: It can be defined as a nature walk for research and examination in large groups, including multiple summits climbing over a long period of time (IDADIK, 2010).

The tracks of the walking types described above are organized are divided into various difficulty levels. These difficulty levels are examined in six different categories. The difficulty levels of the walking tracks are given in Tab. 1 (Sahin, 2010).

Global warming and climate change, which have been a frequent topic in the world agenda in recent years, generally is an issue to be considered in terms of tourism. For this reason, it is necessary to develop and encourage species that are at peace with nature and sustainable. Therefore, it is useful to consider nature walking sport, which does not destroy nature and does not harm ecological balance, as an alternative tourism activity. This activity can have a positive effect in terms of regional development and prevention of migration from the country to town since it is an activity especially in rural areas (Kendir et al., 2019).

Tab. 1: Difficulty Levels of Walking Tracks

Difficulty Level	Explanation
1	The slopes and ups and downs on the track are exceptionally low. Ups are less than 100 m and the trails are wide. Walking time is less than 2 hours. Beginners and those without chronic health problems can participate in walking sports.
2	The slope of the track is low and there are light ups not exceeding 300 m. The most important difference according to the first level is the duration of the walk. Total walking time can take up to 3.5 hours. Anyone who has been hiking before can join.
3	Track slope is higher than the first two levels. Ups on the track do not exceed 500 m and it may be necessary to pass through narrow pathways, dense forest areas and stonier areas. Wet floor transitions are more. Walking time does not exceed 5 hours. Anyone with condition and experience can join the walk.
4	This track is more inclined than the third level. Ups on the track reach 700 m, the trails are distorted and there are almost no trails in the forest areas. More roads are traveled on rocky and stony land. Walking time does not exceed 6.5 hours. Sporty and conditioned people can join the walk.
5	The track has a higher slope than the fourth level. Ups on the track can exceed 1000 m. There are hard, stony, rocky and path free areas. Forest areas are particularly challenging. There are also wet and slippery ground transitions. It is imperative that the walking guide should be very experienced. The walking time is around 8 hours. Sporty, experienced, and conditioned individuals can participate in the march. The tracks are challenging, although they do not contain technical climbs.
6	The tracks of this level are very inclined, with remarkably high ups and downs, with an altitude difference of 1500 m. Tracks that require long-term route tracking are carried out in difficult terrain conditions. Camping stays can be made if necessary. Walking time can exceed 8 hours. Walks require a lot of experience, attention, condition, knowledge, and discipline.

Source: Sahin, 2010

3 The Geographical Structure of Tokat Province

Geographically Tokat is located in the Central Black Sea region of Turkey. Samsun and Ordu are located in the north of Tokat, Yozgat and Sivas in the south and east, and Amasya in the west. With an area of 10.072 km2 in terms of surface area, Tokat has 11 districts. Total population of the province is 612,747 as of year-end in 2019 (TUIK, 2020). The altitude of Tokat center is 623 meters on average. However, altitude varies between 200 and 2200 m across the city (Governorate of Tokat, 2020).

39 % of the lands owned by Tokat are covered with forests, 38 % are covered with cultivated lands and 12 % are covered with meadows and pastures. Tokat is one of the few forest areas of Turkey. Almost all of the mountains at the north and south are woodland. Due to its geography, it is possible to see a great variety of plants and trees.

In the low altitude Niksar and Erbaa regions of Tokat, there are maquis species such as olives, carob, and sandalwood. In this region, Mediterranean climate trees such as red pine and Lebanese cedar can be seen as well. Trees such as oak, scotch pine, larch, fir, and hornbeam are found in Tokat center, Almus, Resadiye, Turhal and Zile districts. Steppe and degraded forests can be seen in Artova, Yesilyurt and Sulusaray districts, which are at a higher altitude and close to Central Anatolia. In addition to this, endemic species such as hazelnut, cranberry, rosehip, blackberry, raspberry grown wildly in the valley sizes are grown throughout the province. On the other hand, Madimak is one of the most known endemic weed species of Tokat and it is frequently found in high altitude areas. (Governorate of Tokat, 2020).

There is an opportunity to see and examine the plants and trees listed above for individuals who are trekking on different routes throughout the city. Therefore, the city of Tokat has the potential to attract the attention of people who are considering joining nature with its rich flora and fauna.

4 Method

This study has been carried out to determine the inventory of trekking routes that can be evaluated within the scope of alternative tourism activities in Tokat. The research conducted in accordance with the study is a qualitative analysis. In the research, primary and secondary data were used. In order to determine the trekking routes of Tokat province, as the primary data, an interview was made with the managers and members (club chairman, guides and professional athletes) of the TODEX (Tokat Nature and Extreme Sports Club) using a "semi-structured interview" method.

The reason for using the semi-structured interview technique in the research is that it gives the researcher the opportunity to obtain more systematic and comparable information. Besides, this technique provides the opportunity to ask additional questions or to make changes to already existing questions according to the course of the interview (Yıldırım and Şimşek, 2005).

In the meantime, the required field work was carried out to determine the routes with the relevant members of TODEX. The secondary data were obtained by scanning academic literature concerning alternative tourism, trekking, recreation and sports.

In the light of the research findings, the trekking routes, having different difficulty levels determined in Tokat province, were detected with the help of GPS and transferred to the "wikiloc" program and maps were prepared using this program. Within the research, the map and information of the 8 most important routes recommended by TODEX have been presented. On the other hand, the routes on Map 9 and Map 10, which were identified by the researchers themselves, were also presented visually in the study since they have great importance. In addition to these, the other routes identified by TODEX and expressed as important are also given in tables.

Map 1: Ballica Village Trekking Route.
Source: Wikiloc/Ballıca Köyü

The walking route on Map 1 starts from Derecayli Village of Pazar district of Tokat and ends in Ballica Village famous for its cave. The altitude of 722 meters at the beginning of the approximately 5.74 km track reaches 1083 meters at the end of the route. The difficulty level of the track is 4.

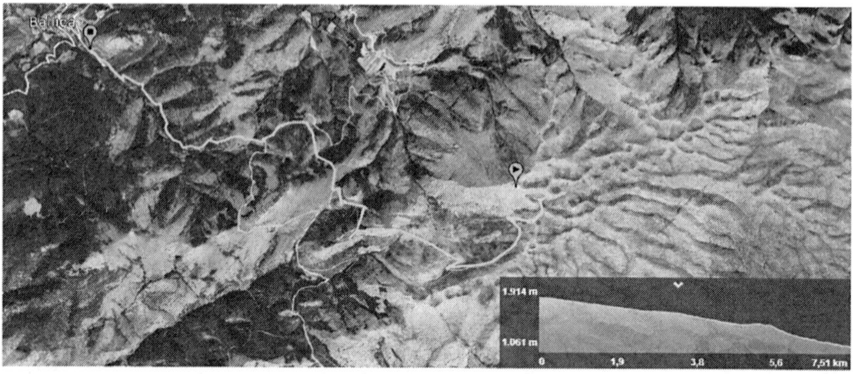

Map 2: Akdag – Ballica Cave Trekking Route.
Source: Wikiloc/Akdağ

The route on Map 2 starts from the summit of Akdag, one of the highest mountains of Tokat, and ends in Ballica Cave. The course, which has a route of approximately 7.5 km, descends from an altitude of 1914 meters to an altitude of 1061 meters. The difficulty level of the track is 2.

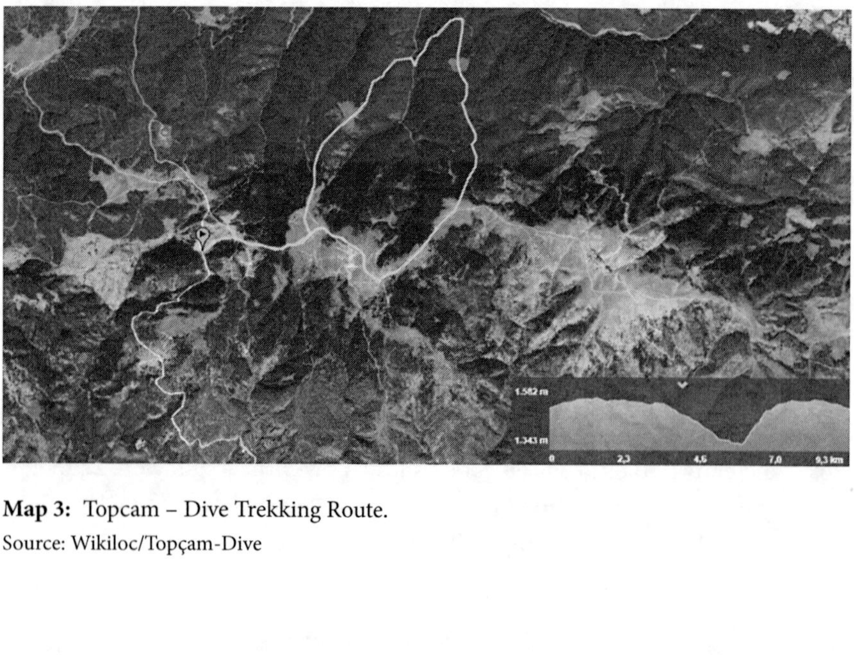

Map 3: Topcam – Dive Trekking Route.
Source: Wikiloc/Topçam-Dive

The walking route in Map 3 is located in Gunevi (Dive) Village and Topcam Plateau location in the north of Camdere (Ilbis) Village of Tokat center. The course continues for approximately 9.3 km on a route ranging from 1343 to 1582 meters' altitude. The difficulty level of the track is 4.

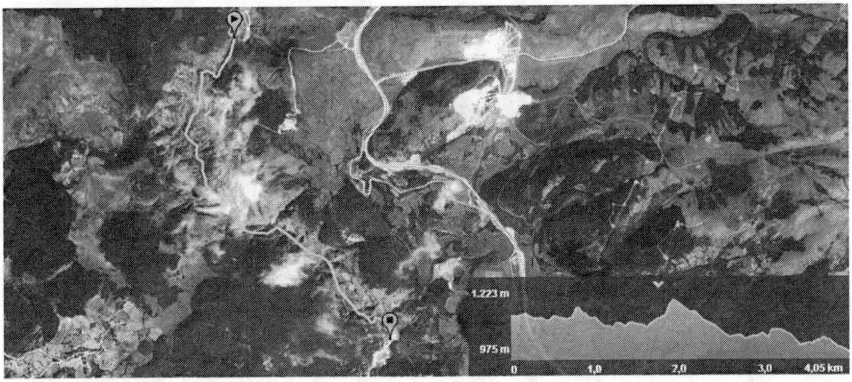

Map 4: Zinav Canyon Trekking Route.
Source: Wikiloc/Zinav Kanyonu

The route on Map 4 has a length of 4 kilometers along the Zinav Canyon in the Resadiye district of Tokat. The track, which initially has an altitude of 1148 meters, goes up to 1223 meters and then ends at an altitude of 975 meters, offering a bumpy feature. The difficulty level of the walking route, which has a very tiring route, is 5.

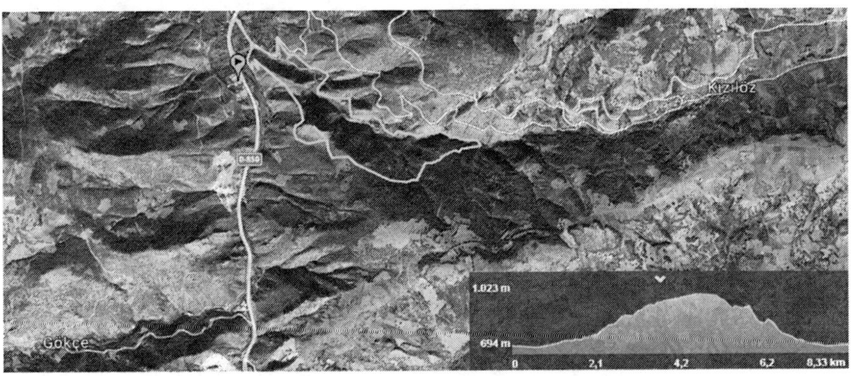

Map 5: Geyras Castle Trekking Route.
Source: Wikiloc/Geyras Kalesi

The route on Map 5 starts from the Geyras district of Tokat center and continues toward Kızıloz Village. The walking route, which also includes Geyras Castle, returns from the west of Kızıloz Village, and goes back to Geyras. The altitude, which is 694 meters at the beginning of the course, goes up to 1023 meters in the middle of the course. The course, which has an 8.3 km route, has a difficulty level of 3.

Map 6: Almus Tufantepe Trekking Route.
Source: Wikiloc/Tufantepe

The walking route on Map 6 starts from the Tufantepe location near the Almus district of Tokat. The route continues toward the surrounding hills and returns to Tufantepe by drawing an arc. The length of the track is about 8.5 km. The altitude, which is 812 meters at the beginning of the course, goes up to 1054 meters. The track has a rough feature and the difficulty level is 3.

Map 7: Almus Catak Plateau – Karadere Trekking Route.
Source: Wikiloc/Almus Çatak Yaylası

The walking route on Map 7 starts from Catak Plateau in Almus district and ends in Karadere Village. The altitude, which is 1486 meters at the start of the track, goes up to 1820 meters in the middle of the track and at the end of the track decreases to 922 meters. The route, which has a quite bumpy route, is 9.8 km long. The difficulty level of the track is 5.

Map 8: Çamiçi-Niksar Trekking Route.
Source: Wikiloc/Niksar Çamiçi Yaylası

The route on Map 8 starts from the camping and caravan area of Çamiçi Plateau in Niksar district and ends in the city center of Niksar. The total length of the track is 6.9 km. The altitude, which is 1151 meters at the start, drops to 397 meters at the end of the track. The difficulty level of the track, which continues completely in the form of downhill, is 2.

Map 9: Guzelbeyli Plateau Trekking Route.
Source: Wikiloc/Zile Güzelbeyli Köyü

The trekking route on Map 9 covers the village and plateau of Güzelbeyli in Zile district. The altitude, which is 995 meters at the beginning of the course, rises to 1371 meters in the middle of the course and drops to around 1000 meters at the end of the course. The course, which has a length of about 10.5 km, has a difficulty level of 3.

Map 10: Beşören Village – İncesu Plateau Trekking Route.
Source: Wikiloc/Beşören Köyü

The route presented on Map 10 includes Beşören Village and İncesu Plateau, which are located in the center of Tokat. The altitude, which is 1539 meters at the beginning of the course, goes up to 1749 meters and then falls continuously to 1166 meters. The course, which has a length of about 8.5 km, has a difficulty level of 4.

Tab. 2: Other Trekking Routes in Tokat Province and Their Difficulty Levels

Name of Routes	Length (km)	Altitude (meter)	Difficulty Level
Geyras – Kemer District Route	10,1	640–742	1
Almus Trekking Route	5	814–981	2
Kocacık – Aydınca – Yayladalı Route	10,5	653–896	1
Aluç Village Trekking Route	5,1	901–1360	3
Kızıliniş Trekking Route	9,8	702–1264	4
Akarçay Route	13,5	875–1598	4
Çamlıbel – Bedirkale Route	7,9	1170–1358	3
Topçam-3 Route	10,5	1342–1583	3
Pazar Route	4,4	676–1035	3
Pınarlı – Bakır Madeni Route	8,6	714–959	3
Bakımlı Village – Almus Route	11	810–1715	4
Pazar – Dereçay Trekking Route	6,1	635–1063	3
Niksar Gerit Village Route	10,1	1032–1292	3
Pazar Değirmen – Waterfall Route	12	620–1057	3
Gümenek – Ballıdere – Aluç Route	18,2	598–1152	5
Akdağ – Tahtoba Route	12,9	1078–1900	5
Topçam Short Route	5,5	1319–1551	4
Topçam Long Route	15,6	1101–1564	5
Hasanbaba – Gıjgıj Hill Route	14,7	591–963	4
Geyras Hill Trekking Route	11,6	697–1251	5

Source: TODEX, 2020

In Tab. 2 above, other trekking routes in Tokat Province can be obseved. All the routes in this table were determined by TODEX (Tokat Nature and Extrem Sports Club) through experience. In the table, the lengths, altitudes, and difficulty levels of these routes can also be seen. The routes whose visuals are given through maps are the most experienced routes. What's more: the routes in Table 2 are also important in terms of creating an inventory.

5 Conclusion and Suggestions

Turkey, thanks to its historical and natural beauties, provides a wide selection of opportunities in terms of tourism activities that can be done in the nature. In this sense, Tokat province, which is one of Turkey's naturally unspoilt regions, also has a lot of resources. Through this study, it has been aimed to determine the trekking routes within Tokat province in the scope of alternative tourism via inventory study.

For the purpose of the study, data were obtained by the semi-structured interview method and field study conducted by the researchers together with TODEX. The highest point in these data is that the trekking routes are generally concentrated in forests and rural areas far from the settlements. Especially during the COVID-19 pandemic, people who act more individually and suitable to social distance have been preferring activities such as trekking, camping and cycling in domestic tourism (Higgins-Desbiolles, 2020; Vos, 2020). This situation creates great opportunities for Tokat province. Therefore, the decision makers in the province have important responsibilities in terms of evaluating new opportunities. From this point of view, it is necessary to carry out informative studies related to trekking routes, to provide security and create direction signs on the routes.

The fact that the rural and forest areas' being very important within the scope of alternative tourism and recreational activities such as trekking and camping is a frequently stated subject in academic studies (Bramwell, 1994; Lane, 1994; Artuğer and Kendir, 2013). The province of Tokat, which is an important forest region of Turkey, embodies the most appropriate potential for trekking with its flora and fauna. What's more, the reality that there are many untouched rural areas makes Tokat one step ahead of its rivals. However, Tokat has a general lack of publicity. Effective studies should be carried out on promotional activities. In addition to these, the completion of ongoing transportation investments such as airports and double highways in Tokat is an important issue in terms of the access of tourists to the region.

As a result, this research is limited to Tokat province. In other provinces and regions some similar studies can be done. That the research covers only trekking routes is another limitation. The academics who will carry out researches on this subject can focus on resources such as winter trekking and history-culture routes. On the other hand, initially the decision makers in the province and all of the stakeholders related to tourism, nature and sports, can benefit from the findings and results of this research.

References

Apollo, M. & Rettinger, R. (2019). Mountaineering in Cuba: Improvement of True Accessibility as an Opportunity for Regional Development of Communities Outside the Tourism Enclaves. *Current Issues in Tourism*, 22(15): 1797–1804, DOI: 10.1080/13683500.2018.1446920.

Artuğer, S. & Kendir, H. (2013). Agritourist Motivations: The Case of Turkey. *International Journal of Business and Management*. 8(21): 63–70. DOI: 10.5539/ijbm.v8n21p63.

Bramwell, B. (1994). Rural Tourism and Sustainable Rural Tourism. *Journal of Sustainable Tourism,* 2(1–2): 1–6. DOI:10.1080/09669589409510679.

Gürsoy, D., Chi, C. G. & Dyer, P. (2010). Locals' Attitudes toward Mass and Alternative Tourism: The case of Sunshine Coast, Australia. *Journal of Travel Research*, 49(3), 381–394.

Higgins-Desbiolles, F. (2020). Socialising Tourism for Social and Ecological Justice after COVID-19. *Tourism Geographies,* 1–14. DOI: 10.1080/14616688.2020.1757748.

IDADIK (Izmir Mountaineering and Nature Sports Specialization Club) Regulation, (2010), URL: http://idadik.org/Source/idadik_yonetmelik_2012.pdf, (Access Date: 22.07.2020).

Kendir, H., Arslan, E. & Asan, H. (2019). Determination of Winter Trekking Routes Within the Scope of Winter Tourism Potential in Sivas Province. *Journal of Recreation and Tourism Research,* 6(3): 294–305. DOI: 10.31771/jrtr.2019.32.

Küçükaslan, N. (2007). *Özel İlgi Turizmi*. Bursa: Ekin Basım Yayın Dağıtım.

Lane, B. (1994). What is rural tourism? *Journal of Sustainable Tourism*, 2(1–2): 7–21. DOI: 10.1080/09669589409510680.

Nepal, S. K. (2008). Tourism-Induced Rural Energy Consumption in the Annapurna Region of Nepal. *Tourism Management*, 29: 89–100, DOI: 10.1016/j.tourman.2007.03.024.

Nyaupane, G. P., Lew, A. A. & Tatsugawa, K. (2014). Perceptions of Trekking Tourism and Social and Environmental Change in Nepal's Himalayas. *Tourism Geographies*, 16(3): 415–437, DOI: 10.1080/14616688.2014.942233.

Selim, S. (2019). En Uygun Güzergâh Algoritması ile Doğa Yürüyüşü Rotalarının Modellenmesi. *Süleyman Demirel Üniversitesi Fen Bilimleri Enstitüsü Dergisi*, 23(2): 617–624, DOI: http://doi.org/10.19113/sdufenbed.567296.

Seraphin, H. & Dosquet, F. (2020). Mountain Tourism and Second Home Tourism as Post COVID-19 Lockdown Placebo? *Worldwide Hospitality and Tourism Themes*, DOI: 10.1108/WHATT-05-2020-0027.

Şahin, K. Z. (2010). *Doğada Yapılan Yürüyüşler Trekking & Hiking*. İstanbul: Dijital Sanat Yayınları.

Tekin, Ö. (2017). Ekoturizm Açısından Konya İli Doğa Yürüyüşü Rotaları Üzerine Bir Araştırma. *MANAS Sosyal Araştırmalar Dergisi*, 6(3): 377–400.

TODEX. (2020). *Tokat Nature and Extrem Sports Club*, URL: https://tr.wikiloc.com/wikiloc/user.do?id=2010255&from=20&to=30, (Access Date: 21.07.2020).

Tokat Governorship. (2020). *Tokat'ta Toprak, Tarım, Su, Coğrafya, Turizm ve Dahası. . .* http://www.tokat.gov.tr/tokatta-tarim-toprak-ve-turizm, (Access Date: 22.07.2020).

Turkish Mountaineering Federation, (2017). *What are the differences of hiking, trekking and mountaineering?* URL: https://drive.google.com/file/d/18ownbW7Jdqdb6lAEs_ovxGTVXyMxbIl-/view, (Access Date: 22.07.2020).

TÜİK. (2020). *Adrese Dayalı Nüfus Kayıt Sistemi.* https://biruni.tuik.gov.tr/medas/?kn=95&locale=tr Access Date: 22.07.2020.

Upadhayaya, P. K. (2018). Sustainable Management of Trekking Trails for the Adventure Tourism in Mountains: A Study of Nepal's Great Himalaya Trails. *Journal of Tourism & Adventure*, 1(1): 1–31.

Vos, D. J. (2020). The Effect of COVID-19 and Subsequent Social Distancing on Travel Behavior. *Transportation Research Interdisciplinary Perspectives*, 5: 1–3. DOI: 10.1016/j.trip.2020.100121.

Wikiloc/Akdağ, (2020). *Akdag – Ballica Cave Trekking Route,* URL: https://tr.wikiloc.com/gezi-yuruyus-rotalari/agdag-ballica-magarasi-doga-yuruyusu-kisa-rota-20374495, (Access Date: 21.07.2020).

Wikiloc/Almus Çatak Yaylası, (2020). *Almus Catak Plateau – Karadere Trekking Route,* URL: https://tr.wikiloc.com/gezi-yuruyus-rotalari/catak-karadere-29225770, (Access Date: 21.07.2020).

Wikiloc/Ballıca Köyü, (2020). *Ballica Village Trekking Route,* URL: https://tr.wikiloc.com/gezi-yuruyus-rotalari/ballica-koyu-doga-yuruyusu-28514957, (Access Date: 21.07.2020).

Wikiloc/Beşören Köyü, (2020). *Beşören Village – İncesu Plateau Trekking Route,* URL: https://tr.wikiloc.com/gezi-yuruyus-rotalari/besoren-incesu-yaylasi-parkuru-53197330, (Access Date: 22.07.2020).

Wikiloc/Geyras Kalesi, (2020). *Geyras Castle Trekking Route,* URL: https://tr.wikiloc.com/gezi-yuruyus-rotalari/geyras-kalesi-doga-yuruyusu-3uncu-rota-17163763, (Access Date: 21.07.2020).

Wikiloc/Niksar Çamiçi Yaylası, (2020). *Çamiçi - Niksar Trekking Route,* URL: https://tr.wikiloc.com/gezi-yuruyus-rotalari/camici-niksar-15641342, (Access Date: 21.07.2020).

Wikiloc/Topçam-Dive, (2020). *Topcam - Dive Trekking Route,* URL: https://tr.wikiloc.com/gezi-yuruyus-rotalari/topcam-dive-29429150, (Access Date: 21.07.2020).

Wikiloc/Tufantepe, (2020). *Almus Tufantepe Trekking Route,* URL: https://tr.wikiloc.com/gezi-yuruyus-rotalari/almus-tufantepe-1-20952293, (Access Date: 21.07.2020).

Wikiloc/Zile Güzelbeyli Köyü, (2020). *Guzelbeyli Plateau Trekking Route,* URL: https://tr.wikiloc.com/gezi-yuruyus-rotalari/guzelbeyli-yaylasi-parkuru-53196225, (Access Date: 22.07.2020).

Wikiloc/Zinav Kanyonu, (2020). *Zinav Canyon Trekking Route,* URL: https://tr.wikiloc.com/kanyonculuk-rotalari/zinav-2-18759276, (Access Date: 21.07.2020).

Yıldırım, A. & Şimşek, H. (2005). *Sosyal bilimlerde nitel araştırma yöntemleri.* Ankara: Seçkin Yayıncılık.

Olca SEZEN DOĞANCILI & Ercan KARAÇAR

10 Pregnant Tourism

1 Introduction

Although pregnancy is named as a "physiological hypercalciuric status" (Akyüz and Bayındır, 2013), it is defined by Turkish Language Association as "having a baby in the womb (woman or animal), being heavy with a child, pregnant, with two lives and monthly"

Pregnancy affects not only the woman herself, but her family, her immediate environment, even the whole society and the population of a country. Because of this feature, it is frequently mentioned within family planning and it is counted in the category of disabled people along with old age and disability (Aslan & Güneş, 2014, p. 13). Having entered into a sociologically and psychologically complex process due to different biological changes during pregnancy, birth and puerperium, the woman goes through a number of processes such as adaptation to the social environment and getting ready for the role of motherhood (Mermer, Bilge, Yücel and Çeber, 2010, p. 72).

Childbirth, which is generally referred to as a happy event (Işık, Akçınar, and Kadıoğlu, 2010, p. 63), brings a responsibility and panic feeling for pregnant women and they have a lot of needs during prenatal, delivery and postpartum periods (Tor Kadıoğlu and Şahin, 2017, p. 983). It's known that physical activity in pregnancy has a positive effect on both maternal health and on the health of the infant (Özdemir et al., 2017, p. 100). It has many benefits such as maintaining the ideal body weight, providing blood pressure control, relaxing the stretched muscles, relieving back aches and headaches, performing breathing exercises, providing short-term and easy childbirth and reducing the cramps of varicose legs (Kutlu et al., 1996, p. 293). With the aim of gaining these advantages, it is recommended to carry out exercises that will benefit both the mother and the baby in the mid-tempo, rather than aerobics, accompanied by a certain guide (Şimşek, Kaya, and Özeskici, 1996, p. 294). Exercise which is defined as physical activities that stimulate the person's breathing and blood circulation, seems to improve the physical conditions of pregnant women and increase their fitness by doing it twice or three times a week (Kramer & McDonald, 2006, p. 2). In this context, according to the level of sports done by pregnant women during pre-pregnancy periods, exercise suggestions change as shown in the table;

Tab. 1: Exercise Tips for Pregnant Women

Previously Inactive	Moderately Active	Sportswoman
15 minutes of exercise, start with 3 days a week	30 minutes of exercise, start with 4 days a week	30 minutes of exercise, start with 4 days a week
Aerobic exercise, 65 % -75 % of maximum heart rate	Aerobic exercise, 65 % -85 % of maximum heart rate	Aerobic exercise, 75 % -85 % of maximum heart rate
Walking	Running	The continuation of previous sports should be discussed with the coach/coach/doctor and adjusted according to the previous activity level.
Swimming	Cycling/Spinning bike	The program must be changed in the third trimester.
Aerobics	Aerobics	Depending on the method of birth, she can return to the races within 4–6 months.
Stationary bike	Swimming	Avoid trauma and avoid stopping sports after the first trimester.
The goal is 30 minutes, 4 days a week	The goal is 30 minutes, 5 days a week	The goal is 30–40 minutes, 6–7 days a week

Source: Olson et al., 2009, p. 152.

While the individuals who are inactive before pregnancy should do exercises such as walking, swimming, stationary bikes for 15 minutes and 3 days a week, the women with moderate activity should do exercises for 30 minutes and 4 days of week. In these exercises, it is recommended that women who are athletes of jogging, cycling, and swimming should start exercising for 30 minutes, with a heart rate of 75–85 %, and exercise for up to 6–7 days per week, consisting of 30–40 minutes per day until the birth process.

Since some changes are observed in the female body with the formation of a baby during pregnancy, the pregnant women should also pay attention to feeding themselves and their babies throughout this period. While mentally and physically healthy individuals could be raised through adequate and balanced nutrition, insufficient and unbalanced nutrition may result in premature birth, babies with low-weight, stillbirth, death in the first days of birth (Aydın, 2014, p. 3–4), and chronic diseases can also be encountered as a result of this. Therefore, providing healthy nutrition in order to meet the needs in woman's body should also support the mother and the baby until the birth (Geyik, 2020, p. 3). Besides, it is observed that the foods and supplement vitamins taken by the pregnant women affect the nutritional competence of the mother during both pregnancy and breastfeeding period (Picciano, 2003, p. 1997). Nutrition, which directly affects breast milk quality, may appear as nutritional deficiencies in people who follow vegetarian and vegan diets (Penney and Miller, 2008, p. 37).

In pregnancy, if the psycho-social mental health of mother is negatively affected, mother, baby, family and even the whole society are affected negatively as parallel to this case. The psycho-social health of a pregnant woman and her attachment to the baby positively affect her future life in terms of physical and mental aspects (Bekmezci and Özkan, 2016, p. 56). In this context, the healthcare personnel, the spouse of the pregnant woman and all the other family members should provide support to the pregnant woman with the aim of preventing the occurrence of undesirable situations by detecting the problems of the pregnant beforehand (Yanıkkerem, Altıparmak, and Karadeniz, 2006, p. 42). When exposed to severe stress during pregnancy, problems such as premature birth and babies with low-weight can be encountered, and stress is shown as the main cause of 30 % of premature births (Akın, 2018, p. 9). When we look at the psychological problems during pregnancy, it is seen that these are emotional turmoil, adulthood, dissatisfaction, emotional imbalances in the stage of maturation (Leifer, 1977, p. 55), chronic stress and depression due to fear of premature birth and miscarriage (Schetter, 2011, p. 531). Within this scope, adequately satisfying the crises that occur during pregnancy is of great importance for the mental health of both the sand babies (Slade, Cohen L, Sadler L, and Miler, 2009, p. 22).

In order to reduce the problems experienced by women during pregnancy and raise healthier individuals, the entire healthcare team, especially nurses, have a major duty. They need to take on the roles such as giving prenatal information and in addition to physical care, they also take active roles in observation, support, assistance, continuing education and counseling (Yanıkkerem, Altıparmak, and Karadeniz, 2006, p. 36).

2 Pregnant Tourism

Pregnant tourism can be expressed as the total of relationships formed by the participation of pregnant women in tourism activities. In this sense, experts recommend that pregnant women should be cautious against malaria while traveling not only for their own health but also the health of the baby (Jaegera et al., 2015; Kaser et al., 2015). Since there is a high risk of many serious conditions such as premature birth, miscarriage and stillbirth because of malaria virus (Centers for Disease Contro and Prevention (CDC)), the pregnant women, who plan to go on holiday, are suggested to make holiday preferences by considering the possibility of exposure to malairia virus due to mosquitos (McGready, Ashleya, and Nosten, 2004, p. 128). Besides, there is also the risk of transmission of many diseases such as measles, mumps, chicken pox, typhoid, rabies, hepatitis (Carroll and Anderson, 2001, p. 397–403).

Pregnant women who are at risk of getting infections and illness during travel (Carroll and Williams, 2008, p. 259), should consider several factors such as general health status, gestational age, risks that may occur during pregnancy, travel type and travel duration, destination and status of health services at the destination when determining their holiday program (Cooper, 2006, p. 196). What's more, the pregnancy month should be taken into account when traveling since during certain pregnancy months flying is permitted with a doctor's report (Centers for Disease Contro and Prevention (CDC)). The risks that may occur during travels are classified as follows;

When the table is analyzed in general, some information on pregnancy related to some diseases such as asthma, lung, heart, anemia and kidney can be observed. In order not to be exposed to these diseases, it is necessary to carefully examine the centers that could be visited for holiday purposes. Carroll & Anderson (2001) stated that the checklist should be looked through for the travel of pregnant women;

Pregnant women, when making holiday decisions, should do a general assessment in terms of a number of risks such as, pregnancy status, medical risk and travel risk. Against these risks by taking into account the state of vaccines and measures to reduce the risks in possible contaminations, the risks and benefits should be compared detailedly. If the risks are higher, the travel has to be postponed, but if the benefits are higher, the travel decision can be taken.

3 Pregnancy in Hospitality Industry

A special massage service is offered for the pregnant women at Double Tree Hotel. As part of the relaxing massage, moisturizing the skin that expands with the growth of the baby and massage some special applications that will make the mother feel peaceful (Double Tree Hotel). At Dedeman Hotel, a sleep menu has been prepared for pregnant women, and a maternity support bed, various pillows and quilts options are offered (Dedeman Hotel). While on holiday during pregnancy, Mirabell Hotel offers wellness services for pregnant women. Thanks to wellness, the happiness and enjoyment of the mothers is defined as main objective and there are various spa facilities for both moms and dads. In addition to

Tab. 2: Possible Risks during Pregnancy in International Travels

Obstetric risk factors	General medical risk factors	Travel destinations to potentially dangerous places
Risk of miscarriage	The history of thromboembolic disease	High altitudes
Cervical problems (cervix)	Pulmonary hypertension	Endemic regions with ongoing outbreaks of life-threatening food or insectic infections
Ectopic pregnancy history (the condition of organ's being elsewhere rather tahn the usual place)	Severe asthma or other chronic lung disease, Symptomatic heart valve disease	The regions in which Chloroquine resistant Plasmodium falciparum malaria is endemic
The history of preterm labor or premature rupture of membranes (membrane rupture)	Cardiomyopathy (heart muscle disorder)	Areas where live virus vaccines are required or recommended
The history of placental abnormalities	Hypertension Diabetes	
Abortion or vaginal bleeding threat during current pregnancy	Renal failure	
Multiple pregnancy in the current pregnancy	Severe anemia or hemoglobinopathy	
Fetal growth abnormalities	Chronic organ system dysfunction requiring frequent medical intervention	
A history of toxemia, hypertension or diabetes with any pregnancy, being 35 years and older or under 15 years of age		

Source: Cooper, 2006, p. 197.

the special demands that the mother will need, there are also special offers for personalized health packages and birth preparations (Mirabell Hotel).

Spa Sport Hotel Zuiver offers several applications for a healthy pregnancy. After the first three months, it states that sauna has a dehumidifying effect and it optimizes blood circulation, as well as helping to remove fluid accumulation

Tab. 3: Checklist for Pregnant Tourists

Control Factors
Pretravel risk assessment
Stage of gestation
Obstetric risk factors
Medical risk factors
Destination risk considerations
– Due to infectious disease
– Cloroquine-resistant Plasmodium falciaprium malaria
– Outbreak of disease requiring a live virüs vaccine
– Outbreak of disease for which no vaccine is available but has a high risk of maternal and fetal morbidity and mortality
* due to food water exposure *due to insect exposure
– Due to environment
*Sports or exercise risk
*Altitude
*Open water
– Lack of access to care
Travel Risk Assessment
– Mode of travel
– Medical services available during transit and at destination
– Medical insurance and evacuation coverage
– Review emergency signs and symptoms for which care should be sought
*Severe vaginal bleeding, abdominal pain, contractions, proteinuria, severe headache with visual change, severe eduma and/or accelerated weight gain
*Rupture of membranes
Recommendations
– Immunizations to reflect actual risk of disease and probable benefit
– Medications-review safety during pregnancy
– Preventative measures to decrease:
*Gestation- related risks
*Typical problems
*Mode of travel risks
*Food related risk
*Water related risk
*Insect related risk
*Environment related risk
*Sexual behavior risk
Medical kit adaptations
Postpone travel if risks outweigh benefit

Source: Carroll and Anderson, 2001, p. 393.

during pregnancy and the crack marks around the navel area with the application of the leg gel. What's more, it provides a relaxation for the body with pregnancy massage and cold shower options (Spa Sport Hotel Zuiver). At Parco San Marco Hotel a relaxing SPA holiday for the health of the babies and mothers and prenatal pain relief massage services are offered. The hotel offers a refreshment opportunity for pregnant women via its private beach, indoor and outdoor swimming pools and face care options and it also offers breakfast and various meal services for a healthy life. In addition, thanks to its close location to Lake Lugano, it provides many activities such as walking by the lake, getting fresh air and doing some light exercises (Parco San Marco Hotel).

Elite World Hotel proposes to make a baby shower to increase the morale of the mother and to help her by making a call to pregnant women and their spouses and relatives. Thus, they serve to make the mothers happy by preparing the party hall with the concepts and equipment that the mothers will love (Elite World Hotel).

3.1 Pregnancy in Travel Business

ETS offers additional services such as cleaning in rooms, extra pillows, coverlets for not only people who have food allergies or diabetes but also for pregnant women. Doctor and ambulance service are also provided in case of emergencies in order for pregnant guests to have a comfortable holiday (ETS Official Web Site). As for Jolly it defined the factors that pregnant women should pay attention while traveling. Besides mentioning about the positive effects of physical activities do by swimming in the pool, it gives some recommendations on sunbathing, as well (Jolly Official Web Site).

A number of holiday tips are given for pregnant individuals in the website prepared by MNG Tourism. By making suggestions for the pregnant women on the hotel's location, hygiene, protection from the sun, more fluid consumption and clothing selection, it is stated that the pregnants can make their holidays better (MNG Tourism).

3.2 Pregnancy in Transportation Business

According to Obstetrician and Gynecologist Surgeon Nigar Güler, women can travel in all periods of pregnancy, but in the first three months, due to the problems such as nausea, vomiting, miscarriage, and in the last three months because of a growing belly the journey can be quite difficult. When traveling for a long time, Obstetrician and Gynecologist Surgeon Nigar Güler makes several suggestions (Private Life Hospital);

- Plenty of water should be consumed while traveling,
- The seat should be comfortable and the seat belt should be worn,
- There should be no movement restrictions and frequent breaks should be given during travel,
- Comfortable clothes and shoes should be worn,
- A travel pillow should be available,
- In case of foot swelling, feet should be kept high and it is necessary to walk during breaks.

Another opinion about traveling during pregnancy has been put forward by Obstetrician and Gynecologist Surgeon Cengizhan Kolata. Kolata suggests that the air balance of the travelling environment should be arranged during bus and train travels since such travels last for a long time and they cannot have breaks whenever needed. If airways are preferred in travels, it may be necessary to move every two hours and to maintain blood circulation at certain intervals by stretching legs, as cabin pressure and oxygen depletion may be encountered (Central Hospital). Expressing that air travel is safe for pregnant individuals, in their studies Roggelin and Cramer (2014) stated that elevation increases blood pressure, heart rate and respiratory rate, but this case, does not cause vital changes in the health of the fetus. However, it is stated that since shipping from the sea is risky in case of emergencies starting from the 28th week of pregnancy, cruise ship companies do not accept the pregnant women for travel during these periods (Roggelin and Cramer, 2014, p. 234).

It is necessary to avoid long-term road travel during pregnancy. However, if traveling cannot be avoided, regular stretching and movement exercises are required to be done. In addition to flexing and twisting the feet, regular fluid consumption and natural energy intake such as fruit and nuts are also essential (NHS Web Site).

Within the context of the "Regulation on the Rights of Passengers Traveling by the Railroad" in the rail transport, some arrangements have been proposed for the convenience of the travel of pregnant women and to benefit them sufficiently (Official Gazette). As for the airline transportation, Turkish Airlines' policy regarding pregnant passengers states that they need a doctor's permit, especially on long-term flights, but the data on the doctor's report according to singular and multiple pregnancy are as follows;

The flight rules of Turkish Airlines about the case of getting doctor's permission change according to different months of pregnancy and these rules are the same for the Pegasus (Pegasus Web Site) Sun Express (Sun Express Web Site) and Anadolu Jet companies. While the rules are the same for all the companies

Tab. 4: The Conditions Permitted to Fly According to Pregnancy Week

	Pregnancy Weeks			
	Weeks 1 - 27	**Weeks 28 - 31**	**Weeks 32 - 35**	**After 36 weeks**
I'm pregnant with a single baby	You can fly	Doctor's report required	Doctor's report required	You can not fly
I am pregnant with more than one baby			You can not fly	

Source: Turkish Airlines Web Site.

above, the pregnant passengers at Anadolu Jet Company are listed under the heading of "Passengers with Special Features" and it is emphasized that the doctor's report is valid for 7 days (Anadolu Jet Web Site). Delta Airlines states that they do not request any documents and that the decision regarding the travel restriction should be made with the doctor (Delta Air Lines). British Airways, in addition to the same flight rules, state that they want a health report indicating that their health status doesn't create an obstruction to fly (British Airways Web Site), while Emirates Airways have listed what should be included in the health reports (Emirates, 2020);

- Whether pregnancy is single or multiple
- Estimated date of birth without pregnancy complications
- The proper dates for the mother to travel
- Confirming that the mother's health is good
- It is necessary to verify that there is no evident situation that prevents the mother from flying.

According to TUI notifications, since the flights of pregnant individuals longer than five hours increase the risk of thrombosis and blood clots; along with wearing compression stockings, some other elements such as fluid consumption, seat belt use and calf exercises are recommended (TUI Web Site). ASL Airlines France does not make any restrictions until the eighth month, while they requests doctor's report after the eighth month (ASL Airlines France).

4 Conclusion and Suggestions

Although pregnancy is a turning point in women's life, it is a social process that concerns both family members and the society. In this context, for the health of both mother and baby, it is necessary to adopt a lifestyle that requires adequate and balanced nutrition, provides a stress-free routine and activities through exercises.

Pregnant tourism that comes out of the participation of women in tourism activities due to the reasons such as women's feelings of happiness, health and motivation needs, also includes special practices for pregnant women. It has been observed that there are not enough investment and applications planned for pregnant women in the sectors such as accommodation, travel, transportation, which are the sub-dynamics of the tourism sector. While there are activities such as SPA, massage and sauna for the comfort and happiness of pregnant individuals in general, there are very few applications in the travel industry. In the travel sector, by giving priority to being comfortable during the travel, wearing a proper clothing, enough fluid intake and moving at regular intervals for blood circulation is included.

In the transportation sector, airline companies have flight conditions for pregnant individuals, and it is generally seen that there are no obstacles until the 28th week of pregnancy. In this context, in order to create special holiday packages and charm for pregnant individuals in accommodation businesses, using the elements such as payment facilities, price discounts, etc. is recommended. In the travel industry, it should be ensured that pregnant individuals feel valued by creating customized package tours for pregnant individuals and creating special interest groups covering pregnant individuals. In the transportation sector, it is recommended to facilitate transportation for pregnant women by making plans concerning road, rail and sea travel of them. Special plans should be made in the tourism sector in order for the pregnancy situation, which is an important period for women, to create unforgettable holiday memories both for mothers and fathers.

References

Akın, Ö. (2018). Gebelik Stresini Değerlendirme Ölçeği'nin Türkçe Geçerlik ve Güvenirlik Çalışması. Ordu Üniversitesi Sağlık Bilimleri Enstitüsü Yüksek Lisans Tezi.

Akyüz, G., & Bayındır, Ö. (2013). Gebelik İle İlişkili Osteoporoz. *Türk Fiziksel Tıp ve Rehabilitasyon Dergisi* (59), 145–150.

Anadolu Jet Resmi Web Sitesi: https://www.anadolujet.com/tr/kurallar/ozelligi-olan-yolcular Access Date: 2020, 01 10.

ASL Airlines France: https://www.aslairlines.fr/en/travellers-information/prepare-your-trip/pregnancy/ Access Date: 2020, 01 10.

Aslan, E., & Güneş, G. (2014). Avrupalı Seçkin Destinasyonlar (Eden) Kapsamında Engelsiz Şehirler ve Turizm "Ulusal Yazılı Medyada Engelsizlik Üzerine Bir Tarama". *15. Ulusal Turizm Kongresi* (s. 1–15). Ankara: Gazi Üniversitresi.

Aydın, E. K. (2014). Son Trimestirdeki Primipar ve Multipar Gebelerin Beslenme ve Kilo Durumlarının Değerlendirilmesi. Erzurum: Atatürk Üniversitesi Sağlık Bilimleri Enstitüsü Yüksek Lisans Tezi.

Bekmezci, H., & Özkan, H. (2016). Gebelikte Psikososyal Sağlık Bakım, Prenatal Bağlanma ve Ebe-Hemşirenin Sorumlulukları. *Uluslararası Hakemli Kadın Hastalıkları ve Anne Çocuk Sağlığı Dergisi* (8), 50–62.

British Airways Resmi Web Sitesi: https://www.britishairways.com/en-ca/information/travel-assistance/medical-conditions-and-pregnancy Access Date: 2020, 01 10.

Carroll, I. D., & Anderson, S. (2001). Women's Health and Travel. J. N. Zuckermane içinde, *Principles and Practice of Travel Medicine* (s. 463–502). UK: John Wilwy &Sons Ltd.

Carroll, I. D., & Williams, D. C. (2008). Pre-travel vaccination and medical prophylaxis in the pregnant traveler. *Travel Medicine and Infectious Disease, 6*, 259–275.

Centers for Disease Contro and Prevention (CDC). (2019, 08 25). https://wwwnc.cdc.gov/travel/page/pregnant-travelers Access Date: 2020, 01 10.

Central Hospital. (2020, 05 10). http://centralhospital.com/Haberler/hamilelikte-seyahat-hakkinda-bilinmesi-gerekenler/ Access Date: 2020, 01 10.

Cooper, M. C. (2006). The pregnant traveller. *Travel Medicine and Infectious Disease, 4*, 196–201.

Dedeman Hotel: URL: https://www.dedeman.com/TR/7-Oteller/287-Park-Dedeman-Elazig/406-Hizmetler/ Access Date: 2020, 01 10.

Delta Air Lines: https://www.delta.com/us/en/children-infant-travel/infant-travel Access Date: 2020, 01 10.

Double Tree Hotel: https://doubletree3.hilton.com/resources/media/dt/LPLDLDI/en_US/pdf/en_LPLDLDI_Spa_Menu_May _2017.pdf Access Date: 2020, 01 10.

Elite World Hotel. URL: https://www.eliteworldhotels.com.tr/blog-en/entertainment/how-is-a-baby-shower-organized.15.64.aspx Access Date: 2020, 01 10.

Emirates: URL: https://www.kidsonaplane.com/emirates-flying-with-kids-policy/ Access Date: 2020, 01 10.

ETS Resmi Web Sitesi: https://www.etstur.com/Ozel-Saglik-Kosullarina-Uygun-Tesisler Access Date: 2020, 01 10.

Geyik, P. (2020). İstanbul: Sağlık Bilimleri Üniversitesi Hamidiye Sağlık Bilimleri Enstitüsü Yükseklisans tezi.

Işık, M. T., Akçınar, M., & Kadıoğlu, S. (2010). Mersin ilinde gebelik, doğum ve loğusalık dönemlerinde anneye ve yenidoğana yönelik geleneksel uygulamalar. *Uluslararası İnsan Bilimleri Dergisi*, 63–84.

Jaegera, V. K., Tschudi, N., Rüegg, R., Hatz, C., & Bühler, S. (2015). The elderly, the young and the pregnant traveler – A retrospective data analysis from a large Swiss Travel Center with a special focus on malaria prophylaxis and yellow fever vaccination. *Travel Medicine and Infections Disease, 13*(6), 475–484.

Jolly Official Website: https://blog.jollytur.com/hamilelerin-tatilde-dikkat-etmesi-gerekenler/ Access Date: 2020, 01 10.

Kaser, A. K., Arguin, P. M., Chiodini, P. L., Smith, V., Delmont, J., Jime´nez, B. C., Schlagenhauf, P. (2015). Imported malaria in pregnant women: A retrospective pooled analysis. *Travel Medicine and Infectious Disease, 13*(4), 300–310.

Kramer, M. S., & McDonald, S. W. (2006). Aerobic exercise for women during pregnancy. *Cochrane Systematic Review, 2*, 1–45.

Kutlu, M., Şimşek, M., Kaya, N., & Özekici, Ü. (1996). Hamilelik Süresince Egzersizin Pozitif ve Negatif Boyutları. *Turkiye Klinikleri J Gynecol Obst, 6*(4), 290–295.

Leifer, M. (1977). Psychological changes accompanying pregnancy and motherhood. *Genetic Psychology Monographs, 95*(1), 55–96.

McGready, R., Ashleya, E., & Nosten, F. (2004). Malaria and the pregnant traveller. *Travel Medicine and Infectious Disease, 2*, 127–142.

Mermer, G., Bilge, A., Yücel, U., & Çeber, E. (2010). Gebelik ve Doğum Sonrası Dönemde Sosyal Destek Algısı Düzeylerinin İncelenmesi. *Psikiyatri Hemşireliği Dergisi, 1*(2), 71–76.

Mirabelle Hotel: URL: https://www.mirabell.it/en/5-stars-hotel-spa-south-tyrol/wellness-pregnancy/ Access Date: 2020, 01 10

MNG Turizm: https://blog.mngturizm.com/hamileler-icin-tatili-kolaylastiracak-5-oneri/ Access Date: 2020, 01 10.

NHS: https://www.nhs.uk/conditions/pregnancy-and-baby/travel-pregnant/?tabname=im-pregnant Access Date: 2020, 01 10.

Olson, D., Sikka, R. S., Hayman, J., Novak, M., & Stavig, C. (2009). Exercise in Pregnancy. *Current Sports Medicine Reports, 8*(3), 147–153.

Özdemir, Ö. Ç., Sürmeli, M., Özel, A., Yavuz, İ. E., Topçuoğlu, A., & Ankaralı, H. (2017). Gebelerde Fiziksel Aktivite ve Yaşam Kalitesi Arasındaki İlişki. *Anadolu Kliniği Tıp Bilimeri Dergisi, 22*(2), 95–102.

Özel Hayat Hastanesi: https://www.ozelhayathastanesi.com.tr/gebelikte-seyahat-etmek Access Date: 2020, 01 10.

Parco San Marco: URL: https://www.parco-san-marco.com/en/wikiparco/hotel-parco-san-marco/well-being-for-baby-and-yourelaxing-spa-breaks-for-pregnant-women/ Access Date: 2020, 01 10.

Pegasus: https://www.flypgs.com Access Date: 2020, 01 10.

Penney, D., & Miller, K. (2008). Nutritional Counseling for Vegetarians During Pregnancy and Lactation. *Journal of Midwifery & Women's Health, 53*(1), 37–44.

Picciano, M. F. (2003). Pregnancy and Lactation: Physiological Adjustments, Nutritional Requirements and the Role of Dietary Supplements. *The Journal of Nutrition, 133*(6), 1997–2002.

Resmi Gazete: https://www.resmigazete.gov.tr/eskiler/2019/03/20190308-6.htm Access Date: 2020, 01 10.

Roggelin, L., & Cramer, J. P. (2014). Malaria Prevention in the Pregnant Traveller: A Review. *Travel Medicine and Infectious Disease, 12*, 229–236.

Schetter, C. D. (2011). Psychological Science on Pregnancy: Stress Processes, Biopsychosocial Models, and Emerging Research Issues. *Annual Review of Psychology, 62*, 531–558.

Slade, A., Cohen L. J., Sadler L. S., & Miler, M. (2009). The Psychology and Psychopathology of Pregnancy. J. Charles H. Zeanah içinde, *Handbook of Infant Mental Health Third Edition* (s. 22–39). New York: The Guilford Press.

Spa Sport Hotel Zuiver: URL:https://zuiveramsterdam.nl/en/spa/sauna-and-pregnancy Access Date: 2020, 01 10.

Sun Express Web sitesi: https://www.sunexpress.com/tr/bilgi/yolcu-bilgilendirme/hamile-yolcular/ Access Date: 2020, 01 10.

Şimşek, M. K., Kaya, N., & Özeskici, Ü. (1996). Hamilelik Süresince Egzersizin Pozitif ve Negatif Boyutlan. *Hekimler Birliği Vakfı Türkiye Klinikleri, 6*(4), 290–295.

Tor Kadıoğlu, C., & Şahin, A. (2017). Hamile Kadınların Sosyal Medya Kullanımı ve Satın Alma Kararları İle İlişkisi:. *Süleyman Demirel Üniversitesi İktisadi ve İdari Bilimler Fakültesi Dergisi, 22*(4), 981–991.

TUI Web Sitesi: https://www.tui.co.uk/destinations/discover/family/safe-flying-during-pregnancy Access Date: 2020, 01 10.

Türk Hava Yolları: https://www.turkishairlines.com/tr-at/bilgi-edin/hamile-yolcular/index.html Access Date: 2020, 01 10.

Yanıkkerem, E., Altıparmak, S., & Karadeniz, G. (2006). Gebelikte Yaşanan Fiziksel Sağlık Sorunlarının İncelenmesi. *Aile ve Toplum, 3*(10), 35–42.

Mehmet Mert PASLI

11 The Case of Dark Tourism in Turkey

1 Introduction

Global competition makes it necessary to be different and revolutionary in the tourism industry. Destinations and businesses should strive to create new and different types of tourism in order to attract tourists in the competitive current environment. It has become imperative for destinations whose economy depends on tourism to diversify. Hence, creating new trends, diversity and product range have become essential. (Aydın and Çoban, 2019: 267).

There is a tendency in tourism toward other alternative types besides mass tourism. This is in order to eliminate the imbalance among the regions and reduce the concentration of tourism in certain destinations. Various considerations such as increasing competition and variance in consumer demands play an effective role. Thus, dark tourism, which is a type of alternative tourism, has been an attraction for tourist in recent years. This covers tourist sites, in areas where deaths, terror and dismal events have occurred

Dark tourism is conceptualized with different names in domestic and foreign literature. The origin of the clashes in this conceptualization can be differences in feelings. For instance, while one tourist can have feelings of sorrow at the visiting of a certain spot, another tourist can feel pride or hatred. The key point is that historical, social, political or environmental events which took place were dark events. Therefore, it can be said that the attribution of the visiting spots can steer to different conceptualizations (Birdir, Dalgıç, Güler and Kayaalp, 2015: 13). Some definitions of the dark tourism are below;

Dark tourism may be defined as "*Visits to places where tragedies or historically noteworthy death has occurred and that continue to impact our lives*" (Tarlow, 2005: 48).

Dark tourism is a term given to a type of tourism that involves travelling to places such as those where torture, genocides and natural disasters took place and where museums and memorials mark those events (Kılıç and Akyurt, 2011: 213).

Dark tourism, in essence, was born because people want to travel to places where suffering occurred by natural reasons or by the actions of other people, (Yıldız, Yıldız and Aytemiz, 2015: 403).

Dark tourism is a type of a particular interest tourism and can be done throughout the year and the purpose of it is to bring various emotions such as pain, death, remembrance, fear and apprehension to the forefront. Thus, there are numerous types of dark tourism sites: prisons, cemeteries, natural disaster areas (tsunami, volcano eruption, etc.), centers of technological disasters (atomic bomb, nuclear disasters, etc.), genocide sites, torture, executions, suicide and euthanasia centers and battlefields are among the dark tourism destinations. (Akyurt Kurnaz, 2019: 235)

In general, every type of tourism can have varieties. Dark tourism can also change depending on places visited and events which took place. Examples are shown below (Alili, 2017: 41–42);

- Battlefields
- Museums and Genocides destinations
- Slum Tourism
- Euthanasia destinations
- Destinations where tragic crime and events took place
- Doomsday Tourism
- Paranormal tourism
- Terror tourism

People who want to absorb death and its related feeling want to visit the places where these feelings are commercialized and remember the lost lives. People do not forget the others who died on behalf of themselves and visit the places where they suffered, died or were killed (Hıdır and Somuncu, 2018: 152).

Stone suggests in his study, that in the classification of dark tourism, sites should be classified according to the degree of impact or savagery they have on society or the world (Fig. 1). Therefore, when the conditions of the regions are considered, classification can be made according to the sadness and feelings they give (Çelik, 2017: 45).

It is obvious that visitors have different purposes when they visit grief destinations. Some visitors plan to grieve for their ancestors while some others aim to live those moments described by guides whereas some visitors might find it antipathic and distressing and become indifferent. Some visit these places once a week or month or a year while some others once in their lifetime or never. People who don't want to face death find these reactions scary but also it depends on jingoistic feelings, kindredship, ethnic similarities or empathy level of the person (Özçelik, 2012: 23).

Numerous battlefields are visited by tourists. Visiting battlefields can give the feeling of pride and sorrow (Akyurt Kurnaz and Kılıç, 2020: 30).

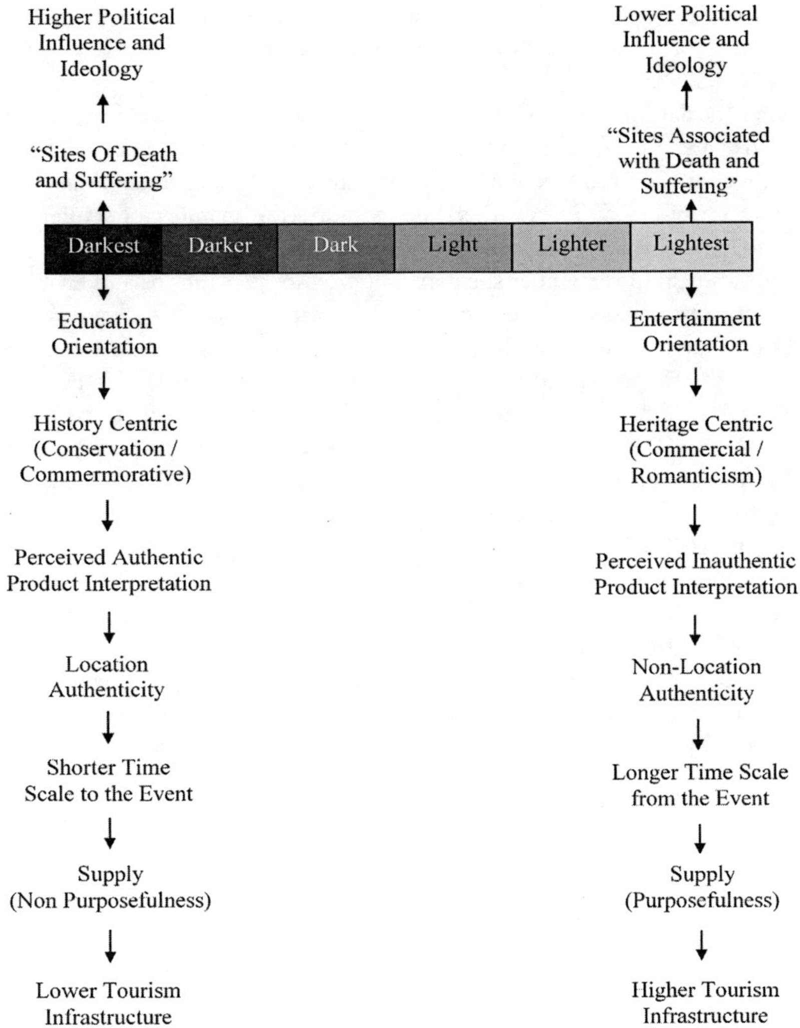

Fig. 1: A Dark Tourism Spectrum: Perceived Product Features of Dark Tourism Within a "Darkest-Lightest" Framework of Supply.

Source: Stone, 2006: 151

Tourists who visit these kind of destinations learn about the culture and the history of those nations when they take part in commemorative events. (Pekerşen, 2020: 34).

Since the day Turks set foot in Anatolia, they established a variety of states and have struggled for existence in this land. Turkey has overcome many difficulties in the region and lived through the empire, the war of independence and world wars. In every city of Turkey, there are memorials in memory of soldiers who lost their lives because of the war of independence as well as terrorism. These memorials help to remember the pain and grief for generations and enable the society to forge a bond between the past and present.

Dark tourism has been defined by many different writers with different expressions. There have been several studies conducted on Dark Tourism in Turkey. Alaeddinoğlu and Aliağaoğlu (2007), Kılıç and Akyurt (2011), Chief Commander (Başkomutan) Historical National Park, Birdir, Dalgıç, Güler and Kayaalp (2015), Gelibolu Peninsula Historical Park, Aksakallı, Selçuk and Işık (2017), Nene Hatun National Historical Park, Akyurt Kurnaz, Kurnaz and Tatlıoğlu (2017), Ulucanlar Prison, Yirik and Seyitoğlu (2014), "Sarıkamış", Güncü and Güneş (2017), Erzincan, Çetinsöz, (2017), Belli and Güneren Özdemir (2018), Sinop Prison, have carried out researches which associated diversified destinations with dark tourism.

This study shows that the visits conducted within the scope of dark tourism consist of monuments, battlefields, dungeons, museums which are converted from prisons where periods of agony hosted, museums and memorials which are created as consequences of natural disasters and symbolic tombs. The examples of the Gallipoli Martyred Submariners Monument, the Monument of Dumlupınar Submarine Martyrs, the Dungeons of Seven Tower, the Historical Sinop Fortress Prison, the Ulucanlar Prison, the Adapazarı Earthquake and Culture Museum, the Anıtkabir Atatürk and the War of Independence Museum, the Memory of July 15 Museum, the Gallipoli Peninsula and the Soma Mine Disaster were given in this context. Other than above given examples Turkey has potential of many dark tourism destinations in various cities.

Memorial for the Crew of the Martyred Gallipoli Submarine

The monument was erected on 19th of May 2006, in honor of two-hundred and seventy-four sailors who were martyred in the Refah, Atılay and Dumlupınar submarine disasters which occurred during various periods of Turkish Maritime history. It has the characteristics of being one of the biggest and most magnificent monuments built on behalf of martyred sailors (www.tsiv.org.tr).

The Monument of the Dumlupinar Submarine Martyrs

This monument is in the Karasu district of the city of Sakarya. It was built in memory of the martyrs of the Dumlupinar Submarine which sank, following a joint NATO military exercise, colliding with a Swedish flagged ship "Naboland" in the Dardanelle straits near Nara Cape on 4th of April 1953. A movie depicting the event was made which included various stories.

The Dungeons of the Seven Towers (Fortress of the Seven Towers)

The dungeons of the seven towers is one of the most important open air museums of Istanbul. The Byzantians named the area "Penta Prikyor", which means 5 towers, named after the number of towers on the city walls, prior to the conquest of Istanbul. When Fatih Sultan Mehmet conquered the city, he added two more towers and the name of this magnificent area was changed to Yedikule which is an example of the integration of Byzantine and Ottoman era structures. From the dark tourism point of view, the most interesting and evocative attractions are the Young Osman Tower and the Bloody Well. The tower is where Osman II (Young Osman), reputed to be a reformist sultan, was assassinated. Hence, the tower is known as "Young Osman Tower". It is known as a very important event in Ottoman history. The site of another important event in Ottoman history is the Bloody Well. It is rumored that the heads of executed prisoners were dumped into this well through the tunnels which led to the Marmara Sea (www.yedikule.org).

Sinop Fortress Prison

Historical Sinop Prison, which used to be described as the Alkatraz of Anatolia, was closed and turned into a museum in 1999. Surrounded by the sea on three sides, this prison has thick and high fortress walls and according to archaeological data, it was built during Hellenistic times. It was renovated and fortified during the Roman, Byzantian, Seljuk Turk and Ottoman eras. The bastions of the inner fortress were used as a shipyard for a long time during the reign of the Seljuk Turks and became a dungeon in 1560, however, officially, turning the inner fortress into a dungeon was acknowledged only in 1887. Another building was added as a juvenile prison in 1939. This historical prison is built from face stone and has a North-South U-shaped structure. The building consists of three parts separated with high courtyard walls and it has 28 wards and several workshops. Some of the most important and famous figures of art and literature in Turkey were imprisoned in Sinop Prison. Although one of the first names to spring to mind is Sabahattin Ali, there are other important names such as Burhan Felek, Refik Halit Karay, Eşber Yağmurdereli. The poems and literary works written by prisoners in this prison have become important works in the history of literature. Sinop Prison has also become a subject for many movies and TV dramas (www.muze.gov.tr).

Sinop Fortress Prison is the most visited place in the city of Sinop. Tab. 1 shows the number of visitors in 2017.

Tab. 1: Number of Visitors to the City of Sinop Museum

Archeology Museum	8.288
Ethnography Museum	5.975
Historical Prison	177.096
TOTAL	**191.359**

Source: sinop.ktb.gov.tr

Ulucanlar Prison

Ulucanlar Central Maximum – Security Prison is situated on the outskirts of the historical town of Ulus, on the top of a hill to the East of Ankara Castle. Its location permits it to have a visual connection with Ankara Castle and the historical town center as well as the new town. According to the 1924 Lorcher Ankara Old City plan, Ulucanlar is located in its current place which was chosen deliberately because it would help the rehabilitation of the prisoners by means of plowing in the surrounding fields. The prison grew larger with annexations. It was a notoriously cold and dark prison where for eighty-one years people were imprisoned in extra ordinarily difficult conditions; hangings were carried out and parents waited outside in the cold for days to be able to have any news from their loved ones. This continued until 2006. Ulucanlar Prison became home to various political figures, poets, journalists and authors who left their mark in Turkish history. Some of these people are; writers such as Necip Fazil Kısakurek, Nazim Hikmet, Ahmet Arif, Fakir Baykurt, Metin Toker. Also, influential figures in the Turkish political history were inmates in this prison; Bülent Ecevit, Muhsin Yazıcıoğlu, Deniz Gezmiş, Yusuf Aslan, Huseyin İnan, Mustafa Pehlivanoglu, Fikri Arıkan, Ali Bülent Orkan were amongst them. The wards are organized and provided with materials which reflect almost all the periods in the history of the prison. Whilst visitors walk around the wards, they can see wax sculptures, belongings of inmates and a documentary which explains the history of the prison.

Altındağ Council opened its doors to visitors on 16 June 2011 as a reminder of those dark days and in the hope of not living them again. Since the opening date, Ulucanlar Prison museum has been visited 1 million times and has been awarded the most visited private museum in Ankara by The Ministry of Culture (www.ulucanlarcezaevimuzesi.com).

Adapazari Earthquake and Culture Museum

Turkey is in a seismic zone because of its geographical position and has encountered many tragedies stemming from earthquakes during which a large number of citizens were lost. There are many earthquake museums all over the world and Turkey has one of them.

The earthquake museum is situated on a 450square meters field where before and after photographs of the Adapazari earthquakes which took place in 1967 and 1999 on display, a vibrating electronic stand which creates an artificial earthquake, information about the technical construction of the buildings which collapsed in the earthquake, seismograph and other related factors available to see and it has exhibition stands, cafe and cinevision areas. The architecture and the interior design of the building were designed as an art gallery to be able to display photographs, paintings and similar objects. The museum opened to visitors in 2004 where simulation of the earthquake which occurred on 17 August 2004 at 03.02am which had a magnitude of 7.4 and lasted 45 seconds can be observed. The memory of people who died in the earthquake is kept alive by placing their names on crystals. Up to now, approximately 1million people have visited the museum (www.kulturportali.gov.tr).

The Mausoleum of Atatürk and Museum of the War of Independence

Anitkabir, is the mausoleum, Anittepe (formerly known as Rasettepe) in Ankara for Atatürk, who was the leader of the Turkish War of Independence and who led the reforms which transformed Turkey into a modern republic and who also was the first President of the Turkish Republic (www.anitkabir.com.tr).

Anitkabir consists of an imposing monument block and the Peace Park which is made up of approximately 48,500 plants and trees gathered from various parts of Turkey and also sent from a variety of countries to represent them. The park is a symbolic "League of Nations" and represents the principle of Atatürk which is "Peace at home, peace in the World" (www.ankara.ktb.gov.tr).

Atatürk and The War of Independence museum consists of four main parts. The first part contains belongings which Atatürk used, items which were given to him by foreign politicians along with belongings donated to the museum by his adopted children Afet Inan, Rukiye Ergin and Sabiha Gokcen. The second part holds panoramic oil paintings portraying the Dardanelle Wars, the battle of Sakarya and The Great Offensive. This section attempts to explain to visitors the difficulties which were faced on the path to the establishment of the Turkish Republic. The third part has galleries where the National Struggle and revolutions are documented and it also contains the tomb chamber of Mustafa Kemal Atatürk. This section narrates, in Turkish and English, the events which took place between 1919 and 1936. The fourth part is the private library of Atatürk where 3,123 books are on display (www.ankara.ktb.gov.tr).

July the 15th Memory Museum

This museum narrates the resistance and struggle against the members of a terrorist organization who attempted a military coup.

The museum, which was built in memory of the martyrs of the attempted coup, has sections such as the history of coups in the world and colonialism. All aspects of the resistance during the night of the putsch are narrated in the museum where memories of the martyrs and veterans are kept alive.

The museum was visited by 450.000 people in the first year when it opened its doors. The number is expected to increase in the coming years (www.iletişim.gov.tr).

Gallipoli Peninsula

As a destination of dark tourism, Gallipoli National Park hosts visitors in every season of the year and shows that tourism is not seasonal (Akyurt Kurnaz, Çeken and Kılıç, 2013: 68). The Dardanelles Campaign was one of the greatest wars in the world. It was between the Allied States and the Ottoman Army during the First World War, during which hundreds of thousands of casualties occurred. While the memorials and martyrdoms reflect one of the most magnificent victories in history, they also remind us of the sorrowful memories of war. The National Park, containing monuments erected as a memorial of more than two hundred and fifty thousand Turkish martyrs and also the burial places of more than two hundred and fifty thousand English, French, Australian and New Zealand soldiers, thus telling the world of the importance of peace. There are 56 local and 35 foreign monuments and memorials of martyrdom within the park (www.canakkale.ktb.gov.tr). There are a myriad of books and art works on this subject and Gallipoli Peninsula is visited by many local and foreign tourists.

Currently, the areas which are not in the status of the National Park are managed by Historical Site of Gallipoli Dardanelles Wars of which reported that average 3 million visitors visited the historical area in 2019.

Soma Disaster

In Soma, Manisa, in 2014, 301 miners lost their lives because of a fire in the coal mine, which was similar to the incident in Chile which took place in 2010. A Mining Martyrdom Monument has been built in memory of the 301 miners in the Soma disaster (www.soma.gov.tr).

2 Conclusion and Suggestions

Dark Tourism appears as one of the most prominent alternative areas of tourism in which there is an increasing interest. Subjects such as cultural heritage, wars, curiosity, fascination with death and interest in history make the destinations of dark tourism appealing. Turkey, like the rest of the world, has examples of this tourism. In this study, some of the examples of dark tourism have been presented. The shared points of these examples are experienced pain, grief, deaths and evoking the feeling of remembrance. From the point of carrying suffering to future generations, these destinations make up the memory of a country or even the world.

In order to resolve the regional imbalance and spread tourism throughout the year, Turkey should look for different types of alternative tourism. Dark tourism accommodates millions of visitors all over the world. Turkey should determine the dark tourism spots, develop and promote them and ensure to captivate tourists' interest. In addition, by updating the web sites of the dark tourism museums and enabling virtual museums tours, these can be made more interesting for tourists.

The academic studies prove that literature, movies and shows increase the interest in destinations. Movies or TV shows which evoke grief, sadness and agony about the dark tourism destinations can inform the audience who are unaware of the events and enlighten the potential visitors

Similar studies should be carried out in Turkey for different destinations and these studies should be compared with other examples in the world and in order to attract more visitors more planning should be achieved with the support of public and private sectors.

References

Aksakallı, Z., Selçuk, G. N. and Işık, C. (2017). Nene Hatun Tarihi Milli Parkı'nın Hüzün Turizmi ve Mabetleşme Yönünden Değerlendirilmesi, *Erzincan Üniversitesi Sosyal Bilimler Enstitüsü Dergisi*, 215–240.

Akyurt Kurnaz, H. (2019). Hüzün Turizmi Kapsamında "Korku Evleri" Üzerine Bir Araştırma, *AVRASYA Uluslararası Araştırmalar Dergisi*, 7(17): 234–248.

Akyurt Kurnaz and Kılıç, B. (2020). *Hüzün Turizmi Alt Disiplinler ve Türkiye Destinasyonlarından Seçmeler*, Nobel: Ankara.

Akyurt Kurnaz, H., Kurnaz, A. and Tatlıoğlu, C. (2017). Hüzün Turizmi Destinasyonlarına Bir Örnek: Ulucanlar Cezaevi, *Avrasya Bilimler Akademisi Sosyal Bilimler Dergisi*, Özel Sayı Aralık 2017, 193–200.

Akyurt Kurnaz, H., Çeken, H. and Kılıç, B. (2013). Hüzün turizmi katılımcılarının seyahat motivasyonlarının belirlenmesi. *İşletme Araştırmaları Dergisi*, 5(2), 57–73.

Alaeddinoğlu, F. and Aliağaoğlu, A. (2007). *Savaş Alanları Turizmine Tipik Bir Örnek: Büyük Taarruz ve Başkomutan Tarihi Milli Parkı*, 18(2): 215–225

Alili, M. (2017). Avrupa'daki ve Türkiye'deki Hüzün Turizmi Destinasyonlarının Karşılaştırması Üzerine Teorik Bir Çalışma, *Uluslararası Global Turizm Araştırmaları Dergisi*, 1(1): 37–50.

Aydın, Ş. and Çoban, G. (2019). Alternatif Turizm Türlerinde Yeni Eğilimler. Edt: Ş. Aydın ve D. Eren. içinde, *Alternatif Turizm.* (267–314). Detay Yayıncılık: Ankara.

Belli, S. and Güneren Özdemir, E. (2018). *Sinop Cezaevi'nin Hüzün Turizmi Çerçevesinde İncelenmesi: TripAdvisor Örneği*, 1. Uluslararası Turizmi Yeni Jenerasyonlar ve Yeni Trendler Konferansı (01–03 Kasım 2018): 117–127.

Birdir, K., Dalgıç, A., Güler, O. and Kayaalp, Y. (2015). Hüzün Turizmi: Gelibolu Yarımadası Tarihi Milli Parkını Ziyaret Eden Yabancı Turistlerin Deneyimleri, *Journal of Tourism and Gastronomy Studies*, 3(4): 12–23.

Çelik, A. (2017). Hüzün Turizmi Üzerine Bir Derleme, *Türk Turizm Araştırmaları Dergisi*, Cilt.1, Sayı.3, ss.41–55.

Çetinsöz, B. C. (2017). *Karanlık (Dark) Alanları Ziyaret Eden Yerli Turistleri Motive Eden Etmenler: Ankara Ulucanlar Cezaevi Müzesine Yönelik Bir İçerik Analizi.* Uluslararası Sosyal Bilimler Sempozyumu, 18–20.

Güncü, A. and Güneş, E. (2017). 1939 Depremi İle Afet Yeri Olan Erzincan'ın Hüzün Turizmi Potansiyeli Üzerine Bir Araştırma, *Turkish Studies*, 12(29): 251–270.

Hıdır, C. and Somuncu, M. (2018). *Auschwitz-Birkenau Anıt ve Müzesi'nin (Polonya) Hüzün Turizmi Kapsamında Değerlendirilmesi*, TÜCAUM 30. Yıl Uluslararası Coğrafya Sempozyumu, 3–6 Ekim 2018: 150–166.

Kılıç, B. and Akyurt, H. (2011). Destinasyon İmajı Oluşturmada Hüzün Turizmi: Afyonkarahisar ve Başkomutan Tarihi Milli Parkı, *Gaziantep Üniversitesi Sosyal Bilimler Dergisi*, 10(1): 209–232.

Özçelik, H. (2012). *Gelibolu-Waterloo Karşılaştırması Yoluyla Türkiye'de Keder Turizmi Potansiyeli Analizi.* (Yayınlanmamış Yüksek Lisans Tezi). Akdeniz Üniversitesi Sosyal Bilimler Enstitüsü. Antalya.

Pekerşen, Y.(2020). *Karanlık Turizm. "Kara Turizmin Köklerini Keşfet".* Detay Yayıncılık: Ankara.

Stone, P. R. (2006). A dark tourism spectrum: Towards a typology of death and macabre related tourist sites, attractions and exhibitions, *Tourism: An International Interdisciplinary Journal*, 54(2): 145–160.

Tarlow, P. E. (2005). *Dark tourism* – The appealing "dark" side of tourism and more, Edited by Marina Novelli, Niche Tourism, Routledge: UK.

Yıldız, Z., Yıldız, S. and Aytemiz, L. (2015). Kara turizm, terör turizmi ve Türkiye potansiyeli. *İnsan ve Toplum Bilimleri Araştırmaları Dergisi*, 4(2), 390–407.

Yirik, Ş. and Seyitoğlu, F. (2014). *Hüzün Turizmi ve Mabetleşme Süreci Kapsamında Sarıkamış,* 15. Ulusal Turizm Kongresi. (s. 754–765). Gazi Üniversitesi Turizm Fakültesi, Ankara, 13–16.11.2014.

https://www.tsiv.org.tr/Home/Detail/5329/gelibolu-sehit-denizalticilar aniti#. Xzz8lugzbIU (Accessed: 07.08.2020).

http://www.yedikule.org/yedikule/yedikule-tarihi/13-yedikule-zindanlar (Accessed: 07.08.2020).

https://muze.gov.tr/muze-detay?DistId=MRK&SectionId=SCE01 (Accessed: 08.08.2020).

https://sinop.ktb.gov.tr/TR-74927/istatistikler.html (Accessed: 08.08.2020).

https://www.ulucanlarcezaevimuzesi.com.tr/tarihce/ (Accessed: 09.08.2020).

https://www.kulturportali.gov.tr/turkiye/sakarya/gezilecekyer/adapazari-deprem-ve-kultur-muzesi (Accessed: 09.08.2020).

https://www.anitkabir.com.tr (Accessed: 09.08.2020).

https://ankara.ktb.gov.tr/TR-152372/anitkabir.html (Accessed: 09.08.2020).

https://ankara.ktb.gov.tr/TR-217305/anitkabir-ataturk-ve-kurtulus-savasi-muzesi.html (Accessed: 09.08.2020).

https://canakkale.ktb.gov.tr/TR-70574/gelibolu-yarimadasi-tarihi-milli-parki-eceabat.html (Accessed: 09.08.2020).

http://www.soma.gov.tr/soma-maden-sehitligi-ve-aniti (Accessed: 11.08.2020).

Alper IŞIN

12 Franchise System and Its Applications in Food and Beverage Business

1 Introduction

In the past few decades, franchising has been shown as one of the most efficient growth strategies for the services and retail industries (Barthelemy, 2008). Franchising businesses especially dominate service sectors such as hotel management, catering enterprises and auto maintenance. Among these the food and beverage industry is considered to be the leading franchise system due to the presence of leading franchising icons such as McDonald's and Kentucky Fried Chicken (Combs and Ketchen, 1999).

Franchising is a business model that gives the franchisor the right to operate under its brand to a third party as the franchisee, in return for an advance fee and ongoing royalties based on sales performance (Barthelemy, 2008). Franchising has become a popular global format for doing business in many industries (Czinkota, Ronkainen and Donath, 2004). Franchising is a strategy applied nationally and internationally to grow a business. (Lederman, 2013).

Since 1945, franchising has shown up as a growing part of our economy. According to the International Franchising Association (IFA) estimates, despite the serious recession in the world economy, the number of franchising businesses and direct franchise-based employees continued to increase significantly, and the contribution of franchising to the gross domestic product continued to increase (Lederman, 2013). According to the association's estimates, all in all, the franchise system has proven to be an efficient and effective expansion strategy in the service industries, contributing over $ 750 billion to the economy (International Franchise Association, 2018).

The tourism sector, one of the most widely accepted sectors of the franchising strategy (Koh, Lee and Boo, 2009), creates additional stakeholder groups such as franchisees of restaurant companies, employees of franchisees, local communities, and governments where they operate (Kim and Lee, 2020). For this reason, the franchise system is very important for food and beverage businesses (restaurants, cafes, bars, etc.). In this research, the relationship of the food and beverage businesses with the franchise system in Turkey and the situation wanted to be explained.

Definition and History of Food and Beverage Business

The food and beverage sector is one of the largest and most important sectors today, which continues to develop even during the times of the greatest economic crises that affect the whole world. One of the main reasons for this is the fact that eating and drinking is a necessity to survive since the day human beings existed, and nutrition is at the most basic step according to the hierarchy of needs as a must, and the history of food and drink goes back to the existence of humans. Therefore, the people and establishments (businesses) that provide services for this basic need go back to very old times in history.

Food and beverage businesses cover many types of businesses in many different areas where different services are provided (Özkoç, Arslan, Kendir and Erdoğan, 2019). Every business shapes its service type according to the wishes and needs of its potential customers (Denizer, 2005). Food and beverage businesses are businesses that aim to gain profit by providing goods and services in order to meet the most basic food and drink needs of individuals in travels and accommodation for different reasons other than their own residences. (Sokmen, 2003). On the other hand, Aktas (2001) defines food and beverage businesses as commercial service businesses that meet the food and drink needs of individuals at certain prices.

In the social context, the beginning of the mass catering service for people and the beginning of civilizations are held together. Regardless of the human reason, they looked for places where they could stay at night and meet their need to eat when they went from one place to another where they were constantly residing. In order to meet this need, facilities such as inns and caravanserais were started to be established along the roads (Denizer, 2005). The activities of people eating together in large groups date back to 12,000 years ago. In Roman and Greek history, banquet events are encountered 2500 years ago. In Europe, it is seen that the activities regarding the operation of private food places coincided with the end of the 1700s (Ninemeier and Hayes, 2005).

The first known form of today's food and beverage businesses are the cafes opened in England toward the mid-1600s. However, the first food and beverage business that serves in today's sense was opened in Paris in 1765 (Denizer, 2005). The first restaurant in America was opened in 1827 under the name "Delmonico". At that time, the food served in this restaurant was both very varied and very expensive. However, in today's worlds "fast food" businesses which offer faster, quicker, and cheaper food have dominated the food and beverage industry in America. (Denizer, 2005). In addition, since the food and beverage industry meets food need which is one of the basic needs of people, has continued to develop in its historical development even in times of crisis, and the number of food and beverage businesses has continued to increase rapidly day by day. This huge industry employs more than 10 million people with more than 835,000 businesses in America alone (Clarke and Chen, 2007).

The Concept of Franchising and Its Historical Development

There are different opinions on the historical development of the franchising system in many sources. According to one opinion, franchising was used in the medieval period to mean that church officials paid money to the government to collect taxes in a particular region (Ulas, 1999). According to another opinion, again in the Middle Ages, the franchising system was defined as the feudal lords to process a certain part of their own land in order to increase production for the people they employed in their service for a certain fee and to allow a certain trade of what they produce in the markets.

Contracts between brewers and bar owners in the 18th century are known as the first examples of franchising (Hall and Dixon, 1988). The franchise system that emerged in Europe in the 18th century developed rapidly in the 1850s in America. Due to the fact that the demand for sewing machines produced by Singer company could not be met, they opened branches in various regions of the country and created the current dealership system (Acheson and Ferris, 1988).

With the industrial revolution, the advancement of technology, the development of transportation and mass media, mass production has emerged. Manufacturers have realized that their success and profitability will increase as their products reach regional markets. The franchise system, which first started with the Singer Company, was followed by the automobile, beverage, and oil sectors (Ulas, 1999).

The concept of franchise has started to be used to express the use of the elements such as emblems, products, symbols, goods, and services of another company as a result of the economic and political changes in the world after the Second World War. Thus, franchising gained the meaning of allowing the use of industrial and intangible goods of a company commercially, by leaving the sales relationship between the manufacturer and the seller in the 1960s. (Arslan, 2006; Gurzumar, 1995; Sogur, 1993; Ulas, 1999).

Since the word franchising does not have an exact local equivalent in the world literature, this word is expressed in the same way in all country languages. The original origin of the word franchising is derived from the English word "to free" which means release (Can, 2012). Franchise is defined as "the formal license granted to sell the product or service of a company in a certain region" (Cowie, 1990).

Franchise's equivalent in the dictionary is concession. Franchising is a long-term and continuous business relationship agreement arising from the privilege of the party holding the concession right of a product or service to the second party to carry out business transactions by providing information support within a certain period, conditions, and limitations. (Tutar, 2010).

There are at least two parties in the franchising system, as franchisor and franchisee. Franchisor is the party that owns a product, service or know-how, a tried, proven, and successful brand, name, and gives the right to sell, distribute or operate them for a certain price (Tutar, 2010). With a shorter expression, it is the party that lets to use its name and brand.

Franchisee, another party in the franchise system, is the real or legal person who pays and uses the right to do business within the scope of the name and system of the parent company, within the conditions specified in the franchising agreement.

Features and Types of Franchise System

In the globalizing system, businesses that want to increase their profitability, promote their brand, and grow more, open up to the world with their own brands without making any financial investment through franchising. The franchise system is one of the most suitable methods for businesses that want to include their brand in the international market with a less risky method.

In other words, the franchising system is a system that protects the profitability of the entrepreneurs from the disadvantages of starting a business from scratch that they know but has no idea how to do (Koc, 2006).

The franchising system can be classified according to the experience of the franchisee and the legal and economic characteristics of the rights provided to the franchisee as well as the location of the contract parties and distribution channels. Accordingly, franchising types can be examined according to the countries of the parties, distribution channels and the nature of the franchising businesses (Alkan, 2015).

Franchising in Regards to the Countries of the Parties

The franchising system is a relationship that begins with a contract signed between two parties: The franchisor and the franchisee. In this system, the two parties can be in the same country or the parties can be in different countries.

Domestic Franchising

It is a franchising system where both sides are in the same country. A business can use the franchise system to increase its profitability and grow. In this way, businesses can bring their brand to distant settlements in the country and increase their awareness.

Domestic franchise agreements are the agreements that made between cities or regions. Since both sides are in the country, all earnings gained remain within the country's borders. The franchisee is granted a narrow zone privilege in franchising between cities, and a wide zone concession in franchising between regions. Granting the wide area concession is similar to the main dealership system (Sogur, 1993).

The concessions granted between regions and cities may create situations that may endanger the brand of the main franchisor from time to time. The fact that franchisees are in different regions may weaken the control of the franchisor and different applications can be seen in different cities. A similar situation was experienced in fast food businesses in the USA in 1955. While some of the franchisee businesses were serving food, some of them were not serving. This has damaged the brand image of the franchisor company by causing the double standard in practice (Alkan, 2015).

One of the basic conditions for being successful in the franchise system is achieving a common standard in goods and services and to maintain quality. People generally prefer the chain of food and beverage that they know in general, in their food and beverage preferences, when they away from the places they live in. In this case, the lack of a standard will damage the image and quality of the brand.

International Franchising

International franchising involves more complex dynamics than local franchising, especially in terms of franchise network configurations. (Alon, 2006). This type of franchising is applicable if the franchisor and the franchisee parties are located in different countries. The franchisor gives the rights and privileges it owns to the franchisee in another country. Thus, it exports the goods or services subject to franchising under the franchising agreement. It also transfers a certain amount stipulated for franchising to another country. (Sogur, 1993).

An entrepreneur can use a variety of strategies to enter a foreign country's market. In the targeted country:

- Direct investment (opening a branch),

- Giving franchise directly to individuals,
- Giving master franchise,

Before deciding which of these strategies to choose, market research should be conducted on business scale, international experience, market potential, investment risk, whether the products or service will be successful in the target country, and the suitability of the management structure of the franchisor, financial resources, competition laws of the country, tax regulations and import controls should be considered (Terpsra, 1983).

Businesses can establish their own production facilities and open branches there by investing directly in that country as a result of the low costs of labor, raw materials, and similar investment factors in the foreign country.

As direct investment provides technology transfer to the target country, the image of the company gets strengthen, earnings are not shared with intermediaries, and the control over the products and services and investment is fully ensured. Goods and services are applied to the region by improving relations with supply sources, customers, and government (Ulaş, 1999).

Another method in the international franchise system is the method that businesses give franchises directly to entrepreneurs in foreign countries. Granting franchises directly to independent entrepreneurs may cause to be faced with some difficulties due to reasons such as financial, consultancy services, information and technology transfer and the legal legislations of the foreign country. In addition, the direct franchising method may be preferred if the target country of the franchisor is geographically, linguistically, and culturally similar to the home country. The advantages of direct franchising are the profits are high due to the fact that the control over the franchise activities are not lost and the income obtained is not shared with third parties.

Franchising in Regards to Distribution Channels

Four types of franchise are defined according to their economic and legal characteristics. These four main franchising types are classified as product and brand franchising and operating system franchising in the literature (Thompson, 1971).

- Manufacturer-Retailer Franchise System
- Manufacturer-Wholesaler Franchise System
- Wholesaler-Retailer Franchise System
- Brand Name Franchise System

In the manufacturer – retailer franchise system, the franchisor is the company that manufactures the product, and the franchisee is the retailer that delivers the product purchased from the manufacturer directly to the final consumer (Kucuk, 2011).

The franchisee tries to create a market where customers can find products produced by the manufacturer. In this system, the franchisee uses the name of the business and follows the procedures of the licensing business (Ulas, 1999). In addition, the franchisee does not have the authority to decide which products or services will be offered. The franchisee is in charge of offering the product or service determined by the manufacturer (Arslan, 2006; Durukan, 2006; Kirca, 1997; Ulas, 1999). As examples of the manufacturer-retailer franchise system, automobile manufacturers Mercedes, BMW, Audi, etc. and petrol companies Shell, BP, etc. can be shown.

Manufacturer-Wholesaler Franchise System

Soft drink companies and brewers can be cited as examples of this system, in which the manufacturer is the franchisor and the wholesaler is the franchisee. In the production of soft drinks, the manufacturer company supplies concentrated syrup to independent bottlers who are wholesalers (Ulas, 1999). The franchisee reaches the final product by diluting this concentrated syrup at the rate that the manufacturer has given it and distributes this product to retailers for local distribution (Hall and Dixon, 1988).

In this type of franchise system, the franchisee wholesaler company ensures that the product is finalized and delivered to the consumer. In this system, the wholesaler company does not have the right to make any changes regarding the content and features of the manufacturer's products. The most well-known international beverage companies (Coca-Cola, Pepsi, 7up) can be cited as an example of this system.

Wholesaler-Retailer Franchise System

In this franchise system, the wholesaler is the franchisor and the retailer is the franchisee. Retailers who will buy a franchise will gather under the leadership of a specific wholesaler and join a chain. Franchisee retailers continue to work without losing their independence by taking advantage of the collective purchasing power of these chain businesses and other chain business transactions. According to the contract, a cooperative is formed that requires member retailers to purchase from it (Ulaş, 1999).

The franchisor will strengthen its own distribution network by arranging retail points to reach the end user for the sale of its products and by predetermining the storage and distribution facilities of its products. At the same time, franchisor will be able to deliver to consumer, which is a heavy burden on the company, with the franchisee retailers. This system is generally applied in car tire companies such as Michelin, Goodyear and for automobile spare parts sales (Yuksel, 2013).

Brand Name Franchise System

The parent company has a well-known brand name that well known to many and grants the franchisee a license to use its brand name by making an agreement with the franchisee. The franchisee shows to retailer their own business methods to be successful. The most well-known hotel businesses such as Hilton, Sheraton, Ritz Carlton, McDonalds, KFC, Vapiano, fast food and restaurant chains, and car rental companies such as Hertz and Avis can be cited as examples of this franchise type (Ulas, 1999).

Franchising in Regards to the Nature of Franchise Businesses Product and Brand Franchise

The basis in this type of franchising is the use of a specific brand or trade name. In addition, the parent company granting the concession has a wide and continuous right to audit on the counterparty, including management and administration. However, in some simple matters, the franchisee may enter the production phase. Automobile and truck dealers, gas stations, soft drink producers are the most typical examples of this type of franchising. As examples of brand name franchising system, hotel chains such as Sheraton, Holiday Inn, restaurant chains such as Mc Donald's, Burger King, KFC, and car rental companies such as Avis, Hertz, and Budget can be shown. (Burgu, 2008).

Enterprises in Turkey prefer to have the franchisee of foreign brands. Due to the fact that trademark protection is easier than protection of business systems, products and brands franchise system is more commonly used in Turkey.

As examples of product and brand franchise from Turkey Mavi Jeans, Cotton, Damat-Tween, Tefal, Colins, and Zara can be shown. In this system, the focus is on marketing, while franchisees try to reach the customer by giving a quality and service standard guarantee like the parent company. The products to be sold are usually produced by the franchisor company (Abdullaev, 2011).

The product and brand franchise is defined as "traditional" or "first generation" franchising in some sources and it is explained under this heading in dealership and distributorship (Stern, Ansary and Coughlan, 1996).

Dealership System

Dealer is defined as a completely independent third person or intermediary between the producer and the consumer, who sells the products purchased from the wholesaler to the customer. (Dogan, 1994). There are some fundamental differences between dealership and franchising systems. When we summarize them:

- While the financial strength, experience, workplace, and location of the entrepreneurs who will get dealership are important in the selection of the dealers, in addition to these features in the franchise system, the personal characteristics of the entrepreneur are also an important variable in obtaining a franchise.
- The control of the dealers is comfortable compared to the franchisees; the dealers can act more freely than the franchisees. (Tuncer, 1992).

Distributorship

Distributor is a legally independent business system that has the right to purchase and sell the products of the manufacturer company in a specific region, maintaining its standards by preserving the image of the brand. The distributor sells by putting the costs and profits on top of the cost of the products purchased from the manufacturer (Akat, 1996).

Compared to franchising, it is seen that the franchisee company is faced with a much stricter control by the parent company than the distributors. Distributors do not pay royalty to the parent company, here the manufacturer companies sell the products they produce to the distributors by adding their profits and earn in this way.

Operating (Management) System Franchise

According to the definition of the American Department of Commerce, the operating system includes franchising, the way of recruiting between the franchisor and the franchisee, the marketing strategy and plan, compliance with standards, quality control and continuous communication relationship. In other words, the case is on the operation of a firm as a whole. A successful retail business gets the sale of the right to operate the same business in a different geography. This type of franchising is a method where the producer provides the goods to more franchisees (Yurdakul, 2003).

The franchise, which is the operating system, is a type of franchise where the franchisor provides the franchisee for the right to use its own brand and sell its own product/service, as well as the entire operating system covering quality standards and training processes. In this system, all activities related to marketing and production takes place between the franchisee and the franchisor, not only in the field of product, service, and brand but also within the operating system as a whole. The places where this system is applied are especially hotels, restaurants, retail stores, rental, and consultancy companies (Arslan, 2006).

As examples of the operating system franchising companies such as McDonald's, Holiday Inn, and Avis can be given. The materials used by a franchised business in the food and beverage sector are purchased from the parent company itself or from the places determined in accordance with the conditions determined by the parent company. The physical layout, positioning and training of the personnel are organized by the parent company (Ulas, 1999).

Conversion Franchising

It is the connection of an independent business operating in a specific sector to a successful parent company in the same sector and transforming the business into a franchisor's sales store. While the franchisor gives its trademark and operating system, the franchisee also presents its own property, professional experience, and knowledge to the franchise chain. The incorporation of existing real estate agents, travel agents and pharmacies into franchise chains are examples of conversion franchises (Sogur, 1993).

While the parent firm gains its expertise and capital from an experienced businessperson's good knowledge of the field in which they operate, the franchisee entrepreneur gains a number of advantages such as advertising, purchasing, research and development with a famous brand and business system. The reasons why experienced people who carry out their business independently in their own sales units switch to the franchise system by transforming business systems are economic partnerships, advertising under the umbrella of a known trademark and providing services with low costs (Groselli, 1989).

Franchise System in Turkey

Franchising in Turkey, used as the dealership system in automobile, white goods and cosmetics sector in the 1970s and 1980s. In the late 1980s, as a result of the removal of bureaucratic obstacles, Turkey has become an attractive market for foreign investors, and foreign companies began to give their rights to use the brand, brand names and business methods to Turkish investors who know better the local market conditions (Gokdeniz and Erdem, 2004: 37).

Fast food, cosmetics, clothing, and transportation are the first sectors in our country to develop with franchising. In 1986, after McDonalds, Kentucky Fried Chicken and Wimpy, have entered the Turkish market. Beauty Shops in cosmetics, Levi's, Benotton, Lee in clothing and Express Cargo and Mail Boxes in transportation are the first businesses that come to mind (Alacaklioglu, 1999).

The National Franchising Association (UFRAD) was established in 1991 in Istanbul with the purposes of: learning and development properly of the franchising in Turkey, providing work in harmony with the parties as an authoritative place against members present in incorrect practices in franchising, and in order to spread the franchising system in the country. Businesses in Turkey give franchise in sectors such as food, service (cleaning, rental cars, computers, cosmetics, real estate, shipping), clothing, and retailing (Ulas, 1999).

According to the UFRAD's data, there are 876 thousand chain stores in Turkey. 24 % of these are foreign and 74 % are Turkish origin brands. Around 50 thousand branches continue to grow with franchise or dealership systems. There are 1,471 franchisor companies in Turkey. The annual volume of the market formed by these brands and their representatives is approximately 35 billion dollars.

Today in Turkey, the number of franchises businesses carry out their activities under the name of certain brands is around 2,500. According to statistics of European Franchise Federation (EFF), Turkey is located as first by the number of franchise in Europe. Turkey is followed by France with 1,375 brand and Germany with 960 brand.

In the food and beverage sector, chain restaurant enterprises prefer the franchising system, which is rapidly spreading in the world, in order to increase their brand, make themselves heard more and increase their profitability. According to international franchise data, franchise businesses hold 65 % of the employment in the food.

2 Conclusion

According to 2017 data of international franchise, the top 5 companies, which have the most franchisee businesses in the world, operate in the food sector. These are listed as McDonald's, KFC, Burger King, Subway, and 7-Eleven. (http://www.franchisedirect.com). In the coming years, it is predicted that franchise investments will be directed toward businesses that produce healthy food, as the franchise preferences in the food sector will change by affecting people's eating habits in this direction.

References

Abdullaev, B. (2011). Franchising Uygulamalarında Franchise Alanların Memnuniyetine Yönelik Bir Araştırma. Yüksek Lisans Tezi, İstanbul Üniversitesi Sosyal Bilimler Enstitüsü, İşletme Anabilim Dalı, İstanbul.

Acheson, K. and Ferris, S. (1988). Retail and Wholesale Trade Services in Canada. Singapore: The Fraser Institute.

Akat, Ö. (1996). Uluslararası Pazarlama Karması ve Yönetimi, 1. Baskı, Bursa: Ekin Kitabevi Yayınları.

Aktaş, A. (2001). Ağırlama Hizmet İşletmelerinde Yiyecek ve İçecek Yönetimi, Antalya: Eren Ofset.

Alacaklıoğlu, A. S. (1999). Yatırım Destekleri Konusunda Soru ve Cevaplar, İstanbul Ticaret Odası İstanbul.

Alkan, P. (2015). Türkiye Gıda Sektöründe Franchising Uygulamaları, Yüksek Lisans Tezi, Uludağ Üniversitesi, Fen bilimleri Enstitüsü, Bursa.

Alon, I. (2006). Market conditions favoring master international franchising. *Multinational Business Review*. 14, pp. 67–82.

Arslan, K. (2006). Kobi'lerde Yönetim ve Pazarlama Aracı Olarak Franchising, İstanbul: İstanbul Ticaret Üniversitesi Yayınları.

Barthélemy, J. (2008). Opportunism, knowledge, and the performance of franchise chains. *Strategic Management Journal*. 29(13), 1451–1463.

Burgu, A. (2008). Franchising Sisteminde Güç ve Çatışma İlişkileri. Yüksek Lisans Tezi, Kocaeli Üniversitesi Sosyal Bilimler Enstitüsü, İşletme Anabilim Dalı, Kocaeli.

Can, E. (2012). Uluslararası işletmecilik: Teori ve uygulamalar, 4. Baskı, İstanbul: Beta Basım.

Clerke, A. and Chen, W. (2007). International Hospitality Management, Butterworth-Heinemann, El-sevier, GB.

Combs, J. G. and Ketchen, D. J. (1999). Can capital scarcity help agency theory explain franchising? Revisiting the capital scarcity hypothesis. *Academy Management Journal*. 42 (2), 196–207.

Cowie, A. P. (1990). Oxford Advanced Learner's Dictionary. Oxford: Oxford University Press.

Czinkota, M., Ronkainen, I. and Donath, B. (2004). Mastering global markets – Strategies for today's trade globalist. Mason, OH: Thomson-South-Western.

Denizer, D. (2005). Konaklama İşletmelerinde Yiyecek ve İçecek Yönetimi, 1. Baskı, Ankara: Detay Yayıncılık.

Doğan, N. (1994). Franchising Bayilik Değildir, Dünya.

Durukan, T. (2006). Uluslararası Pazarlara Giriş Stratejileri Franchising Sistemi ve Türkiye Uygulanması, Ankara: Asil Yayın.

Groselli, J. (1989). Franchising: As a Bridge Between Small And Large Firms, London: Graham & Trotman.

Gökdeniz, A and Erdem, B. (2004). Uluslararasılaşma süreci, sözleşmeye dayalı uluslararasılaşma yolları-(franchısıng (isim hakkı)) ve bir hizmet işletmesi (mc.donald's) üzerinde kavramsal bir araştırma, *Verimlilik Dergisi*, 4.

Gürzumar, O. B. (1995). Franchise Sözleşmeleri ve Bu Sözleşmelerin Temelini Oluşturan "Sistem" lerin Hukuken Korunması, İstanbul: Beta Basım.

Hall, P. and Dixon, R. (1988). Franchising, London: Pitman Publishing.

International Franchise Association. (2018, January 28). Franchise sector expected to outpace U.S. economy-wide job creation in 2018.

Kim, B. and Lee, S. (2020). The impact of material and immaterial sustainability on firm performance: The moderating role of franchising strategy. *Tourism Management*. 77, 1–13.

Kırca, Ç. (1997). Franchise Sözleşmesi, Ankara: Banka ve Ticaret Hukuku Araştırma Enstitüsü.

Koç, M. (2006). Bir Girişimcilik Modeli Olarak Franchising. Yüksek Lisans Tezi, Dokuz Eylül Üniversitesi Sosyal Bilimler Enstitüsü, İşletme Anabilim Dalı, İzmir.

Koh, Y., Lee, S. and Boo, S. (2009). Does franchising help restaurant firm value? *International Journal of Hospitality Management*. 28(2), 289–296.

Küçük, B. (2011). Uluslararası Pazarlara Giriş Stratejilerinde Franchising Sistemi ve Türkiye Uygulaması, Yüksek Lisans Tezi, Fatih Üniversitesi Sosyal Bilimler Enstitüsü, İşletme Anabilim Dalı, İstanbul.

Lederman, H. Y. (2013). Franchising and franchise Law: An introduction. *Michigan Bar Journal*. 92(1), 34–37.

Ninemeier, J. D. and Hayes, D. K. (2005). Restaurant Operations Management: Principles and Practices. Prentice-Hall. New Jersey, USA.

Özkoç, A. G., Arslan, E., Kendir, H. & Erdoğan, T. (2019). Otel İşletmelerinde Yeşil Mutfak Kalitesinin (Y-Mutkal) Ölçülmesi: Nevşehir İlinde Bir Araştırma. *Journal of Tourism and Gastronomy Studies*, 7(3), 2294–2309.

Sökmen, A. (2003). Ağırlama Endüstrisinde Yiyecek ve İçecek Yönetimi, Ankara: Detay Yayıncılık.

Stern, L. W., Ansary, A. E. and Cpughlan, A. T. (1996). Marketing Channels, New Jersey: Prentice Hall.

Şoğur, M. (1993). İşletme Sistemi Olarak Franchising, 2. Baskı, İstanbul: Der Yayınları.

Terpsra, V. (1983). International Marketing, Third Edition, The Dryden Press, New York: CBS College Publishing.

Thompson, D. (1971). Contractual Marketing Systems, Lexington: Heath Lexington Books.

Tuncer, S. (1992). Bir Dağıtım ve Pazarlama Aracı Yöntemi Olarak Franchsising, İSO Dergisi, Sayı: 311.

Tutar, H. (2010). İşletme Yönetimi, Ankara: Seçkin yayıncılık.

Ulaş, D. (1999). Bir Pazarlama ve Dağıtım Yöntemi Olarak Franchising Sistemi, Ankara: Nobel Yayın.

Yurdakul, M. (2003). Franchising Sisteminin KOBİ'lerde Pazarlama İşlevi Bakımından Değerlendirilmesi, Dumlupınar Üniversitesi Sosyal Bilimler Dergisi, 8(2).

Yüksel, H. L. (2013). Franchise Sözleşmesinin Genel Olarak İncelenmesi, İfası ve Sona Ermesi, Yüksek Lisans Tezi, İstanbul Üniversitesi Sosyal Bilimler Enstitüsü, İstanbul.

https://www.franchisedirect.com/top100globalfranchises/rankings, Accessed on 02.08.2020.

Esra ŞAHİN

13 Gastronomy in Tourism Marketing

1 Introduction

The origin of the word gastronomy in almost all sources is shown as the words *gastros* (stomach) and *nomos* (law) in Ancient Greek (Scarpato, 2002). It is known that the word gastronomy, which can be translated as rules regarding the stomach, has a meaning beyond this shallow definition. Having been related to many different disciplines from medicine to chemistry, agriculture to anthropology makes it difficult to define agreed gastronomy. In the literature, different authors have described gastronomy in different ways. For example, Santich (2004) defines gastronomy as the art and science of good eating. Hegarty and O'Mahony (2001) describe gastronomy as a concept that includes the basic ingredients used in the preparation of food, the methods of food storage, preparation and cooking, the types and quantities of food and beverages, popular and unpopular flavors, customs and traditions related to food and beverage serving, tableware, and beliefs about food and beverages. Gillespie (2001) associated the concept with pleasure and described it as a discipline to increase the pleasure from food. Santich (2004) explains gastronomy as providing and guiding suggestions on what to eat or drink in where, when, how and with which combinations based on historical, cultural, and environmental influences. According to Brillat-Savarin, gastronomy should bring pleasure and health together and enjoying the food should not be considered as a sin anymore (Freedman, 2008: 266). When we look at the early definitions of gastronomy, it is seen to be mostly associated with luxurious and flamboyant dishes (Barkat and Vermignon, 2006). When all the definitions are evaluated, gastronomy can be divided into two main categories. The first category can be best associated with pleasure from food and beverages. The second category includes everything related to eating and drinking in a wider sense (Scarpato, 2002). The relationship of gastronomy with pleasure and luxury has led to the perception of gastronomy as an eating and drinking activity that only the elite group enjoyed until the 1980s. Those who had reached a certain moral satisfaction and were economically self-sufficient could travel to France to visit a starred restaurant or to taste a special drink or dish. Beginning in the 1950s and highly popular in the 1970s, the behavior of wondering ethnic food and traveling to taste these foods has become one of the most enjoyable activities of citizens of high-wealth countries (Şahin, 2015). Over time, factors such as the increase in income levels, the interest in quality time-spending, the turn toward experience-oriented activities, the advances in internet and mobile technologies facilitating access to information and the search for differences have also changed the structure of consumer behavior. Eating and drinking, which is the common ground for all humanity have become more and more popular, and different cultures, as well as foods and beverages of these cultures have become a matter of curiosity. This has also affected the tourism marketing activities, and the tendency to move from mass tourism to niche markets has developed. Gastronomic tourism has especially become prominent in the recent years for those who know what they want and seek a more distinctive and unique product.

Due to the increase of both social and academic interest in gastronomy, this section will examine the place and importance of gastronomy in tourism marketing based on secondary data. Firstly, the concept of gastronomic tourism will be explained, tourist motivations and typologies will be evaluated. Then, gastronomic products that have an impact on tourism marketing and their use will be mentioned, and the results and recommendations will be presented.

2 Gastronomic Tourism, Tourist Motivations and Typologies

Perhaps the first person to write the relationship between gastronomy and tourism is the Sicilian Greek Archestratus. Archestratus has travelled through the Ancient Mediterranean Region to find the best food and beverage and took notes about the food and the places he visited. His work named Gastronomia, written in 4th century B.C., is a food and wine guide for the Mediterranean Region. This book provides information on what are the best foods and beverages in the region and where to find them and is also known as the oldest cookbook in the world (Santich, 2004).

The concept of gastronomy, which was not mentioned for a long time after the ancient times, started to become more popular in Europe in the 19th century. It is observed that food and beverages, including wine, were not considered as a product of tourism until the 1950s. Since the 1960s, the spread of cooking shows on television and radio, the increasing number of cookbooks, and the increased interest in products such as wine have made food increasingly popular. Since the 1970s, it became easier for people to travel, and with the growing interest in dishes of different countries, it has become inevitable for gastronomic tourism to develop. The shape of food consumption has begun to change, and besides being a physiological need, it has gained a different meaning with social status and cultural background (Hall and Mitchell, 2005). In the pressure of capitalism and

Post-Fordist consumption, eating and tasting new food has become a psychological and socio-cultural need, besides being a biological one.

These changes have brought about academic interest and research that food can be used as a tourism product (Belisle, 1983; Hobsbawm and Ranger, 1983). Although Lucy Long from Bowling Green University used the term "culinary tourism" in 1998, the beginning of gastronomic tourism in the sense we use today is accepted as Erik Wolf publishing "Culinary Tourism White Paper" in 2001. Erik Wolf then established the International Culinary Tourism Association (ICTA) in 2003, with the opinions and support of the tourism industry. This term did not come to the fore until ICTA was established (Wolf, 2002). Long (2004) defined the culinary tourism as "the intentional exploratory participation in the foodways of another participation including consumption, preparation, and presentation of a food item, cuisine, meal system, or eating style considered to belong to a culinary system not one's own" (Long, 2004: 21). Since then, it has been observed that food and drink motivated activities have been named in different ways such as food tourism, food and wine tourism (Hjalager and Richards, 2002), cuisine travel, culinary travel (Ignatov, 2003), gastronomic tourism (Hall and Mitchell, 2001; Hjalager and Richards, 2002), culinary tourism (Wolf, 2002), tasting tourism (Boniface, 2003), gourmet tourism (Okumuş, Okumuş and Kercher, 2007). Although there are nuances between these concepts, it is seen that they are mainly used to describe food and beverage-oriented activities. In this study, the concept of gastronomic tourism will be used considering that the word gastronomy is more comprehensive for food and beverage-based tourism activities.

The World Tourism Organization (UNWTO) defines gastronomic tourism as "a type of tourism activity which is characterized by the visitor's experience linked with food and related products and activities while travelling" (UNWTO, 2019). Apart from daily routine visits to any restaurant, Hall and Sharples (2003) evaluate the visits made to taste a special food, to experience the products and production processes of a region or to taste the food prepared by a famous chef within the scope of gastronomic tourism. Lin, Pearson and Liping (2011), define gastronomic tourism in the form of traveling with the purpose of tasting food and beverages or acting with that motivation at least in part of the trip. Deneault (2002) considers gastronomic tourism as more than a dining experience and defines it as a cultural discovery movement, which includes a wide variety of culinary and agritourism activities, from food festivals to farm tours, and tasting of regional flavors. When these definitions, as well as others in the literature are evaluated together, it is clearly understood that the main motivation of gastronomic tourism is food and beverages (Hall and Mitchell, 2005). Therefore, it can be said that the distinctive point of food and beverages, which is the subject of gastronomic tourism is whether it is the main or assistive motivation for travel. Foods that are eaten to meet the physiological needs are not of interest in gastronomic tourism (Hall and Mitchell, 2005). In their studies, Quan and Wang (2004) found that in the travel of tourists, food is preferred and consumed as (1) the main attractive factor; (2) a secondary or supporting factor; (3) a part of the daily routine. Those who consider food and beverages as the main attractive or supportive factor in terms of travel motivation make the target market in gastronomic tourism activities.

Therefore, it can be said that the travel motivation is one of the main criteria to determine the gastronomic tourist. Fields (2002) interpreted the categories of tourist motivation by McIntosh et al. (1995) in terms of gastronomy. How gastronomy can be integrated into tourist motivations can be interpreted as follows:

1. Physical motivators: Tourists can be influenced beyond their physical needs with their five senses and can be encouraged to pursue, try, or discover new tastes in the region. Some destinations can make a difference by offering healthy activities or foods that have positive effects on health.
2. Cultural motivators: Motivates tourists to learn and experience the culture of societies other than their own. Food and beverages are one of the most enjoyable ways of discovering and experiencing different cultures.
3. Interpersonal motivators: Helps individuals to socialize and improve their relationships. In the research of Warde and Martens (2000), people stated that eating outside values the social function more than the quality of the food. In this sense, food can be used as a means of communicating and socializing. Activities such as gala dinners and sundowning of touristic organizations or creating mixed groups during meals are examples of food-based socialization activities.
4. Status and prestige motivators: Status and prestige can easily be associated with gastronomy. It is important for many people to eat in a Michelin star restaurant or to be seen in such places. Sharing the gastronomic experience and associated cultural interactions is also seen as prestigious.

Mak, Lumbers, Eves and Chang (2012) examined and suggested that the motivational factors for gastronomy can be categorized into five main dimensions. These are:

- Symbolic
- Obligatory
- Contrast

- Extension
- Pleasure

The symbolic dimension refers to motivations that include symbolic meanings of gastronomy such as status and prestige. Generally, food consumption in tourism is considered as a "symbolic" form of consumption (Mak, Lumbers and Eves, 2012). Gastronomy has many activities that can easily be used as an indicator of status and prestige. The obligatory dimension reflects the relationship between vital needs, health, and nutrition. The contrast dimension refers to the search for difference, apart from the daily routine experiences of the tourist. The extension dimension refers to the motivations to seek food experiences that extend the daily routine of tourists. The pleasure dimension is related to enjoying the food experience.

All these different motivations have led to various categorizations on the typologies of gastronomic tourists. For example, Mitchell and Hall (2003) divided gastronomic tourists into four categories according to their interest in local foods. These are: Gourmet tourists; gastronomy and cuisine tourists; culinary tourists and rural/urban tourists. Ignatov and Smith (2006), on the other hand, gathered gastronomic tourists under three titles with a much simpler approach. These are: Food tourists, wine tourists and food and wine tourists. Hjalager (2002) proposed a relatively more sociological model according to the attitudes and preferences of tourists regarding food and beverage. According to Hjalager (2002), gastronomic tourists can be grouped as follows: (1) Recreational, (2) Existential, (3) Diversionary, (4) Experimental. These tourist typologies explained by Kivela and Crotts (2006) can be summarized as follows.

Recreational tourists are conservative, and they do not like foreign foods. Existential gastronomy tourists want to discover the cuisine of a region beyond the physiological needs and learn about its culture. They are eager to find local restaurants and dishes. They participate in activities, trips and culinary courses related to gastronomy and usually buy local products to take home. Diversionary gastronomy tourists prefer abundant and easily accessible foods. For them, eating is more of a social activity. Experimental gastronomy tourists are interested in new ingredients and new food preparation methods. They are innovative and exploratory people. They consider food and beverages as a part of their social status and prestige. They buy culinary gifts such as kitchenware.

In terms of tourism marketing, knowing the motivations and typologies of gastronomic tourists is important for planning marketing activities. For example, existential and experimental gastronomic tourists constitute an important target market in the typology of Hjalager (2002). Recreational and diversionary tourists, on the other hand, are not the primary target group, but the market can be expanded by adding activities related to gastronomy as a supportive factor in the activities of these tourists.

3 Gastronomic Products in Tourism Marketing

It is observed that destinations are getting increasingly similar with the effect of globalization in international tourism (Ooi, 2004). It is also known that the tourism sector has intense competition in terms of both geographical diversity and the number and quality of businesses. This competitive environment forces destinations to make a difference. Gastronomy and gastronomic products that are included in this concept create an important opportunity for destinations in tourism marketing. The fact that tourists spend nearly 40 % of their budget on food during their travels (Boyne, Hall and Williams, 2003) and that gastronomic tourists spend 24 % more than other tourists makes this market attractive for tourism destination (www.worldfoodtravel.org). Both its ability to create a tourism branch on its own, as well as its product variety makes gastronomy a unique and significant tool in tourism marketing. Especially in the tourism sector where there is intense competition, gastronomic products are one of the areas to successfully achieve destination differentiation. Many qualities such as local gastronomic products, and the ability of these products to grow in certain geographic areas, the variety of food and beverages, the ceremonies of food preparation, cooking and serving, as well as consumption ceremonies, stories about that food, and the meaning they contain, make gastronomic products inimitable. The difficult to imitate and unique structure of gastronomic heritage (Harrington and Ottonbacher, 2010), offers important advantages in creating the destination image and differentiating the destination. The production of the unique food and beverages specific to the region with the skills of local people can help the destination be perceived as different from other destinations by tourists (Haven-Tang and Jones, 2006). Fig. 1 shows the elements that can be used to analyze the gastronomic products and resources of a destination.

All kinds of assets that a destination has and can offer related to food and beverages, can be called gastronomic products. From the grapes grown in vineyard to the wine, and from the factory where this wine is produced to the tasting workshops, many elements are included in the gastronomic product group. It is possible to categorize gastronomic products as natural products (milk, wheat, grape, etc.), products obtained by processing natural products (cheese, wine, pickle, etc.), dishes made from natural products, souvenirs with gastronomy theme, places related to gastronomy (museums, cooking courses, workshops,

Fig. 1: Elements for Analyzing a Destination's Resources, Products and Gastronomy.
Source: UNWTO, 2019

restaurants, factories, shops), intangible assets of gastronomy (rituals, traditions, customs, etc.), and touristic activities based on these.

Various activities that can take place in gastronomic tourism can be listed as follows (Şahin, 2015):

- Visits to the farms where foodstuffs are grown
- Grape harvest events, trips to vineyards
- Visiting local product producers on site
- Cheese making workshops
- Olive oil production and tasting activities
- Visits to local markets
- Participation in local food and beverage festivals
- Tasting of special menus in private venues
- Wine degustation
- Gourmet trips for special occasions (Christmas time market trips and tasting of Christmas food, etc.)
- Visiting private gastronomy venues (gastronomic universities, wine routes, culinary museums, etc.)
- Dining at Michelin star restaurants
- Local cuisine lessons at home in authentic regions
- Luxurious cooking courses
- Gourmet tours that are made to only eat in the best restaurants of the world
- Organizing and participating in cooking competitions

Research shows that cuisine affects the choice of destination (Karim and Chi, 2010). However, not only the products, but also the experiences of the tourists in the places where they receive services affect the choice of destinations. Face to face interaction with the service employee, enjoying the service, not having communication problems (such as providing an English menu (Karim and Chi, 2010)), fulfilling the expectations from the service employee, and performing the service at the desired speed may also affect their evaluations (Richards, 2002).

4 Conclusion and Suggestions

Gastronomy is a value chain that includes many types of activities from food production to distribution, processing to tasting (www.e-unwto.org). The benefits it provides to both the producer and the consumer side lead to increased interest in gastronomic tourism. Although it is claimed that an increasing number of tourists are traveling for gastronomic purposes (Okumuş et al., 2007), according to Long (2004), food and beverages are rarely the main reason for visiting a destination but are often seen as part of the total destination experience. It should be noted, however, that local food and beverages can be the main attraction for tourists seeking to have new and authentic experiences (du Rand and Heath, 2006). For this reason, gastronomic elements need to be successfully integrated into tourism marketing strategies.

In order to achieve successful tourism strategies in terms of gastronomy, first of all, the gastronomic profile of the destination must be created. This profile provides information on the resources of the destination and how to make use of these resources. Analysis of gastronomic tourists is another important subject. It is necessary to make strategic plans by analyzing different data such as how many nights gastrotourists stay, their consuming habits, how they spend, and their demographic features.

The value chain mentioned at the beginning causes gastronomy to be frequently mentioned together with sustainability. Gastronomic tourism is an important resource for the support of local farmers and sustainable agricultural activities, along with the production of local foods. Strong links between tourism and local food can create significant added value, while weak links can be destructive (Hjalager, 2002). If the demands of tourists related to food and beverages can be met from local resources, this will be an important source for the growth of the local economy. However, if local resources are insufficient, it will increase imports, damage local production, and reduce its economic benefits (Richards, 2002). Destinations can increase demand by producing a variety of activities to support the positive impact of gastronomic products on local development. Apart from agrotourism, participation in various activities such as harvesting, which allows the inclusion in the production part of the gastronomy, can increase the pleasure of the tourist from the experience while supporting the local producer. Participating in the production process and having a different and rare experience are among the factors that increase this pleasure. Since local food production is dependent on agriculture, hunting and fishing, proper development of links with tourism can encourage increased local enterprise activities and the "bottom-up" development of community-based tourism initiatives (Richards, 2002).

In addition to sustainability, there has been an increasing interest in food safety and traditional production in the recent years. Increased awareness and health concerns about food risks lead individuals to know the origin of the food and the methods of production. In fact, some farms are known to keep their production areas open to visits and control. Establishing a trust-based relationship also helps to establish marketing strategies for producers and tourism agents. Labelling and quality certificates that guarantee the origin and the methods of production of foods can provide an effect that strengthens the image of the destination (Richards, 2002). It can also serve as a resource for gastronomic tourism.

The increasing interest in gastronomic tourism facilitates the emergence of innovative approaches in tourism marketing. It is seen that even the products that are not cared enough or seen as ordinary by the local people can be commercialized and the tourism potential it creates is satisfying. For example, festivals of products that we have not heard of until recent years, such as the Urla International Artichoke Festival, Datça Almond Flower Festival, or Asparagus Festival have become an important tourism product. The number of food-based festivals is increasing, both due to the emphasis on women's employment and labor, and the tourism potential it creates. Eating and drinking, which is a common denominator for everyone, adds a different vitality to these festivals and is effective in introducing and promoting the local culture. Increasing both the number and quality of such festivals will make a difference in destination marketing.

Festivals are places where local foodstuffs are often sold as souvenirs. However, destinations also offer their own food and beverages as souvenirs, apart from festivals. Researches conducted by EUROTEX crafts tourism project in Greece, Finland, and Portugal (Richards, 1999) indicated that 84 % of tourists buying souvenirs in rural areas buy food or beverages to take home. Relatively cheap and easy to carry foodstuff souvenirs are useful products that almost everyone can enjoy. Such gastrosouvenirs also encourage the production of that product.

For example, buying olive oil as a souvenir from Ayvalık is also important for the continuation of olive cultivation in the region.

Gastronomy plays an important role not only in the promotion of a city or a region but also in the promotion of the whole country. The close relationship of gastronomy with the culture contributes to the promotion of the country through the cuisine. Countries such as Italy, who successfully integrate their cuisine into their national identities, offer their cuisine as a unique touristic product along with their historical and geographical beauties. In addition, gastrodiplomacy, which is a concept that has been mentioned more in recent years, enables us to get acquainted with the cultures of the countries that we have not had much knowledge about until now. Gastrodiplomacy is defined as "using the culinary culture of a country as a tool for creating intercultural interaction" (Chapple-Sokol, 2012). South Korea, which is one of the countries that successfully apply gastrodiplomacy, used kimchi as a country branding tool and even the words kimchi diplomacy was used for gastrodiplomatic activities (Strugar, 2015). Gastrodiplomacy is one of the examples showing how important gastronomic assets are in tourism marketing and destination promotion.

The contribution of the products and assets that will be the source of gastronomy-based tourism activities to both the destination and the economic and social structure of the country is indisputably important. However, it is only possible to benefit from this contribution with a successful tourism marketing strategy. Achieving a gastronomic position in harmony with the general strategy of the destination, combined with its culture, is significant to be successful in the long term. For this reason, destinations should promote their gastronomical assets through various promotional activities, social media, mass communication tools, brochures, airlines and strive to raise awareness. The awareness created should be supported in a way to create continuous demand while preserving local values.

References

Barkat, S. M., and Vermignon, V. (2006). *Gastronomy Tourism: A Comparative Study of Two French Regions: Brittany and La Martinique.* Sustainable Tourism with Special Reference to Islands and Small States Conference. pp. 1–22. Valletta: University of Malta.

Belisle, F. (1983). Tourism and Food Production in the Caribbean. *Annals of Tourism Research*, 10: 497–513.

Boyne, S., Hall, D., & Williams, F. (2003). Policy, Support and Promotion for Food Related Tourism Initiatives: A Marketing Approach to

Regional Development. *Journal of Travel and Tourism Marketing*, 14(3/4): pp. 131–154.

Chapple-Sokol, S. (2013). Culinary Diplomacy: Breaking Bread to Win Hearts and Minds. *The Hague Journal of Diplomacy*, 8: 161–183.

Deneault, M. (2002). *Acquiring a taste for cuisine tourism: A product development strategy*. Ottawa: Canadian Tourism Commission.

du Rand, G. E., and Heath, E. (2006). Towards a Framework for Food Tourism as Element of Destination Marketing. *Current Issues in Tourism*, 9(3): 206–234.

Fields, K. (2002). *Demand for the gastronomy tourism product: motivational factors*. Anne Mette Hjalager & Greg Richards (Eds.) in *Tourism and gastronomy* (pp. 36–50). London: Routledge.

Freedman, P. (2008). *Yemek Damak Tadının Tarihi*. İstanbul: Oğlak Yayınları.

Gillespie, C. (2001). *European gastronomy into the 21st century*. Oxford: Butterworth-Heinemann.

Hall, C. M., and Mitchell, R. (2001). *Wine and food tourism*. Norman Douglas, Ngaire Douglas & Ros Derrett (Eds.) in *Special interest tourism: context and cases* (pp. 307–329). Australia: Wiley &Sons.

Hall, C. M., and Mitchell, R. (2005). *Gastronomic tourism comparing food and wine tourism experiences*. Marina Novelli (Ed.) in Niche tourism contemporary issues, trends and cases (pp.73–88). Oxford: Elsevier Butterworth-Heinemann.

Hall, M., and Sharples, L. (2003). *The consumption of experiences or the experience of consumption? An introduction to the tourism of taste*. C. Michael Hall, Liz Sharples, Richard Mitchell, Niki Macionis, Brock Cambourne (Eds) in *Food tourism around the world: development, management and markets* (pp. 1–24). Oxford: Butterworth-Heinemann.

Harrington, R. J., and Ottenbacher, M. C. (2010). Culinary Tourism – A Case Study of the Gastronomic Capital. *Journal of Culinary Science & Technology*, 8: 14–32.

Haven-Tang, C., and Jones, E. (2006). Using Local Food and Drink to Differentiate Tourism Destinations Through a Sense of Place. *Journal of Culinary Science & Technology*, 4(4): 69–86.

Hegarty, J. A., and O'Mahony, G. B. (2001). Gastronomy: A Phenomenon of Cultural Expressionism and an Aesthetic for Living. *International Journal of Hospitality Management*, 20: 3–13.

Hjalager, A. M. (2002). *A typology of gastronomy tourism*. Anne Mette Hjalager & Greg Richards (Eds.) in *Tourism and gastronomy* (pp. 21–35). London: Routledge.

Hjalager A. M., and Richards, G. (2002). *Still undigested: research issues in tourism and gastronomy.* Anne Mette Hjalager & Greg Richards (Eds.) in *Tourism and gastronomy* (pp. 224–234). London: Routledge.

Hobsbawm, E., and Ranger T. (1983). *The invention of tradition.* Cambridge: Cambridge University Press.

Ignatov, E. (2003). *The Canadian Culinary Tourists: How Well Do We Know Them?* (Master's thesis). University of Waterloo, Ontario, Canada.

Ignatov, E., and Smith, S. (2006). Segmenting Canadian Culinary Tourists. *Current Issues in Tourism*, 9(3): 235–255.

Karim, M. S. A., and Chi, C. G. Q. (2010). Culinary Tourism as a Destination Attraction : An Empirical Examination of Destinations' Food Image. *Journal of Hospitality Marketing & Management*, 19(6): 531–555.

Kivela, J., and Crotts, J. C. (2006). Tourism and Gastronomy: Gastronomy's Influence on How Tourists Experience a Destination. *Journal of Hospitality and Tourism Research*, (30): 354–77.

Lin, Y. C., Pearson, E. T., & Liping, A. (2011). Food as a Form of Destination Identity: A Tourism Destination Brand Perspective. *Tourism and Hospitality Research*, 11(1): 30–48.

Long, L. M. (2004). *Culinary tourism*. Lexington: University Press of Kentucky.

Mak, A. H. N., Lumbers, M., & Eves, A. (2012). Globalization and Food Consumption in Tourism. *Annals of Tourism Research*, 39(1): 171–196.

Mak, A. H., Lumbers, M., Eves, A., & Chang, R. C. (2012). Factors Influencing Tourist Food Consumption. *International Journal of Hospitality Management,* 31(3): 928–936.

Mitchell, R. D., and Hall, C. M. (2003). Seasonality in New Zealand Winery Visitation: An İssue of Demand and Supply. *Journal of Travel & Tourism Marketing,* 14(3–4): 155–173.

Okumus, B., Okumus, F., & McKercher, B. (2007). Incorporating Local and İnternational Cuisines in the Marketing of Tourism Destinations: The Cases of Hong Kong and Turkey. *Tourism Management,* 28(1): 253–261.

Ooi, C.C. (2004).Poetics and Politics of Destination Branding: Denmark. *Scandinavian Journal of Hospitality and Tourism,* 4(2): 107–128.

Quan, S., and Wang, N. (2004). Towards a Structural Model of the Tourist Experience: An İllustration from Food Experiences in Tourism. *Tourism Management*, 25(3): 297–305.

Richards, G. (1999). *Developing and marketing crafts tourism*. Tilburg: Atlas.

Richards, G. (2002).*Gastronomy: An Essential Ingredient in Tourism Production and Consumption?* Anne Mette Hjalager & Greg Richards (Eds.) in *Tourism and gastronomy* (pp. 3–20). London: Routledge.

Santich, B. (2004). The Study of Gastronomy and its Relevance to Hospitality Education and Training. *International Journal of Hospitality Management*, 23: 15–24.

Scarpato, R. (2002). *Gastronomy as a tourist product: the perspective of gastronomy studies*. Anne Mette Hjalager & Greg Richards (Eds.) in *Tourism and gastronomy* (pp. 131–151). London: Routledge.

Strugar, T. (2015). Eastern Gastrodiplomatıc Efforts: Asian Nations as Pioneers in the Use of Cuisine in Cultural Diplomacy. *The Art of Food: Culture and Food Diversity, Gastrodiplomacy* (1–3 July 2015), Barcelona, Spain.

Şahin, E. (2015). *Bir Destinasyon Unsuru Olarak Yerel Gastronominin Marka Şehir Pazarlamasında Etkileri:Gaziantep Örneği*.(Master's thesis). Gaziantep University, Gaziantep, Turkey.

Warde, A., and Martens, L. (2000). *Eating out: social differentiation, consumption and pleasure*. Cambridge: Cambridge University Press.

Wolf, E. (2002). Culinary tourism: A tasty economic proposition. *International Culinary Tourism Task Force*. http://www.culinarytourism.org/faq.php (Accessed: 20.05.2020).

World Tourism Organization (2019). UNWTO Tourism Definitions, UNWTO, Madrid, DOI: https://doi.org/10.18111/9789284420858 (Accessed: 20.05.2020).

Burcu Ayşenur AKBULUT

14 An Evaluation on the Effects of Science and Technology on Gastronomy and the Transformation of Restaurants

1 Introduction

The first phase of the gastronomy field in the modern sense has been the introduction of classical recipes. The development of classical recipes in the field of gastronomy can be expressed as the second phase, and the creation of new and creative recipes using new materials and techniques can be expressed as the third phase (Linden, 2013). The third phase of gastronomy provided the combination of science and cuisine, it has triggered innovations in the field of kitchen technology. It includes many processes that take place in gastronomy, physics and chemistry. Understanding the physical and chemical processes of cooking enables the development of kitchens and restaurants. (Hegarty, 2005). Basically, the invention of the refrigerator and the production of various cooking tools are among the first technological discoveries that affect gastronomy. In the 21st century, innovations in the field of science and technology have been a force in the field of gastronomy, which not only causes the change of equipment, but also the change of food and drink elements. Various currents such as molecular gastronomy, gastrophysics, neurogastronomy, note-by-note cuisine, digital gastronomy and various technologies used in the kitchen (rotary evaporator, gastrovac, sous vide, 3D food printers, robotic chefs, etc.) has affected the process. In restaurants where these products are presented, the use of technology has been increased and virtual reality environments appear in the form of tables using touch and sensor systems, e-menus, and menus with music. The aim of the study is evaluating the scientific-technological revolution in terms of gastronomy and examining the restaurant transformation provided by this revolution with examples. For this purpose, it has been tried to determine how science and technology affect gastronomy, how scientific and technological developments in the field of gastronomy are reflected in restaurants, and examples of restaurants of which use scientific-technological innovations effectively.

2 The Effect of Science and Technology on Gastronomy

Science can be defined as "the collection of systematic information with accepted validity" (Karasar, 2000), while technology can be defined as "the application of science for an industrial or commercial purpose to develop a product or process" (Çengel, 2012). When the development adventure of the modern understanding of science is examined, it is seen that the 16th century is the milestone. With the Renaissance movement in Europe, developments in the field of science have increased and many new technologies have been developed. Technologies of which have developed with science have brought industrial revolutions. The industrial revolutions, which started with the use of water-powered mechanical production benches and the invention of steam-powered machines toward the end of the 18th century, developed as the discovery of electricity and the use of electrical energy in the 20th century. In the 1970s, it continued with a series of applications such as automation provided by electronic and information technologies, digitalization, online business models and digital interfaces (Bağcı, 2018). Nowadays, the scientific and technological revolution resulting from the technological reflections of science has begun to be called "Industry 4.0". This revolution, which expresses a triple structure in the form of the internet of things, the internet of services and cyber-physical systems (Stendustri, 2020) means a series innovations with devices such as devices communicating through sensors, robots and robotic systems, three-dimensional printers, virtual and augmented reality applications, simulation technologies

The effect of developments in science and technology on the field of gastronomy was realized in the form of culinary technologies in the first place. The kitchen equipment production has started to be designed for temperature distribution and control, fast cooking, less energy use, safe and hygienic production (Rodgers, 2007). The change of energy types used in kitchens changes the structure of kitchen equipment day by day. With the use of electrical energy in kitchen equipment in parallel with scientific and technological developments, the use of electrical equipment has become more and more common. The most famous invention as electrical equipment is undoubtedly the refrigerator. Such that the refrigerator was chosen as the first among the twenty most important inventions in the world made in the field of food by the Royal Academy of Science in England in 2012 (Akbulut, 2012). After the Second World War, electric cookers became widespread and they were preferred due to their easy use. Among the kitchen equipment, another kitchen equipment that works with different mechanisms and which is renewed with each new technology is the ovens. Ovens commonly used in kitchens and they work according to three basic heat transfer principles, which are used alone or in multiples. These are in the form of conventional ovens (direct), convection ovens (convection) and radiation/microwave ovens (radiation) (Fisher, 2002; Katsigris and Thomas, 2005). In addition, it is possible to talk about high performance hybrid ovens and induction ovens with new technological developments (Ficher, 2002; Panosonic, 2019). Another new system used in kitchen technologies is induction systems. Induction is an impressive method of heating and is new to use in kitchen equipment which is discovered by Michael Faraday in the 1830s. In this system, heat is provided by creating a magnetic field and currents of which is being converted to heat are created with the resistance of the cooking vessel suitable for induction (Miller et al., 2018). Induction technology has been made suitable not only for stoves but also for ovens (Panasonic, 2019), grill and wok pan equipment (Inoksan, 2019). The auxiliary equipment used in the kitchen is also renewed with different features every day depending on the development of technology.

In the 21st century, when gastronomy and science have started to be mentioned together, kitchen equipment has gained an innovative dimension and started to show similarities with laboratory equipment. It can be said that the most important driving force in the emergence of these innovative kitchen equipment is the studies on science and gastronomy, the rise of molecular gastronomy and molecular cuisine applications.

The importance of using scientific methods in understanding the properties of foods was first realized with the studies of Lavoiser in the 17th century and after half a century with the book "Physiology of Taste" written by Brillat-Savarin. Nicholas Kurti was the scientist who emphasized the connection between gastronomy and science and made important studies in this field (Linden, McClements, and Ubbink, 2008). Hungarian-born Nicholas Kurti (1908–1998) was Professor of Physics at Oxford University. Kurti, who specialized in low temperature physics, made a presentation at the Royal Institute of London in 1969 titled "Physicist in the Kitchen" (Pedersen, Meyer, Nursten and Redzepi, 2006). In this presentation, Kurti mentioned that how science affects cooking and expressed his most well-known statement: "It is a sad reflection that we know better the temperature inside the stars than inside a soufflé" (Linden et al., 2008). Heat is one of the most basic culinary processes that change the physical or chemical structure of food. For example, the knowledge that water boils at 100 ° C can vary according to many variables such as sea level, atmospheric pressure values, the water used and the salt content in it (Barham, 2013). Studies on how this and a number of similar scientific knowledge can affect culinary knowledge have continued. Kurti continued his studies related to cuisine and science, and by the 1990s he gave a series of workshops in Sicily with the French

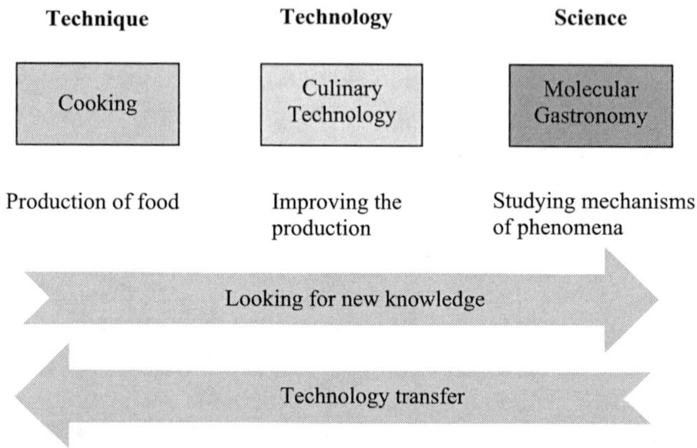

Fig. 1: Gastronomy in Technique, Technology and Science.
Source: This, 2011

chemist Herve This on science-based cuisine. The trainings titled "Molecular and Physical Gastronomy", which continued until the 2000s, were first shortened as "Molecular Gastronomy" in 2004 (Pedersen et al., 2006; Linden et al., 2008).

Molecular gastronomy is the discipline that studies sensory phenomena associated with the physicochemical changes and consumption of edible materials during cooking (This, 2002 as cited in Linden et al., 2008). With the increase in molecular gastronomy studies, this concept has been discussed and the differences between molecular gastronomy and molecular cuisine applications have been revealed. It is possible to define molecular cuisine applications as the application of the phenomena revealed by molecular gastronomy. This (2013) defined molecular cuisine as "producing food by using new tools, materials, methods". In addition, he evaluated the applications in the kitchen in terms of technique, technology and science (This, 2011). This (2011) defined cooking as a technical activity because it includes cooking, making and producing, and kitchen technology as the process of developing cooking technique either by using science or alone. Molecular gastronomy is the discipline of which examines cooking activity and tries to understand the events that occur during cooking (Fig. 1). A distinction similar to Herve This's distinction between molecular gastronomy and molecular cuisine applications is made between the fields of gastronomy and gastrophysics. This distinction between gastronomy and gastrophysics is made similar to the distinction between astrophysics and astronomical science. Just as astronomers observe planets and stars and study how they move; astrophysicists provide the scientific basis for why and how stars came into being (Barham, 2013; Moller, 2013; Mouritsen and Risbo, 2013). The authors fundamentally differentiate gastronomers and gastrophysicists in terms of practice and theoretical science.

Herve This (2011, 2013), the most important name that guides molecular gastronomy, argued that molecular cuisine is only the first step and the new trend is "note-by-note cuisine". Note-by-note cuisine originated from the idea of "cooking by using pure compounds". Note-by-note cuisine is based on the separation of compounds found in plant and animal tissues and recombining them to form combinations that do not exist in nature. With this new trend, new flavors and tastes of which never existed before can be obtained, compounds that cause diseases or should not be consumed by patients can be separated from foods, and personalized menus can be developed by considering their metabolic characteristics and needs (This, 2013; This, 2014; Aksoy and Uner, 2016; Burke and Danaher, 2016).

Cooking basically consists of a whole of physical and chemical processes. Such that the components such as texture, fragmentation, sound, color and smell, which are necessary to ensure sensory quality in the field of gastronomy, are physical and chemical conditions. Linden (2013) stated that the textures of foods are related to mechanic, breaking and fragmentation while chewing is related with sound, color properties are related with light, smell related to volatile release and transfer, and stated that the integration of physics and gastronomy will affect many innovations. Another scientific discipline influenced by science-based culinary studies is neurogastronomy. Shepherd's (2012) book "Neurogastronomy" begins with the question of how the sense of smell creates taste and provides examples of how the human brain perceives eating and drinking. Studies in the field of neurogastronomy have increased in recent years and the results of studies in this area have changed the presentation characteristics of the food served in restaurants. For example, there are a series of studies and technologies, restaurants and menu designs have been developed taking into account the results of these studies. For instance, when people eat alone, their food consumption is less than when they eat with a group (Piqueras-Fiszman, Varela, and Fiszman, 2013); the type of music that is listened while having meal affects the pleasure of the meal (Spence and Piqueras-Fiszman, 2013); the color of the plate has an effect on taste perception (Piqueras-Fiszman, Alcaide, Roura, & Spence, 2012), the atmosphere and light of the environment in which the food is consumed effect the consumers (Spence and Piqueras-Fiszman, 2013).

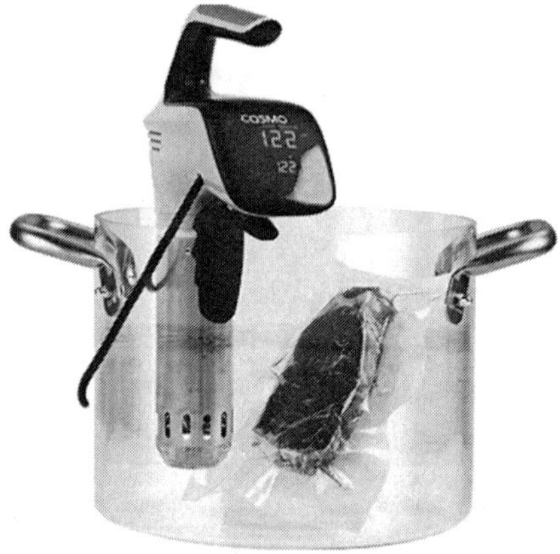

Fig. 2: Sous Vide Machine.
Source: Amazon, 2020

Ruiz, Calvarro, Pulgar, and Roldan (2013) expressed the contribution of science to cooking as developing new recipes, new cooking adjustment and optimization of cooking methods, the use of non-traditional ingredients, and the use of technological and scientific tools which are not belong to kitchens. It is known that many products of which are not used in the kitchen, and even chemicals of which are used in chemistry laboratories, have been suitable to use with food by means of innovative kitchen trends. In molecular kitchen applications, monosodium glutamate, gelling agents, thickeners hydrocolloids, sodium alginate, xanthan gum, gellan gum, carrageen or methylcellulose, lecithin, monoglycerides are commonly used material. (Ruiz, Calvarro, Pulgar and Roldan, 2013). On the other hand, the relationship of science with the kitchen has brought about revolutionary changes in kitchen technologies (Fig. 2).

A series of projects have been carried out to provide the change of kitchen equipment with innovative kitchen trends. One of these projects is the project of Zaragoza University in 2004 which was conducted to develop new kitchen tools based on scientific and technological knowledge, and in order to popularize the use of such tools by chefs through courses and collaborations (García-Segovia et al., 2014). Along with the molecular kitchen applications, many new techniques, especially the sous vide technique, have led to changes in industrial kitchen equipment. It is known that vacuuming technologies and sous vide machine production have increased with sous vide cooking technique. The sous vide technique is a thermostat bath technique, similar to the steam table cooking

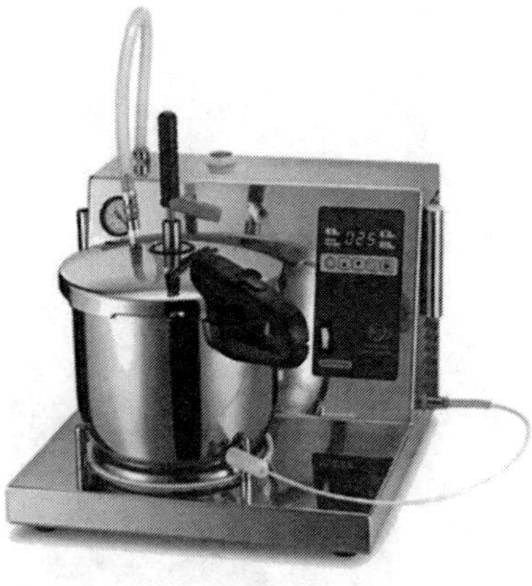

Fig. 3: Gastrovac-Cooked Vide Machine.
Source: García-Segovia et al., 2014

method, but it is more stable, controllable, and gives the same temperature to all of the water (Ruiz et al., 2013; García-Segovia et al., 2014). With this technique, it is ensured that all kinds of food are more delicious by controlling cooking in vacuumed bags and at low temperature (Fig. 3).

In 2003, researchers from the Polytechnic University of Valencia developed Gastrovac in a joint research project with Chef Javier Andrés (Valencia La Sucursal Restaurant) and Chef Sergio Torres (El Rodat Restaurant) with an innovation in cooking methods. Researchers have named the cooking technique performed with Gastrovac as "cooked-vide." This machine, which is similar to a pressure cooker, has a vacuum pump that creates an oxygen-free low-pressure atmosphere. This tool allows food to be cooked at lower than normal temperatures by preserving their texture, nutritional value and color, and reducing the fat content in fried products (García-Segovia et al., 2014).

Another example of innovative kitchen equipment is Rotavap equipment, which is mainly used in chemistry laboratories for solvent evaporation. Its use in kitchens is for concentrating flavors by distilling food-based liquids or in order to transfer them to other foods (Linden, McClements, and Ubbink, 2008). A rotary evaporator, which is used to reveal the essence of aroma, is a kitchen equipment generally used in restaurants with molecular cuisine applications.

A Rotary Evaporator allows extraction of volatile components by rotating the food to be distilled in a bottle (Fig. 4). Rotating the bottle of which contains the product that will be distilled provides a better essence. It also allows the

Fig. 4: A Rotary Evaporator.
Source: Polyscienceculinary, 2019

production of concentrated liquid solutions such as wine reduction or concentrated broths (Ruiz et al., 2013).

Undoubtedly, among the innovative kitchen equipment, the most interesting are those developed with the effect of digital technologies. While 2D food printers can print on specially produced pulp papers for the printer, 3D food printers enable three-dimensional printing of foods (Fukuchi and Jo, 2012; Cbinsights, 2020). 3D (three-dimensional) food printers are generally used in chocolate, dessert, cake making and decoration.

There are developments of which suggest that robotic systems may gradually take place in kitchens. Robotic arms called *Flippy*, which can turn hamburger patties, have been made in America and there are studies for the continuation of the project (Gültekin, 2017). Another example is a robotic chef designed under the name of *Moley Robotic Kitchen* (Moley, 2020). The use of these robotic systems in kitchen applications and their reflection on technology have been increasing in recent years.

3 Transforming Restaurants in the Light of Science and Technology Developments

The word of restaurant is derived from the word of "restauratif" used in pre-revolutionary France. Restaurants initially appear as places that serve soup-type

Fig. 5: Gastronomic Laboratory of Chef Paco Roncero

foods to reduce respiratory ailments and calm the soul. Restaurants, which have changed and differentiated with the phenomenon of eating out, undoubtedly experienced the biggest change with science-based kitchen studies. The intermediary factor that transforms the science of gastronomy into new culinary technology is undoubtedly innovative chefs. Revolutionary innovations in kitchen technologies also affect the structure of restaurants. The best example where the relationship between gastronomy and science has been firmly established is undoubtedly the work of Chef Ferrán Adrià. Ferrán Adrià's elBulli Restaurant, which is operating in Spain, is a 3 Michelin-starred restaurant that has been voted the best restaurant in the world five times (Capdevila et al., 2015). The procedures used by physical and molecular chemists have inspired chefs (Linden, McClements and Ubbink, 2008). The best reflection of this inspiration at the restaurant level is the *elBulliLab* project created by Ferrán Adrià (Capdevila et al., 2015). Adrià and his team established a restaurant laboratory where the restaurant developed new recipes and techniques, and developed innovative menus in this laboratory for six months of a year. Another example of a restaurant laboratory is the gastronomic laboratory (Fig. 5) of Chef Paco Roncero (Spence and Piqueras-Fiszman, 2013).

Innovative activities are encountered in the production, presentation or eating process of the meals (Albors-Garrigos et al., 2013). The Fat Duck Restaurant in England is among the restaurants that benefit from science and technology in terms of both production and presentation. Heston Blumenthal, Chief of Fat Duck Restaurant, is among the innovative chefs and collaborates with scientists. For example, Chef Heston Blumenthal, with the help of Professor Malcolm Povey (Professor of Food Physics at the University of Leeds), combined science and cooking to create the crispy fish paste and tested what determines ideal vitality. Working on a texture analyzer and an acoustic envelope detector, the chef and professor produced the dough that allowed the fish to harden the outside while the inside of the fish gently steamed (Fiszman, 2013; Fat Duck, 2016; Spence and Piqueras-Akbulut and Aksoy, 2016). One of the striking examples is that Chef Blumenthal developed the menu of the sound of the sea with the ability of sound to affect taste as a food component as a result of studies in the field of neurogastronomy (Piqueras-Fiszman et al., 2013)

Chefs who keep science-based cooking develop and use new technologies to create new looks, textures and new flavors in food. Firo Vazquez, the chef of El Olviar Restaurant in Spain, produced aromatic edible paper as a result of the studies he conducted with Murcia University and created an edible menu with different flavors (García-Segovia et al., 2014; Frisbie, 2016; Vazquez, 2012).

With the development of technology, the use of augmented reality and virtual reality in the eating and drinking experience has become important in terms of providing memorable experiences. Inamo Restaurant which is located in London, serves its customers with an augmented reality e-menu through its smart tables (Fig. 6). These smart table menus offer the ability to monitor the menu preparation and games (Echtler and Wimmer 2013; Margetis et al., 2013; Spence and Piqueras-Fiszman, 2013; Aksoy and Akbulut, 2016; Inamo, 2016). One of the other applications related to the use of technology at the table is the "telematic dinner party" (Barden, Comber, Bryan-Kinns, Stockman and Olivier, 2012). Projects have been carried out for the use of telematics applications at tables, which can be expressed as a synthesis of semiconductor, communication, user interface, smart devices, computer technologies and digital technologies (Fig. 6). The telematics banquet application offers the same environment in two different venues by allowing people to intervene in the tables as well as the sound and images of the people on both tables (Comber, Barden, Bryan-Kinns, Stockman and Olivier, 2014).

The most striking example of the application of virtual reality in restaurants is undoubtedly the El Celler de Can Roca Restaurant in Spain (Fig. 7). El Celler

Fig. 6: e-menu of Inamo Restaurant (England)

Fig. 7: El Celler de Can Roca

de Can Roca Restaurant, which ranked first in the "World's 50 Best Restaurants" awards held in London in 2015, combines virtual images and reality to offer its customers different experiences (Spence and Piqueras-Fiszman, 2013; Afroditi Krassa, 2016; World's 50 Best Restaurants, 2016). The restaurant, which changes the atmosphere according to the theme of the meal by using projection on each incoming food and combines the multi-sensory environment with a theatrical presentation, offers unforgettable experiences to the customers. The work called El Somni, which is referred to GastrOpera (combination of gastronomy and opera) by El Celler de Can Roca Restaurant, is a feast opera of which is formed by the combination of contrasting environments where images, art, music, taste and scent are presented and tells about the cyclical journey of human existence. (Fig. 7) (El Somni, 2016).

Innovative and high-end restaurants now blend craft, art and tradition with innovation in harmony while preparing and serving food (Linden et al., 2008). In the future, it will bring many innovations in restaurants' production, serving and post-eating experiences. In the future, it will be possible to see more robotic chefs in restaurants or will be possible to see machines that prepare food production according to nutritional values.

4 Conclusion

The developments in science and technology have significantly affected gastronomy. The relationship of gastronomy with science has led to the birth of innovative culinary trends and this has led to the rise of innovative developments in gastronomy and culinary technologies accordingly. The study has presented the sample applications concerning the use of gastronomy with technology, how they have shaped restaurants, what the future gastronomy trends can be, and sample applications for the use of digital technologies in gastronomy. The scientists, chefs and restaurants samples that give a scientific view to the kitchen reveal the dimension of the relationship between science and gastronomy. It is possible to list a series of examples such as the use of augmented reality and virtual reality applications in restaurants, e-menus, edible menus, 3D printers, robot chefs. The scientific information on the results of the use of new technologies in the kitchen is very limited. Actually, the use of equipment and additives, which are used in chemistry laboratories, in the kitchen may have negative consequences for human health. This situation has not been fully evaluated and scientifically analyzed, as well. For future studies, it is highly recommended to evaluate the effects of science and technology on the kitchen in terms of health.

References

AfroditiKrassa. (2016, Mart 13). Fine Dining. *The Salmon*, 1–31. paulliebrandt:http://www.paulliebrandt.com/files/TheSalmonby%20 AfroditiKrassa%20Sept2014.pdf (Date accessed:27.07.2020).

Aksoy, M. & Akbulut, B. A. (2017). Restoranlardaki teknolojik yeniliklerin deneyim pazarlaması açısından değerlendirilmesi, ICCHT 2017- International Cangress on Cultural Herigate and Tourism, Konya, Turkey, 1–10.

Aksoy, M. & Üner, E. H. (2016). Rafine mutfağın doğuşu ve rafine mutfağı şekillendiren yenilikçi mutfak akımlarının yiyecek içecek işletmelerine etkileri, *Gazi Üniversitesi Sosyal Bilimler Dergisi*, 3 (6), 1–17.

Barden, P., Comber, R., Green, D., Jackson, D., Ladha, C., Bartindale, T., Olivier, P. (2012). Telematic dinner party: designing for togetherness through play and performance. *Proceedings of the Designing Interactive Systems Conference* (s. 38–47). Newcastle (UK): ACM.

Barham, P. (2013). Physics in the kitchen. *Flavour*, 2(1), 1–5. doi:10.1186/2044-7248-2-5

Brillat-Savarin, J. A. (2016). *Lezzetin Fizyolojisi: Ya Da Yüce Mutfak Üzerine Düşünceler*, (Çev: Heval Bucak) Oğlak Yayıncılık: İstanbul.

Burke, R. ve Danaher, P. (2016). Note by Note: A New Revolution in Cooking. *Dublin Gastronomy Symposium, Food and Revolution, 31 May-1June 2016*, Dublin.

Çengel, Y. (2012). Bilim ve Teknoloji, *Bilim ve Teknik Dergisi*, Ekim 2012, 50–53.

Dünyanın En İyi 50 Restoranı. (2016, Mart 10). theworlds50best: http://www.theworlds50best.com/list/1-50-winners (Date accessed: 20.07.2020).

Echtler, F. & Wimmer, R. (2013). The Interactive dining table. *Human Factors in Computing Systems Conference*. Paris: ACM.

El Somni. (2016, Mart 15). http://www.elsomni.cat/en/el-somni/ (Date accessed: 20.07.2020).

Fisher, D. (2002). *Commercial Cooking Appliance Technology Assessment*, Fisher- Nichel, inc., Food Service Technology Center, [URL: https://fishnick.com/equipment/techassessment/Appliance_Tech_Assessment.pdf] (Date accessed: 10.07.2020).

Frisbie, R. (2016, Mart 16). *The Latest Culinary Craze: Edible Books in Spain.* travelingboy: http://travelingboy.com/archive-travel-frisbie-edible_books.html#null (Date accessed: 20.07.2020).

García-Segovia, P., Garrido, M. D., Vercet, A., Arboleya, J. C., Fiszman, S., Martínez-Monzo, J., Laguarda, S., Palacios, J., Ruiz, V. (2014). Molecular Gastronomy in Spain, *Journal of Culinary Science & Technology*, 12(4), 279–293.

García-Segoviaa, P., Garridob, M. D., Vercetc, A., Arboleyad, J. C., Fiszmane, S., Martínez-Monzoa, J., Ruiz, J. (2014). Molecular gastronomy in Spain. *Journal of Culinary Science & Technology,* 4(12), 279–293.

Hegarty, J. A. (2005). Developing "Subject Fields" in Culinary Arts, Science, and Gastronomy. Journal of Culinary Science & Technology, 4(1), 5–13. doi:10.1300/J385v04n01_02

Hegarty, J. A. (2008). Developing "Subject Fields" in Culinary Arts, Science, and Gastronomy. doi:10.1300/J385v04n01_02

Inamo Restoran. (2016, Mart 12). Inamo restaurant: http://www.inamo-restaurant.com/ (Date accessed: 15.07.2020).

İnoksan (2019). *İndüksiyonlu Pişiriciler,* [URL: https://www.inoksan.com/Urun-Kategorileri/INDUKSIYONLU-PISIRICILER/65/82] (Date accessed: 20.07.2020).

Karasar, N. (2000). *Bilimsel Araştırma Yöntemi.* Nobel Yayın Dağıtım: Ankara.

Katsigris, C. & Thomas, C. (2005). *Design and Equipment for Restaurants and Foodservice: A Management View*. 3th ed., Hoboken, NJ: John Wiley & Sons, Inc.

Linden, E. (2013). Integration of gastronomy and physics for innovation. doi:10.1186/2044-7248-2-11

Linden, E. v., McClements, D. J. & Ubbink, J. (2008). Molecular gastronomy: A food fad or an ınterface for science-based cooking? *Food Biophysics* 3, 246–254.

Miller, B., Rama, M. & Adamson, E. (2018). *Yemek Pişirmenin Temelleri.* Nobel Yayınları, İstanbul

Mouritsen, O. G. & Risbo, J. (2013). Gastrophysics-do we need it?, *Flavour*, 2(3) doi:10.1186/2044-7248-2-3

Møller, P. (2013). Gastrophysics in the brain and body. *Flavour*, 2(1), 1–8. doi:10.1186/2044-7248-2-8

Panaconic (2019). *Panasonic Countertop Induction Oven,* [URL: https://www.youtube.com/watch?v=4EZkut-HBD0] (Date accessed: 10.07.2020).

Pedersen, T., Meyer, C., Nursten, H. & Redzepi, R. (2006). Gastronomy: the ultimate flavour science? *Developments in Food Science*, 43(C), 611–616. doi:10.1016/S0167-4501(06)80144-5

Piqueras-Fiszman, B., Alcaide, J., Roura, E. & Spence, C. (2012). Is it the plate or is it the food? Assessing the influence of the color (black or white) and shape of the plate on the perception of the food placed on it. Food Quality and Preference, 24(1), 205–208. doi:10.1016/j.foodqual.2011.08.011

Piqueras-Fiszman, B., Alcaide, J., Roura, E. & Spence, C. (2012). Is it the plate or is it the food? Assessing the influence of the color (black or white) and shape of the plate on the perception of the food placed on it. *Food Quality and Preference* 24, 205–208.

Piqueras-Fiszman, B., Varela, P. & Fiszman, S. (2013). How Does the Science of Physical and Sensory Properties Contribute to Gastronomy and Culinary Art? *Journal of Culinary Science & Technology*, 11(1), 96–109.

Polyscienceculinary (2019). Rotary Vacuum Evaporatıon System, [URL: https://polyscienceculinary.com/products/rotary-vacuum-evaporation-system-120v] (Date accessed: 07.07.2020).

Rodgers, S. (2007). Innovation In Food Service Technology and Its Strategic Role, *Hospitality Management*, 26, 899–912.

Ruiz, J., Calvarro, J., Sánchez del Pulgar, J. & Roldán, M. (2013). Science and Technology for New Culinary Techniques. Journal of Culinary Science and Technology. *Taylor & Francis Group*. doi:10.1080/15428052.2013.755422

Spence, C. (2012). Book Review: 'Neurogastronomy: How the Brain Creates Flavor and Why It Matters' by Gordon M. Shepherd. *Flavour*, 1(1), 21. doi:10.1186/2044-7248-1-21

Spence, C. & Piqueras-Fiszman, B. (2013). Technology at the Dining Table. *Flavour*(2), 2–13.

Stendustri (2020). https://www.stendustri.com.tr/ (Date accessed: 11.07.2020).

This, H. (2002, 4 Ocak). Molecular gastronomy. Angewandte Chemie – International Edition.

This, H. (2011). Molecular Gastronomy in France. *Journal of Culinary Science*, 9(3), 140–149. doi:10.1080/15428052.2011.600212

This, H. (2013). Molecular Gastronomy is a Scientific Discipline, and Note by Note Cuisine is the Next Culinary Trend. *Flavour*, 2(1), 1. doi:10.1186/2044-7248-2-1

Vázquez, F. (2012, Ocak 12). Interview with Chef Firo Vázquez. (OliveOilTimes, Röportaj Yapan) http://www.oliveoiltimes.com/olivarama/interview-with-chef-firo-vazquez/23981/2 (Date accessed: 12.07.2020).

Ayşe SÜNNETÇİOĞLU

15 Eating Out Behaviors of Tourists

1 Introduction

Eating out behaviors of tourists will be examined. The purpose of writing this section is to contribute to the understanding of tourists by considering their eating out behaviors within a conceptual framework. Accordingly, primarily tourism and the phenomenon of eating out will be discussed. Then, a conceptual framework will be tried to be presented by addressing the behaviors exhibited by tourists within the scope of eating out purposes, information seeking, food selection, restaurant selection, meal experience, complaint-satisfaction, and behavioral intentions.

2 Tourism and the Phenomenon of Eating Out

For millions of tourists worldwide, it is stated that food plays an important role (González, Curtis, Washburn & Shirsat 2020, p. 2). Food consumption, which is an integral part of the tourism sector, is one of the basic needs of tourists (Guan & Jones, 2014, p. 417). Tourism can be considered as one of the factors that bring out the phenomenon of eating out (Akarcay & Sugur, 2014, p. 265).

Eating out is defined as "consuming the food and beverages prepared and served outside the home by a catering business by paying their fee" (Özdemir, 2010, p. 219). Eating out is limited to food and drinks consumed in catering establishments and examined on a commercial basis (Özdemir, 2018, p. 47). When people have a chance to choose, each answer to questions such as; why they choose which food, how they eat, who they eat with and where they eat, is within the scope of the study of different branches of science such as; Sociology, anthropology, history, tourism, nutrition and business (Altaş & Varnacı Uzun, 2017, p. 4436).

The phenomenon of eating out within the scope of tourism is defined as "consuming the food and beverages prepared and served by commercial enterprises operating in the destination by paying the fee" (Sünnetcioğlu, 2018, p. 76). In this context, each answer to questions such as; why tourists choose which food, how they eat, who they eat with, and where they eat can be evaluated within the scope of the study of tourism.

3 Tourists' Eating Out Behaviors

3.1 Eating Purposes of Tourists

It is observed that the purpose of eating for consumers and tourists differs (Yılmaz & Gültekin, 2016, p. 217). It is stated that deciding to eat out is related to the purposes of eating out (Özdemir, 2010, p. 220).

It can be stated that the phenomenon of eating out has two basic dimensions. While in the first of these dimensions; ease, hunger, and economic efficiency are emphasized, in the second dimension; social interaction, mood improvement, fun, happiness, status, and prestige are emphasized (Cullen, 1994, p. 7–8; Özdemir, 2010, p. 221). In another study conducted; the goals of consumers to eat in restaurants are determined as psychological, sociological, economic, and imperative (Bekar & Gumus Donmez, 2016, p. 5).

It is stated that tourists' eating behaviors have five dimensions: symbolic, imperative, contrast, extension, and pleasure (Mak, Lumbers, Eves & Chang, 2012). The imperative dimension of the eating behavior of tourists emphasizes to physiological needs; symbolic dimension to cultural goals, gaining status and prestige; the dimension of contrast to escaping from the routine of daily life; extension size, feeling the comfort of being at home, seeking diversity in routine; the pleasure dimension establishing new relationships and strengthening social ties with the smell, taste and visual image of the food. In the context of international tourism, food consumption is expressed as a tourist paradox, and this paradox is expressed to change between the imperative and the symbolic dimension of food consumption in a foreign tourist destination (Mak, Lumbers, Eves & Chang, 2017, p. 1).

By considering the food and beverage consumption motives of the tourists (Kim, Eves & Scarles, 2009; Kim, Goh & Yuan, 2010; Kim &Eves, 2012; Mak et al., 2012; Mak et al., 2017; Mak, 2018) it is possible to list the purposes of eating as follows (Sparks, Bowen & Klag, 2003, p. 9; Sparks, Bowen & Klag, 2004, p. 32; Quan & Wang, 2004a, p. 301–302; Kim et al., 2009, p. 427);

- Meeting physiological need
- Healthy eating
- Convenience (not having to wait and cook it yourself)
- Learning the destination culture
- Gaining intellectual accumulation
- Having a touristic experience (exploring, eating new and different foods that they cannot cook)

- Gaining status and prestige (showing local food experiences to friends and other people)
- Showing their lifestyle
- Getting away from daily routine
- Feeling the comfort of being at home
- Having sensory pleasure (taste, appearance, smell, presentation of the food, nice food or feeling like treating yourself, etc.)
- Having new relationships and strengthening social ties (developing and strengthening social relationships, having fun with the close environment, getting the opportunity to meet new people)
- Searching for novelty
- Searching for variety

In a study conducted, the most important reason determined as 'convenience' (in other words not having to wait and cook yourself). In addition, the participants attached great importance to eating healthy food (Sparks et al., 2004, p. 32).

3.2 The Information Search of Tourists

Even for a variety of purposes, a consumer who has felt the need to eat out and, in this context, has decided to eat out, has to decide what to eat and where to eat. To make this decision, tourists need information (Özdemir, 2018, p. 50).

The services offered by restaurant businesses are abstract and it is not possible to experience them before purchasing. Since it is not possible to experience a service before it is used, it is stated that potential customers perceive more risks when deciding whether to use a service or not (Selvi, 2009). In addition, although foods has been experienced before, the first experience in a restaurant can be described as the first experience of that food. Because the services are experience. Therefore it is very difficult to evaluate them before purchasing. For this reason, there is a need to search for information from various sources (Pedraja & Yagüe, 2001, p. 316).

The search for information in the context of restaurants takes place in two ways: the search for internal information and the search for external information. The search for internal information refers to the application of information obtained from previous experience and stored in memory. In the case of insufficient internal knowledge or lack of previous experience with restaurant services, external information seeking takes place. External knowledge seeking behavior occurs in two ways, active and passive. In the search for passive information, the consumer does not make great efforts to collect information and rather pays regard to the information that has been received in some way. In the pursuit of active information, consumers strive to access information. For example, visiting restaurants before eating, reading restaurant menus (Pedraja & Yagüe, 2001, p. 316; Özdemir, 2018, p. 52).

Sources of information search in the choice of food restaurant while on holiday are determined as; advertisements, the reputation of the chefs, food magazines, food guides, local newspapers. Additionally, from mouth to mouth communication has also been determined as playing an important role in restaurant selection. The suggestions of family, friends, local business, and local people are among the important information sources for tourists (Sparks et al., 2003, p. 12).

While most of the tourists are on holiday, they search for information in restaurant selection until they find something nice. Then, from mouth to mouth suggestions from friends and/or relatives, suggestions from local or businesses in the region, and past experiences are among the resources used to search for information (Sparks et al., 2004, p. 15). In another study by Stephen, Julia & Stephen (2005, p. 158) as a source of information for tourists; they choose their restaurant through their previous experiences (in other words they choose a restaurant they have previously eaten), the advice of a friend or relative, or simply returning to a restaurant they have passed, and returning to a restaurant they had previously found attractive.

In restaurant selection, e-comments can also be considered as an important source of information and an important factor affecting tourists' restaurant selection (Law, To & Goh, 2008, p. 351). The difference between WOM (word of mouth communication) and E- WOM (electronic word of mouth communication) results from the use of an electronic word of mouth communication tool (Joeng & Jang, 2011: p. 357). For example, blogs of gastronomy (Wang, 2011). It is possible to say that electronic from mouth to mouth communication is an important source of information in food selection as well as restaurant selection. Potential gastronomy blog readers use the recommendations and knowledge of other gastronomes to make decisions (Wang, 2011, p. 503).

3.3 Food Selections of Tourists

When the researches about food consumption of tourists have been examined, it has been observed that they generally show four approaches; consuming their own cuisine, consuming local foods of arrived destination, consuming another country's cuisine at the destination and having no preference about consuming their own cuisine or any local food (Telfer & Wall, 2000, p. 440; Quan & Wang,

2004, p. 301; Chang, Kivela & Mak, 2010, p. 1002–1003; Vu, Li, Law & Zhang, 2019, p. 155–156).

When the food selection behaviors of the tourists are examined, it is seen that the tourists from different nationalities show different behaviors. However, it can be seen that tourists of the same nationality also exhibit different food preferences. However, this does not constitute an obstacle to making an overall assessment of the food preferences of tourists according to nationalities. As Hofstede (1983, p. 78) stated; the characterization of a national culture does not mean that every individual within that culture is programmed mentally in the same way, it is a kind of average belief and value pattern of national culture.

It can be said that the local food preference of German, Iranian, Saudi Arabian, Iraqi, Russian, British, French, American, Ukrainian, and Dutch tourists is low. (Sunnetcioglu, 2018, p. 156). Some tourist groups such as Italian, Indian, and French prefer their own cuisine (Vu et al., 2019, p. 164). While Japanese tourists are perceived as a group of tourists who avoid local food consumption in one study (Pizam & Sussman, 1995), in another study they are perceived as the most adventurous tourist in food choice (Pizam & Jeong, 1996). In another study, while some of the Chinese tourists are willing to consume local food, some others prefer Chinese food during their travels. Another part is that they do not show a particular preference in food choice, but they prefer to adapt to the food choice of the group they belong to (Chang et al., 2010, p. 1003–1004).

Tourists prefer their own kitchen meals for reasons such as feeling at home and missing their country; on the other hand, they see local food as a tool to increase their intellectual knowledge and learn culture. They also see local food selection as an indicator of status and prestige, a way to increase cultural capital and means of enriching culinary knowledge (Chang et al., 2010, p. 1003–1004).

Factors affecting the choice of eating during eating out are classified as the characteristics of the environment, the individual, and the food (Randall & Sanjur, 1981). Based on this classification, Mak et al., (2012, p. 930) classified the factors affecting the food choices of the tourists as the characteristics of the food and the destination.

The characteristics of tourist that affect their food choices have been determined as culture and religion, socio-demographic factors (age, gender, income, etc.), food personality traits (food neophobia, food neophilia, search for diversity, etc.) and motivational factors. The features of the foods in the destination have been considered as the sensory properties of the foods (taste, smell, appearance, etc.), the ingredients used in foods, the preparation and cooking methods, food or cuisine type (national, regional and local dishes), food availability (e.g., types available, variety, alternatives), price, value and quality. As for the destination characteristics, they are, gastronomic image/identity (e.g., perceived gastronomic image of the destination), marketing communications (e.g., Internet, guidebooks, TV programs), contextual influences (e.g., time, place, with whom), service encounter (e.g., expected service quality), servicescape (e.g., physical elements in a consumption setting's built environment), seasonality (e.g., season, temperature of the destination) (Mak, et al., 2012, p. 930).

3.4 The Restaurant Selections of Tourists

It is possible to say that the restaurants in the destinations are important in the choice of holiday destinations for tourists (Gyimothy, Rassing & Wanhill, 2000, p. 373; Sparks et al., 2003, p. 9; Sparks et al., 2004, p. 27).

It has been examined that while the consumers are making a restaurant selection, what kind of qualities are important for them and which of these features are given more importance by the consumers (Özdemir, 2018, p. 53). It has been revealed that each of the consumers of different age groups, different genders, different educational levels, and the ones who eat at restaurants with different frequencies gives priority to a different criteria (Kim et al. 2010; Harrington et al. 2011; Albayrak, 2014, p. 197; Yüksekbilgili, 2014, p. 6358).

Just like other consumers, tourists also consider some qualities of the restaurants when deciding on their restaurant selection. Regardless of their order of importance, these qualities can be stated as follows (Yüksel & Yüksel, 2002a, p. 322; Yüksel & Yüksel, 2002b, p. 59; Sparks et al., 2003, p. 11; Sparks et al., 2004, p. 11; Law et al., 2008, p. 351; Özdemir, 2010, p. 224; Harrington et al., 2011; Albayrak, 2014, p. 196; Yüksekbilgili, 2014, p. 6358; Yılmaz & Gültekin, 2016, p. 221; Araz, 2019, p. 57);

- Atmosphere,
- Food/menu variety,
- The meals' being of good quality and healthy,
- Service quality,
- Price (worth buying),
- Location,
- Hygiene,
- The quality of ingredients and services,
- The behaviors of the employees and their appearance,
- Service (speed, presentation)
- The nutritional content of meals,
- Comfort,
- The image of the business,

- Getting the money's worth that you've paid
- Having a car park opportunity,
- A non-smoking area,
- The visibility of the food preparation area
- Decoration,
- Working hours,
- Being crowded,
- The presentation of the menu outside the restaurant,
- Including local dishes on the menu,
- Word of mouth communication,
- Advertising,
- New experiences

3.5 The Meal Experience of Tourists

The concept of eating does not only refer to the consumption of food to take the nutrients and energy that the human body needs, and the process related to physiological and biological actions such as chewing, swallowing and digestion (Özdemir, 2018, p. 59). Eating is considered by tourists as an important part of the overall holiday experiences (Sparks et al., 2004, p. 32). Restaurants are generally assumed to be in the business of selling food only. Restaurants are, however, primarily retailers of 'food service experiences' (Yüksel & Yüksel, 2002b, p. 54).

According to tourists eating is not only a physiological need. At the same time, for them, it can be considered as a way of learning from the destination culture, having fun with the immediate surroundings, getting the opportunity to meet new people, experiencing sensory (taste, appearance, smell, presentation, good food) and emotional (pampering yourself) aspects and having a new touristic experience by getting away from the daily routine.

If food consumption in tourism takes place as a supportive touristic experience, it becomes an extension of their daily eating routine and as a result, tourists prefer the foods they are familiar with. However, if food consumption takes place as an intense touristic experience, it becomes the opposite of daily eating routine behavior; and so they tend to consume new and different foods. In order for the food consumption to turn into an intense touristic experience; it has to become a motivating factor. This tourism type is stated as gastronomic tourism (Quan & Wang, 2004a: 302).

Experience can be defined as cognitive sensory and other experiences as a result of the interaction with services and goods that are consumed (Aho, 2001; Grundey, 2008, p. 135). Eating out is recognized as an experience by consumers (Johns & Kivela, 2001; Pettersson & Fjellström, 2007) and there are some components that make up this experience (Andersson & Mossberg, 2004). When the restaurant qualities that are taken into account by tourists and restaurant selection are considered, it can be uttered that the elements that make up the meal experience of tourists are similar to those that make up the meal experience of the consumers. It's possible to list these elements as follows (Anderson & Mossberg, 2004; Hansen, Jensen & Gustafsson, 2004; Gustafsson et al., 2006; Pettersson & Fjellström, 2007; Özdemir, 2010);

- Food and beverages (core products)
- The interior design of the restaurants (color, furniture, service sets, etc.)
- Social encounter (customer-customer and customer-service staff interaction)
- Being accompanied (with whom to eat, formal and informal relationships between the people who eat)
- Restaurant atmosphere
- Service
- Payment

3.6 Satisfaction, Complaint, and Behavioral Intentions of Tourists

It is stated that satisfaction, complaints and behavioral intentions come to the fore depending on the evaluations of the consumers after the meal experience (Özdemir, 2018, p. 63). Service quality, product quality, menu diversity, hygiene, location, crowd, working hours, noise, service speed, price and value, children's activities and atmosphere are the factors that tourists consider in their evaluation of their experiences (Yuksel & Yüksel, 2002b, p. 62). In a study conducted on Chinese tourists, it was determined that food quality (visual/olfactory impressions, ingredients, and palate/touch distinction) has a significant effect on tourist satisfaction (Nam, Shim & Jeong, 2017, p. 68). Study examining complaints about Top 50 Restaurant List of the World restaurants on the TripAdvisor site, it has been determined that the complaints are about fairness of price and originality, food quality, service quality and atmosphere. It was determined that the most complained issue was the high prices and not meeting the expectations (Şengül, Çavuş & Taşkın, 2018, p. 324). In another study investigating hotel restaurants, it was revealed that tourists made their complaints about restaurant through word-of-mouth communication and not also ceasing to patronize the restaurant, but also warn people not to patronize the restaurant (Heung & Lam, 2003, p. 285).

It can be said that the behavioral intentions of tourists can be measured by variables such as intention to revisit destination, recommending that restaurant to acquaintances or other people (WOM), making positive & negative comments about restaurants online (E-WOM), loyalty, intention to visit and that tourists exhibit these behaviors (Heung & Lam, 2003; Sparks et al., 2003; Lee et al., 2005; Stephen et al., 2005; Wu et al., 2017; Şengül et al., 2018; Nguyen, Dang & Ngo, 2019).

4 Conclusion

In this section, food consumption behaviors of tourists are examined in the context of eating out. Thus, it is attempted to establish a theoretical framework. As a result, the behaviors of tourists in the context of food consumption are revealed with a holistic perspective. Variables related to tourists' eating out behaviors are revealed. These variables have been identified as tourists' purposes to eat out, search for information, food choices, restaurant choices, dining experiences, complaint-satisfaction, and behavioral intentions after meal experience. By considering these variables, it is possible to examine food consumption behaviors of tourists in a holistic way. In addition, it is possible to examine the relationships among the variables of tourists' eating out behaviors.

References

Aho, S. K. (2001). Towards a general theory of touristic experiences: modeling experience process in tourism. *Tourism Review*, 56 (3, 4), 33–37.

Akarçay, E. & Suğur, N. (2015). Dışarıda Yemek: Eskişehir'de Yeni Orta Sınıfın Fast-Food Yeme-İçme Örüntüleri. *Sosyoloji Araştırmaları Dergisi*, 18 (1), 1–29.

Albayrak, A. (2014). Müşterilerin Restoran Seçimlerini Etkileyen Faktörler: İstanbul Örneği. *Anatolia: Turizm Araştırmaları Dergisi*, 25 (2), 190–201.

Altaş, A. & Varnacı Uzun, F. (2017). Üniversite Öğrencilerinin Yemek Tercihleri: Aksaray Üniversitesi Öğrencileri Üzerine Bir Çalışma. *Journal of Human Sciences*, 14 (4), 4435–4452.

Andersson, T. D. & Mossberg, L. (2004). The dining experience: do restaurants satisfy customer needs? *Food Service Technology*, 4, 171–177.

Araz, U. (2019). *Kapadokya Bölgesine Gelen Turistlerin Restoran Seçim Ölçütleri Üzerine Bir Araştırma (Yayınlanmamış yüksek lisans tezi)*, Nevşehir Hacı Bektaş Veli Üniversitesi, Nevşehir.

Batra, A. (2008). Foreign tourists' motivation and ınformation source(s) ınfluencing their preference for eating out at ethnic restaurants in bangkok. *International Journal of Hospitality & Tourism Administration*, 9(1), 1–17.

Bekar, A. & Gümüş Dönmez, F. (2016). Tüketicilerin Dışarıda Yemek Yeme Nedenlerine İlişkin Bir Değerlendirme. *Social Sciences,* 11(1), 1–15.

Chang, R., Kivela, J. & Mak, A. (2010). Food preferences of Chinese tourists. *Annals of Tourism Research,* 37(4), 989–1011.

Cullen, P. (1994). Time, tastes and technology: the economic evolution of eating out. *British Food Journal,* 96 (10), 4–9.

González, A. F., Curtis, C., Washburn, I. J. & Shirsat, R. A. (2020). Factors in tourists' food decision processes: a US-based case study. *Journal of Tourism Analysis: Revista de Análisis Turístico,* 27 (1), 2–19.

Grundey, D. (2008). Experiential marketing vs. traditional marketing: creating rational and emotional liaisons with consumers. *The Romanian Economic Journal,* 11(29), 133–151.

Guan, J. ve Jones, D. L. (2014). The contribution of local cuisine to destination attractiveness: an analysis involving Chinese tourists' heterogeneous preferences. *Asia Pacific Journal of Tourism Research,* 20(4), 416–434.

Gustafsson, I., Öström, A., Johansson J. & Mossberg, L. (2006). The five aspects meal model: a tool for developing meal services in restaurants. *Journal of Foodservice,* 17, 84–89.

Gyimothy, S., Rassing, R. S. & Wanhill, S. (2000). Marketing works: a study of the restaurants on Bornholm, Denmark. *International Journal of Contemporary Hospitality Management,* 12(6), 371–379.

Hansen, K. V., Jensen, Q. ve Gustafsson I. (2004). Payment- an undervalued part of the meal experience. *Food Service Technology,* 4, 85–91.

Harrington, R. J., Ottenbacher, M. C. & Kendall, K. W. (2011). Fine-dining restaurant selection: direct and moderating effects of customer attributes. *Journal of Foodservice Business Research,* 14(3), 272–289.

Heung, C. S. V. & Lam, T. (2003). Customer complaint behaviour towards hotel restaurant service. *International Journal of Contemporary Hospitality Management,* 15(5), 283–289.

Hofstede, G. (1983). The cultural relativity of organizational practices and theories. *Journal of International Business Studies,* 14(2), 75–89.

Jeong, E. ve Jang, S. (2011). Restaurant experiences triggering positive electronic word-of-mouth (e-WOM) motivations. *International Journal of Hospitality Management,* 30, 356–366.

Johns, N. & Kivela J. (2001). Perceptions of the first time restaurant customer. *Food Service Technology,* 1, 5–11.

Kim, Y. G. & Eves, A. (2012). Construction and validation of a scale to measure tourist motivation to consume local food. *Tourism Management,* 33, 1458–1467.

Kim, Y. G., Eves, A. & Scarles, C. (2009). Building a model of local food consumption on trips and holidays: a grounded theory approach. *International Journal of Hospitality Management*, 28, 423–431.

Kim, Y. H., Goh, B. K. & Yuan, J. (2010). Development of a multi-dimensional scale for measuring food tourist motivations. *Journal of Quality Assurance in Hospitality & Tourism*, 11(1), 56–71.

Law, R., To, T. & Goh, C. (2008). How do mainland Chinese travelers choose restaurants in Hong Kong? *International Journal of Hospitality Management*, 27, 346–354.

Mak, A., Lumbers, M., Eves, A. & Chang, R. C. Y. (2012). Factors influencing tourist food consumption. *International Journal of Hospitality Management*, 31(3), 928–936.

Mak, A., Lumbers, M., Eves, A. & Chang, R. C. Y. (2017). The effects of food-related personality traits on tourist food consumption motivations. *Asia Pacific Journal of Tourism Research*, 22(1), 1–20.

Mak, A. H. N. (2018). Motivations underlying tourist food consumption. In C. Pforr, & I. Phau (Eds.), *Food, wine and china: a tourism perspective* (pp. 72–95). Abingdon: Routledge.

Nam, M. J., Shim, C. & Jeon, C. (2017). The effect of food quality on behavioral intention in Korean restaurants. *International Journal of Tourism and Hospitality Research*, 31(10), 59–72.

Nguyen, H. M., Dang, L. A. T. & Ngo, T. T. (2019). The effect of local foods on tourists' recommendations and revisit intentions: The case in Ho Chi Minh City, Vietnam. *Journal of Asian Finance, Economics and Business*, 6(3), 215–223.

Özdemir, B. (2010). Dışarıda Yemek Yeme Olgusu: Kuramsal Bir Model Önerisi. *Anatolia: Turizm Araştırmaları Dergisi*, 21(2), 218–232.

Özdemir, B. (2018). *Yemek deneyimi*. In Kurtuluş Karamustafa (Ed.), *Yiyecek ve içecek yönetimi* (pp. 47–74), Ankara: Detay Yayıncılık.

Pedraja, M. ve Yagüe, J. (2001). What information do customers use when choosing a restaurant. *International Journal of Contemporary Hospitality Management*, 13(6), 316–331.

Pettersson, A. & Fjellström, C. (2007). Restaurants as friends of the family: functions of restaurant visits in everyday life. *Journal of Foodservice*, 18(6), 207–217.

Pizam, A. & Jeong, G. H. (1996). Cross-cultural tourist behavior perceptions of Korean tour-guides. *Tourism Management*, 17(4), 277–286.

Pizam, A. & Sussmann, S. (1995). Does nationality affect tourist behavior?. *Annals of Tourism Research*, 22(4), 901–917.

Quan, S. & Wang, N. (2004). Towards a structural model of the tourist experience: an illustration from food experiences in tourism. *Tourism Management*, 25, 294–305.

Randall, E. & Sanjur, D. (1981). Food preferences—their conceptualization and relationship to consumption. *Ecology of Food and Nutrition,* 11(3), 151–161.

Selvi, S. M. (2009). *Turizm İşletmelerinin Pazarlanmasında Fiziksel Kanıt (Plysical Evidence).* In Cevdet Avcıkurt, Şehnaz Demirkol, Burhanettin Zengin (Eds.), *Turizm İşletmelerinin Pazarlanmasında 7P ve 7C* (215–236). İstanbul: Değişim Yayınları.

Sparks, B. (2007). Planning a wine vacation? factors that help to predict tourist behavioral intentions. *Tourism Management*, 28(5), 1180–1192.

Sparks, B., Bowen, J. & Klag, S. (2003). Restaurants and the Tourist Market. *International Journal of Contemporary Hospitality Management,* 15(1), 6–13.

Sparks, B., Bowen, J. & Klag, S. (2004). *Restaurants as a contrıbutor to a tourıst destınatıon's attractıveness phase three (pt ı) consumer report*, https://sustain.pata.net/wp-content/uploads/2015/02/Sparks_RestaurantsPh3-PtI.pdf, Erişim Tarihi: 21.07.2020.

Stephen, W. L., Julia, E. B. & Stephen, T. L. (2005). Tourists' use of restaurant webpages: is the internet a critical marketing tool? *Journal of Vacation Marketing*. 11(2), 155–157.

Sünnetçioğlu, A. (2018). *Seyahat Yaşam Tarzı, Kültürel Duyarlılık Ve Yeni Yemekleri Deneme Korkusunun Yerel Yemek Tercihine Etkisi* (Yayınlanmamış doktora tezi), Çanakkale Onsekiz Mart Üniversitesi, Çanakkale.

Şengül, S., Çavuş, O. & Taşkın, E. (2018). Analysis of e-complaints for restaurant enterprises in the world's 50 best restaurant list. *Journal of Tourism and Gastronomy Studies*, 6(1), pp. 320–334.

Telfer, D. J. & Wall, G. (2000). Strengthening backward economic linkages: local food purchasing by three Indonesian hotels. *Tourism Geographies,* 2(4), 421–447.

Vu, Q. H., Li, G., Law, R. & Zhang, Y. (2019). Exploring tourist dining preferences based on restaurant reviews. *Journal of Travel Research*, 58(1), 149–167.

Wang, Y. H. (2011). Exploring the factors of gastronomy blogs influencing readers' intention to taste. *International Journal of Hospitality Management,* 30(3), 503–504.

Yılmaz, G. & Gültekin, S. (2016). *Consumers and Tourists' Restaurant Selections.* In Cevdet Avcıkurt, Mihaela S. Dinu, Nejdet Hacıoğlu, Recep Efe, Abdullah Soykan, Nuray Tetik (Eds.) *Global Issues and Trends in Tourism* (217–230). Sofia: St. Kliment Ohridski University Press.

Yüksekbilgili Z. (2014). Restoran Seçim Ölçütleri Üzerine Bir Araştırma. *Journal of Yaşar University*, 9(36), 6353–6360.

Yüksel, A. & Yüksel, F. (2002a). Market segmentation based on tourists' dining preferences. *Journal of Hospitality & Tourism Research*, 26(4), 315–331.

Yüksel, A. & Yüksel, F. (2002b). Measurement of tourist satisfaction with restaurant services: A segment based approach. *Journal of Vacation Marketing*, 9(1), 52–68.

Barış DEMİRCİ, Hilal AYVALI, Halil CODA & Ahmet ÜNVER

16 Boza as a Gastronomic Value in Anatolia

Introduction

The boza of which etymological origin has been examined in two views, is derived from the Persian word "büze", which means maize and soured, dark drink (Yıldırım, 2008). Other than the Anatolia, it is a traditional beverage which obtained by grain fermentation, and it has similar products in the Balkans, Caucasus, Turkic Republics, Egypt, Iran and Africa (Birer, 1987, Meric, 2010; Levent and Cavuldak, 2017). The second view that defends the Turkish etymological origin of boza is grounded on the samples of boza usage in different dialects of Turkish (Uighur, Chagatai, Chuvash, Yakut, etc.) and the fact that revealed by Ali Sir Nevai's (1441–1501) work named Muhakemetu'l Lugateyn, in which he studied Turkish and Persian comparatively, Persians took some words such as "kımız, bahsum, boza, sürme, etc." from Turkish (Ali Sir Nevai, 1941; Isıtman, 1941; Gabain, 1988). Since boza, which is similar to traditional beer (without hops), is included in the classification of fermented beverages obtained with the same raw materials and similar techniques, theorists again show two different approaches to the discovery of boza. In order to understand the approaches, it is useful to consider the studies of beer and boza production processes.

Traditional Beer

Information obtained from tablet and source research shows that the prerequisite of beer, which was first discovered in Sumer, was growing grain and making bread (Bottero, 2018). The 5000s B.C. were the years when barley-wheat (triticum dicoccum) cultivation started in the lands between the Euphrates and Tigris, and both bread and beer were produced from grains. In the Epic of Gilgamesh, beer begins with drying the malt obtained by germinating the soaked barley, grinding it with a mortar and pestle or grinding it with a millstone and turning it into a homogeneous dough with water and beer bread. The mixture prepared from malt, water and beer bread is left to ferment and ferment in a closed container. After the mixture swells and foaming is observed, sweetening and flavoring substances (honey, spices, etc.) are added to this mixture. The mixture prepared in vats with holes in the bottom turns into a kind of fermented barley water, beer, and the liquid that drips from the holes in the bottom of the trough to the bottom container (Bamfort, 2003; Eren, 2005; Maden, 2013).

Traditional Boza

The ground grains (wheat, barley, corn, etc.) are soaked and boiled with the addition of water, and then cooked until it becomes mushy and separates from the outer skin of the grain.

The grain gruel is passed through the process of filtering in order to filter the residues that may reduce the quality of the desired final texture, such as rough texture, crust, and membrane. Sugar or honey and a little boza or bread is added to the creamy grain gruel, and its brim is covered with a thin cover (cheesecloth) and fermentation is applied. Since the fermentation period is the time to determine the aroma, taste and alcohol ratio of the boza, the boza gets covered and taken to the storage area before souring and foaming appear (Turan, 2007). In the light of this information, while it is possible to say that boza is the coeval and even the origin of the beer of which formula was changed with the addition of hops; the reason of the theoretical differences regarding the discovery of boza is based on the studies regarding to the food and beverage culture of the currently consumed and preferred geographies (geographies where Turks lived and live) and the fact that the grain gets grounded and boiled during the production of boza.

The first theory for the discovery of boza is in the direction of Anatolia and Gobeklitepe, which are the ancestors of wheat, grain cultivation and agriculture in human history, are the ancestors of bread and beer and therefore boza. Six large cooking pots with 160 liters of liquid capacity in which gray-black sticky residue traces which are the traces of calcium oxalate that emerged during the soaking and fermenting of the grain discovered in the Gobeklitepe excavation area are considered as indicators of early agricultural period related to the production-consumption of fermented grain beverages for Gobeklitepe and Anatolia by the researchers (Heun et al., 1997; McGovern, 2009). In this view, which defends that the source of boza is Mesopotamia, it is stated that its spread took place on the course of Sumer, Anatolia, Egypt, Africa, Asia, China, Iran, Afghanistan, Caucasia and then the Balkans (Kramer and Noah 1995).

According to the other theory, boza is the traditional drink of the Turks. The most important data supporting the theory is that while boza was produced and consumed according to its current low alcohol state (1 %) in the geographies where Turks live, the alcohol ratio of the beer produced in Sumer is quite high compared to boza (Çakır, 2019; Kaşgarli Mahmut, 1941; Kaya, 1994; Sami,

1987). In the scientific data that Sait Maden added to the preface of his Epic of Gilgamesh translation, Central Asia was flooded as far as Eastern Anatolia due to the glacier melting during the Great Flood, the Turks who were living there came to Mesopotamia and started a civilization, and they carried the tastes they consumed, agricultural and grain culture to Mesopotamia (Çiğ, 2008). In the light of historical data, it is not clear which civilization has been discovered before and from whom to whom it is transferred, but boza has managed to become one of the cultural ambassador flavors of today, without spoiling its traditional recipe, by transferring it from culture to culture, especially in geographies where Turkish populations live.

Food and beverage recipes are important and deep resources by which societies and cultures can convey their "selves", as well as being an indicator of nutritional diets, accessible materials and cooking techniques. The philosophy of the Turkish traveler Evliya Çelebi, who lived in the 17th century, "If history is water, culture is the land irrigated by it" (Yerasimos, 2018) should be interpreted as nutritional prescriptions, if food is water, culture is the land irrigated by it. In this context, boza is the transmitter of civilizations that carry the accumulation of centuries behind them, it is a drinkable culture. Since gastronomic tourism is also the tourism of the taste, food culture and the story in its content (Hjalager, 2002), food and beverage recipes should be transferred to those who taste and experience, based on the history, meaning and culture they contain. Tourism activities and food and beverage phenomenon are considered as a whole. Exploring different flavors and experiencing food and beverages belonging to different cultures are among the first reasons for some tourists to choose the relevant destination (Yüncü, 2016). In this sense, boza can be made an element of touristic attraction by being promoted in the international arena due to its nutritious properties, its content that has preserved its standard recipe for centuries, and its gastronomic cultural value.

The Adventure of Boza in Anatolia

Since the seriousness of written transmission and written record tradition in Anatolia was rarely seen outside the Hittite period, especially in the Turkish population, the transference continues through verbal (stories, poems, epics and sayings), written sources about boza are often encountered in periods when recording was once again important (Görkem, 2009). The written work in which boza has been featured for the first time was written in the 14th century and the principles of its preparation were written in the 15th century, and the production of boza obtained from wheat as well as rice and barley boza was described in the work (Turan, 2007). According to the expressions of Evliya Çelebi (17th century), it is known that two different boza, sour and sweet, were produced in the Ottoman Empire. While sour boza is a drink close to the traditional beer recipe, high in alcohol, and has been banned by ulama at various times due to its intoxicating nature, sweet boza is a drink obtained by lactic acid fermentation before the alcohol rate rises, and it is a drink that people of all ages enjoy. Besides of community, Matbah-ı Amire records show that boza was frequently preferred by the Sultan and Statesmen, boza expenses are also found in the kitchen records of the Prince Sanjaks (Turhan, 2007). According to the information in Seyahatname, there were 300 boza houses in Istanbul in 1635, there were also boza houses in Ankara, Bursa, Erzurum, Kirikkale, Edirne, Tosya, Rumelia, Crimea and Egypt and it has been a beverage preferred by the locals in many provinces in summer and winter (Yerasimos, 2014). Along with the banned taverns during the reign of Sultan Murat IV (17th century), boza makers increased their sour boza production and started to produce intoxicating boza like opium boza. It is understood that, as a result of this change, which was reflected in Turkish with the expression "the şıra producer's witness is boza producer" (which expresses the meaning of "the blind leading the blind"), wine was produced with the name of şıra and beer with the name of boza. When the boza houses ceased to be a place of chitchat and took on the atmosphere of a pub, the sale of sweet boza turned into street sales, and the boza culture evolved into a culture of street flavors (Koç, 2007; Ünsal, 2007). Hacı Sadık Bey, who immigrated to Istanbul from Prizren during the heavy immigration to the Ottoman Empire in 1870 after the '93 War, experimented with a new technique instead of the juicy, brown and sourish boza produced, by producing yellowish, thick and sweetish boza. Since it has been deteriorated after a while because of storing in oak barrels, He has been able to preserve its flavor by preserving the boza in marble cubes. By establishing the world's first official boza trade house in 1876 with the brand name Vefa Bozacısı, he standardized the boza and the members of the same family brought the boza to industrial production in 1918. The recipe for the boza culture that has existed since the Republic Period is this standard recipe belonging to Hacı Sadık Bey. (İgüs, 2016). The reasons why the people love boza: Besides its taste, boza is a nutritious, easily accessible and cheap beverage that can be produced from all kinds of grains.

Production and Nutritive Properties of Boza

In the Turkish Food Codex (TGK) article TS 9778, boza is described as: "A product prepared by cooking millet, rice, wheat, corn and similar grains cleaned of their foreign materials by adding drinking water to one or more of their crushes or flours and subjecting them to alcohol and lactic acid fermentation in accordance with the technique by adding white sugar". The first fermentation that takes place in Boza fermentation is alcohol fermentation, which provides the development of yeast and the release of carbon dioxide, which is seen to swell. The second fermentation is the lactic acid fermentation carried out by the lactic acid bacteria that shape the acidic structure of boza. As in the traditional recipe, in order to obtain boza, first the grain (usually in the form of millet semolina) is grounded, boiled and cooled, then sugar is added and left to fermentation at room temperature and then cooled to storage and preservation. Again, as stated in the TGK, the alcohol ratio should not be over 2 %, it should contain 10 % sugar, the acidity should not exceed 0.5 % and the dry matter rate should be at least 20 %. Boza, which contains bacteria from the Lactobacillus family, which is vital for the intestinal flora and immune system, also contains yeasts such as candida and saccharomyces, while lactic acid bacteria show a dense spread compared to yeasts. Boza, which is rich in B group vitamins due to the grains it contains, can also be called liquid bread thanks to its beneficial effects on health, (Levent and Cavuldak, 2017).

The Touristic Importance of Boza

While tourism dynamism takes place in many ways today, one of the mentioned types is auditory tourism. Auditory tourism value is related to the cultural heritage of a destination that has made a sound on the streets. For this reason, it can be said that the rich historical cultures of ancient cities have tourism potential that feeds auditory memories. It can be seen in the memoirs of a writer that he wrote about the connection he established with the voices rising from the streets of the city he lived in and entering from his window and his various memories, and these sounds often revive his beautiful memories. In this context, in a study aimed at achieving the auditory heritage of Istanbul, it is stated that great importance is attached to the voices of boza sellers' shouting in the streets in winter evenings (Yelmi, 2016, s. 308). In the study, these voices are named as part of a deep culture and indicated as a different attraction for tourists.

With the recent healthy nutrition trend, gastro tourists explore the cultural qualities of foods beyond taste, test their health benefits and travel to find and taste ingredients that enrich food (Fabbri and Crosby, 2016, p. 2). Boza, which is widely used in various methods in the Balkans, Iran, Caucasus, Central Asia, Africa and Anatolia, has a high potential in terms of beneficial intestinal bacteria it contains, as it can be included in various routes for healthy eating tours. Because it can be produced from various raw materials and can be passed through different fermentation processes, this increases its potential (Novak, Kovacs and Czipa, 2016, p. 67). For example, it draws attention with the types of cocoa included in Bulgaria and black colored in Egypt with an alcohol content of up to 7 %, with their different benefits (Arıcı and Dağlıoğlu, 2002, p. 39).

Conclusion and Suggestions

As a result, boza is the drink of chitchat, meeting and entertainment culture that has existed in these lands for centuries, with cinnamon, cloves and roasted chickpeas. It is the bread of the military camps, breast milk for newborns, heat in winter and cool in summer. The reason why boza is a conversation and entertainment culture that makes people waiting its seller at meals every night, rather than a drink, is the fact that boza sellers on the street at night are welcomed home by the households, while sipping boza with roasted chickpeas and cinnamon, they tell their stories, the places they see and visit, they set up poems, and they fight chitchat with the household. Actually, boza is not just a taste, it is the ambassador of cultures of which adventure has passed over the centuries, and it is a very important value of gastronomy and cultural tourism. Promotion of this value has a great importance for tourism. For this reason, boza can be added to the gastronomy routes that will be created. Because gastronomy routes are transforming the gastronomic heritage within the food and beverage culture of a destination into an attraction element in gastronomy tourism.

Gastronomy museums play an important role in conveying a product or culture to visitors. These museums convey the history of the products in their theme, their changes from the past to the present, the old and new equipment used in food production, the effects of the products on humanity and other objects related to the products. Many museums have sections where the mentioned products can be tasted and sold. In addition to this, various professional trainings are also provided in some gastronomy museums (Sezen, 2018). “Boza Museum” can be added to these museums with many examples in the world and Turkey. This museum will play a key role in promoting boza and ensuring its sustainability, beyond being a point to be visited especially for local and foreign tourists who are interested in gastronomy.

Food and beverage festivals, which are an integral part of culture, are known as important in promoting a destination, creating a touristic attraction element that will bring tourists to a destination, and in the development of tourism in a destination (Yatkın and Tolga, 2010). In this sense, boza festivals, which will be organized internationally, are important in terms of promoting boza, which is considered as a Turkish drink within the scope of gastro diplomacy, as well as its social, economic and cultural effects on the destination.

References

Ali Şir Nevai. (1941). *Muhakemetu'l Lugateyn*, (Çevirmen: Işıtman, İ.), TDK, Ankara.

Arıcı, M. & Dağlıoğlu, O. (2002). Boza: A Lactic Acid Fermented Cereal Beverage as a Traditional Turkish Food. *Food Reviews International*, 18 (1), pp. 39–48.

Bamfort, C. (2003). *Beer: Tap Into the Art and Science of Brewing*, New York: Oxford University Press.

Birer, S. (1987). Boza Yapımı ve Özellikleri, *Gıda* 12 (5), pp. 341–344.

Bottero, J. (2018). *Gılgameş Destanı; Ölmek İstemeyen Büyük İnsan*, (Çeviren: Suda, O.), İstanbul: Yapı Kredi Yayınları.

Çakır, C. (2019). *Tarihi, Etimolojisi ve Edebiyatı ile Boza Kitabı*, Türk Dünyası Araştırmaları Vakfı, İstanbul.

Çığ, M. İ. (2008). *Sümerlilerde Tufan Tufanda Türkler*, Ankara: Kaynak Yayınları.

Eren, E. (2005). *Geçmişten Günümüze Anadolu'da Bira*, İstanbul: Tarih Vakfı Yurt Yayınları.

Fabbri, A. D. & Crosby, G. A. (2016). A Review of the Impact of Preparation and Cooking on the Nutritional Quality of Vegetables and Legumes. *International Journal of Gastronomy and Food Science*, 3 (2016), pp. 2–11.

Gabain, A. V. (1988). *Codex Cumanicus'un Dili – Tarihi Türk Şiveleri*, (Çeviren: Akalın, M.), Ankara: Türk Kültürünü Araştırma Enstitüsü Yayınları: 73.

Görkem, İ. (2009). Dünden Bugüne Türk Sözel Edebiyatı: Değişim ve Dönüşüm, *A.Ü. Türkiyat Araştırmaları Enstitüsü Dergisi*, 39 (2009), pp. 411–422.

Heun, M., Schäfer-Pregl, R., Klawan, D., Castagna, R., Accerbi, M., Borghi, B. & Salamini, F. (1997). Site of Einkorn Wheat Domestication Identified by DNA Fingerprinting. *Science* 278 (5341), pp. 1312–1314.

Hjalager, A. M. (2002). A Typology of Gastronomy Tourism. Anne-Mette Hjalager & Greg Richards (Ed.), in *Tourism and Gastronomy*, Londra: Routledge.

Işıtman, İ. R. (1941). Büyük Şairin 500. Doğum Yıldönümü Dolayısıyla; Muhakemetu'l Lugateyn, Ali Şir Nevai İncelemesi, TDK, Ankara, 11 (2), pp. 115–121.

İgüs, E. (2016). Balkanlar'dan Anadolu'ya Boza ve Türleri ile Türkiye'deki Balkan Kökenli Bozacılar, *Karabük Üniversitesi Sosyal Bilimler Dergisi*, 6 (1), pp. 101–111.

Kaşgarlı Mahmut. (1941). *Divanu Lugati't – Türk*, (Çeviren: Atalay, B.), Ankara: Alaadin Kıral Basımevi.

Kaya, C. (1994). *Uygurca Altun Yaruk* (Giriş, Metin ve Dizin), Atatürk Kültür, Dil ve Tarih Yüksek Kurumu, Ankara: TDK Yayınları: 607.

Koç, Ü. (2007). *Klasik Dönem Osmanlı Ülkesinde Boza, Acısıyla Tatlısıyla Boza*. T.C Kültür ve Turizm Bakanlığı, Ankara.

Kramer, N. (1995). *Tarih Sümer'de Başlar*, (Çeviren: Çığ, M. İ.) Ankara: Türk Tarih Kurumu Yayınları: X (11a).

Levent, H. & Cavuldak, Ö. A. (2017). Geleneksel Fermente Bir İçecek: Boza. *Akademik Gıda*, 15 (3), 300–307.

Maden, S. (2013). *Gılgameş Destanı*, İstanbul: Türkiye İş Bankası Kültür Yayınları.

McGovern, P. E. (2009). *Uncorking the Past. The Quest for Wine, Beer, and Other Alcoholic Beverages*. Berkeley/Los Angeles/London: University of California Press.

Meriç, A. (2010). *Trakya Bölgesinde Üretilen Bozaların Bazı Fizikokimyasal ve Mikrobiyolojik Özellikleri*. (Yayımlanmamış Yüksek Lisans Tezi), Namık Kemal Üniversitesi, Tekirdağ, Türkiye.

Novák, A., Kovacs, B. D. & Czipa, N. (2016). Boza beltartalmi paramétereinek változása az alapanyagok függvényében. *Agrártudományi Közlemények*, pp. 67–70.

Sâlephur, C. (1381). Farsça Türkçe Genel Sözlük. Tebriz.

Sezen, T. S. (2010). *Gastronomi Müzeleri*. Atilla Akbaba & Neslihan Çetinkaya (Ed). İçinde *Gastronomi ve Yiyecek Turizmi*. Ankara: Detay Yayıncılık.

Turan, N. A. (2007). Acısıyla Tatlısıyla Boza, Kültür ve Turizm Bakanlığı Yayınları, Ankara.

Ünsal, A. (2007). Şıra, Boza ve Kimi Şerbetlere Dair, Acısıyla Tatlısıyla Boza. T.C Kültür ve Turizm Bakanlığı, Ankara.

Yatkın, Ö. ve Tolga, Ö. (2010). *Gastronomi Konulu Festivaller*. Atilla Akbaba & Neslihan Çetinkaya (Ed). *Gastronomi ve Yiyecek Turizmi*. Ankara: Detay Yayıncılık.

Yelmi, P. (2016). Protecting Contemporary Cultural Soundscapes as Intangible Cultural Heritage: Sounds of Istanbul. *International Journal of Heritage Studies*, 22 (4), pp. 302–311

Yerasimos, M. (2014). *500 Yıllık Osmanlı Mutfağı* (13. Baskı). İstanbul: Boyut Yayın Grubu.

http://www.vefa.com.tr/index.php?dil=tr&sayfa=uretim Erişim Tarihi: 01.07.2020.

Yıldırım, N. (2008). Fars Mitoloji Sözlüğü, İstanbul: Kabalcı Yayınları.

Yüncü, H. R. (2010). *Sürdürülebilir Turizm Açısından Gastronomi Turizmi ve Perşembe Yaylası.* 10. Aybastı Kabataş Kurultayı: Yerel Değerler ve Yayla Turizmi içinde Şengel, S. (Edt.) Ankara: Detay Yayıncılık.

Emrah KESKİN & Bülent YORGANCI

17 Memorable Food Tourism

1 Introduction

Food has an effect that increases the motivation in the travel process as well as being one of the most important elements that reveals satisfying senses and ensures the satisfaction of the tourists in their destination experiences (Hall, 2003; Quan ve Wang, 2004). Food and beverages, which are the main building blocks of culinary tourism, are an important resource for culinary tourism and are also an element that reflects the values of a country or region to tourists. For this reason, culinary tourism adds meaning to the travels of visitors (Telfer ve Wall, 1996).

According to Boyne et al. (2003), food and beverage has an important power in ensuring the continuity of the travel of especially gastronomic tourists. According to Wolf (2002), the destination cuisine allows gastronomic tourists to experience unforgettable experiences while enjoying the unique taste of the dishes during their local food experiences and also to make various comments about the dishes of the local cuisines they have experienced. Luxury hotels, famous restaurants, world metropolises, unique ethnic cuisines of exotic peoples and seaside resorts are places where the destination image of the food is exhibited. For this reason, kitchen products are important in the promotion of destinations (Dann, 1996; Noguchi, 1992).

While operations strive to make their guests experience memorable tourism, they should also pay attention to their having memorable dining experience. The presentation of the dishes and the quality of the food and the service provided contribute to the guests' memorable dining experience (Di Pietro, Robin & Campell, 2014; Jang & Namkung, 2009). The value that the multidimensional features of the restaurants make customers feel in their dining experiences provides unforgettable experiences (Tsaur and Lo, 2020). According to Chandralal et al. (2014), the food, space design or extra entertainment elements prepared by some restaurants specially for the important days of the customer are effective in having a memorable experience.

2 Memorable Food Experience

According to Kim et al. (2010), the memorable tourism experience is the recall of those which are remembered with positive feelings from all the events that take place during the touristic experience. In the formation of these experiences, hedonism, spaciousness, local culture, meaningfulness, knowledge, participation and innovation are effective. Memorable tourism experiences cover all the activities that tourists carry out during their travels (Quan &Wang, 2004). In addition, the memorable tourism experience theory is important because the memories of tourists about their experiences are influential in their next travel preferences (Chandralal et al. 2014; Kozak, 2001; Lehto, O'Leary & Morrison, 2004; Marschall, 2012; Mazursky 1989). According to Zhong et al. (2017), positive feelings such as happiness, excitement, hedonism, relief, and emotional experiences are experiences that are easy to remember from memorable tourism experiences. Factors such as services provided by tourism workers, expectations of tourists, influences, remembering, social relations, imagining nostalgia and feeling of freedom are the elements that affect emotions in memorable tourism experience (Tung & Ritchie, 2011a; Tung & Ritchie, 2011b). According to Morgan (2006), the moments of surprise of tourists, shared experiences, the difference of local elements and the positive value attributed to the experience are effective in having unforgettable experiences.

Elements such as photographs taken in unique places and places where tourists travel and souvenirs not found everywhere are experiences that travels have to be made to be remembered. This kind of factors makes the travel experience unforgettable (Hannabus, 1999; Marschall, 2012). According to Morgan and Xu (2009), the fact that tourists meet the local culture and communicate with the local people of the destination region and participate in the social activities involving the local people ensure that they have a memorable tourism experience. The culinary traditions of the destination areas, various dishes, different presentation styles and the attraction, fantasy and multidimensional effect of the general food culture show that the dining experiences are effective in the memorable tourist experiences (Sajna, 2005; Sparks et al. 2003).

In memorable experiences, the fact that the tourists have positive feelings toward the destination food image makes their experience happy. The positive emotions are affected by many factors, including the content of the food, its presentation and the characteristics of the environment in which it is consumed (Seyitoğlu & Ivanov, 2020; Solunoglu & Yayla, 2020). People who are happy with their food experiences share their happiness on social media by taking photos of the food at the time of experience. Thus, experience is recommended to others (Zhu et al. 2019). According to the study of Gregorash (2018), in memorable experiences in tourism, people first remember where they do the activities, with whom they do these activities and what activities they do and secondly, what foods they experience. Therefore, in strengthening the image of destination and in being visited by tourists, the experience of eating is very important.

3 Food Tourism

The concept of culinary tourism was first defined by Long (1998) and it was stated that the main purpose of travel in culinary tourism is to experience and interpret the dishes of traditional and exotic cuisines. Since food consumed in unique kitchens in culinary tourism is different from ordinary foods consumed in daily life, it provides memorable dining experiences (Hall & Mitchell, 2001; Long, 2004; Ottenbacher & Harrington, 2010; Jamie et al., 2017). Activities such as farm visits by tourists, participation in festivals, local food market visits, orchard visits, local food experiences sold at rural roadside stalls, wine production site visits, watching chefs' shows and visiting fishermen are included in culinary tourism (Long, 2006; Wolf, 2002). According to Karim and Chi (2010), food, ambience, menu, menu variety, accessibility, service staff, food tours, cooking courses and food festivals in the region contribute to the reflection of the destination food image (Lai et al., 2017). Kitchen products of a region are elements that increase the excitement and curiosity of tourists.

Wine and beer experiences of tourists, physical elements of places to experience, regional wine excursions, wine-related stories of food, hospitality of local people, local cookbooks and magazines are among the attractions of culinary tourist (Ignatov & Smith, 2006). Hjalager (2003) divided the tourists into four groups according to their attitudes toward tourism and food: existential gastronomic tourist, recreational gastronomic tourist, tourist seeking difference and experiential gastronomic tourist. Existential gastronomic tourists are tourists who are open to learning and experiencing new foods, experience cultural food and drink, love to learn more and experience organic food from rural areas. Experiential gastronomic tourists are tourists who experience food and beverage during their holidays, pay attention to menus, go to interestingly designed cafes, experience various cooking methods, read current food magazines and follow the developments. Recreational gastronomic tourists are tourists who do not like their food experiences and consume the food they consume in daily life. The imitation gastronomy tourists are the gastronomic tourists who do not take time for their food experiences during the holiday process and buy popular and familiar menus by choosing chain restaurants. This type of gastronomic tourists prefer standard shuttle services, it is important for them that the amount of food is high, not the quality of the food. These tourists rely on brochures and travel agencies in their destination preferences. (Kivela and Crotts, 2006). Of all these, gastronomic tourists who are open to experience and have existential features are the types of tourists who enjoy culinary tourism and tend to visit such destinations.

According to Piqueras-Fiszman and Jaeger (2015), experiencing homemade-looking, traditional or newly discovered food types makes food experiences unforgettable. Local products, food taste, originality, innovation, food-borne togetherness, social interaction, hospitality and service are the effective factors in realizing a memorable dining experience (Sthapit, 2018).

Societies' unique dishes and eating styles provide local cuisine culture. The inclusion of local culinary cultures in tourism makes the destination a center of food culture experience center culture open to the world. (Horng & Tsai, 2010). Local cuisines are not only an influential factor in the marketing of the destination, but also a motivation factor for tourists' intention to visit. Local foods, which motivate tourists, also provide recognition of the food heritage of a destination (Hjalager, 2002). It is a pleasure for many tourists in time for tourists to experience the local cuisine of the destination where they travel and comment on the taste of local cuisine. This pleasure contributes to the preference of the destination by many visitors (Stone et al. 2018).

In order for local cuisines to be in the attention of tourists, they must have some attraction and factors that will affect the visits of tourists. Local cuisines, which tourists consider unique, are kitchens with such attraction and may be the focus of visits (Cohen, 2000). In order to contribute more to tourism and create a more qualified local food heritage in places where local foods are associated with tourism, local people should be provided with higher quality products from agriculture and local products should be created.

In this way, both local products can be strengthened and the gastronomy image of the region can be increased (Boyne et al. 2003; Fox, 2007; Hall & Sharples, 2003).The fact that local food and beverage operations, which focus on tourism, use local products to provide unique dining experiences to tourists, contributes to the continuation of interest in local cuisines (Cohen and Avieli, 2004). For this reason, destinations that aim to have a memorable culinary experience should pay attention to culinary cultures.

4 Conclusion and Suggestions

Enjoying the food experiences made during touristic travels and going on trips to taste food is an indicator of how important the food is for tourism (Stone et al. 2017). The fact that the food image of the destination visited by the tourists is positive has an effect on the satisfaction of the tourists with their travels (Oliver, 1980). For this reason, the culinary heritage, local dishes and traditional environment of a region are very important in the satisfaction of tourists after the experience (Kim et al., 2012).

Local cuisines are an important factor in the satisfaction of the travel experience (Lopez-Guzmận & Sậnchesz-Canizares, 2012). The fact that the destination has a good food image is effective in making tourists satisfied with their experiences (Quing-Chi et al., 2013). According to Peŝtek and Cinjarevic (2014), the positive effect of food and beverages that are effective in carrying out touristic activities on tourists contributes to their revisiting the destination. According to Ab Karim et al. (2009) and Quing-Chi et al. (2013), the food image of the destination area is effective in revisiting the same area and making recommendations to others about the visit.

According to Tsai (2016) and Ali et al. (2014), local products in tourism provide unforgettable experiences. Some tourists check the food images of the regions in the travel region selections and expect to enjoy the food experiences they will make during their travels. This situation shows how important the food experiences are in the choice of destination area for tourists (Tikkanen, 2007). Especially in food experiences carried out in gastronomic tourism, local foods enable tourists to enjoy the experience and have memorable experiences by revealing the local food, service, food and beverage services and the relationship with the region (Björk and Räisänen, 2014).

Since there is a positive interaction between the remembered experiences and the decisions to be made in the future (Hoch & Deighton, 1989), what happened in the touristic experiences is remembered in the process after the experience and becomes effective in the next experience decisions (Miao et al., 2014). According to Harrington et al. (2012), while the quality of the food in the touristic experiences ensures that the quality experiences are made, it also helps to remember these experiences, and it is effective in the formation of the intention again.

Governments need some sanctions to make food experiences better and more active. It is necessary to raise the awareness of the public about local foods by the government and to provide certain trainings, to train qualified culinary staff who will offer good recalled food experiences, to evaluate the food tourism resources and to increase the promotional activities of the foods used in tourism. If these strategies are realized, sustainable food tourism will be developed. In addition to these strategies, unique tastes, local recipes and food photos shared on websites keep food tourism alive and create intention to realize quality and unique dining experiences in humans (Horng and Tsai, 2010).

Food, travel, festival, local products that provide memorable tourism experiences and attract the attention of customers can be handled in detail, and elements that may create a lack of interest can be removed from the experience process, or items that will increase the excitement of individual or group participation that will provide more interest and add excitement to the experience can be increased. At the time of experience, more attention can be paid to the points that touch the emotions of individuals. For example, the fact that the experience is high quality, diverse, unique and full of activities that people can show their energies can provide memorableness and intention to recommend. Especially in food tourism, memorable food experiences can be felt more intensely by paying attention to the quality of local products, preserving their originality, increasing the promotion of food festivities and by paying more attention to the naturalness of people's relationship with other elements of the environment they are experiencing.

References

Ab Karim, S., Chua, B. L. & Salleh, H. (2009). Malaysia as a culinary tourism destination: international tourists' perspective, *Journal of Tourism, Hospitality and Culinary Arts,* 1(3), 63–78.

Ali, F. K., Hussain ve N. & Ragavan, A. (2014). Memorable customer experience: examining the effects of customers experience on memories & loyalty in Malaysian Resort Hotels, *Procedia-Social and Behavioral Sciences,* 144, 273–279.

Björk, P. & Kauppinen-Räisänen, H. (2014). Culinary-gastronomic tourism – a search for local food experiences. *Nutrition & Food Science*, 44(4), 294–309.

Boyne, S., Hall, D. & Williams, F. (2003). Policy, support and promotion for food-related tourism initiatives, *Journal of Travel & Tourism Marketing,* 14(3–4), 131–154.

Chandralal, L., Rindfleish, J. & Valenzuela, F. (2014). An application of travel blog narratives to explore memorable tourism experiences. *Asia Pacific Journal of Tourism Research*, 20(6), 680–693.

Chi, C. G.-Q., Chua, B. L., Othman, M. & Karim, S. A. (2013). Investigating the structural relationships between food image, food satisfaction, culinary quality, and behavioral intentions: the case of Malaysia. *International Journal of Hospitality & Tourism Administration*, 14(2), 99–120.

Cohen, E. (2000). *The Commercialized Crafts of Thailand*. London: Curzon Press and Hawaii University Press.

Cohen, E. & Avieli, N. (2004). Food in tourism: attraction and impediment. *Annals of Tourism Research,* 31(4), 755–778.

Dann, G. (1996). *The Language of Tourism: A Sociolinguistic Perspective*, Oxford: CAB International.

Di Pietro, R. B. & Campbell, J. (2014). The influence of servicescape and local food attributes on pleasure and revisit intention in an upscale-casual dining restaurant, *Hospitality, Review*, 31(4), 1–35.

Fox, R. (2007). Reinventing the gastronomic identity of Croatian tourist destinations. *International Journal of Hospitality Management*, 26(3), 546–559.

Gregorash, B. J. (2018). Understanding authenticity within gastronomic experiences. *Authenticity & Tourism,* 24, 145–163.

Hall, C. M. & Mitchell, R. (2001). *Wine and food tourism*. In N. Douglas & R. Derrett (Eds.), Special interest tourism: Context and cases (s. 307–329). Brisbane: John Wiley.

Hall, C. M., Sharples, E., Mitchell, R., Macionis, N. & Cambourne, B. (2003). *Food Tourism around the World: Development, Management and Markets*, Oxford: Butterworth Heinemann.

Hall, M. & Sharples, L. (2003). The consumption of experiences or the experience of consumption? An introduction to the tourism of taste. In

Hannabus, S. (1999). Postmodernism and the heritage experience. *Library Management,* 20(5), 295–392.

Harrington, R. J., Ottenbacher, M. C., Staggs, A. & Powell, F. A. (2012). Generation Y consumers: Key restaurant attributes affecting positive and negative experiences. *Journal of Hospitality & Tourism Research*, 36(4), 431–449.

Hjalager, A. M. (2002). A typology of gastronomy tourism. In A. Hjalager, A. ve Richards, G. (Eds.), *Tourism and Gastronomy* (s. 21–35), London, United Kingdom: Routledge.

Hjalager, A. M. (2003). What do tourists eat and why? Towards a sociology of gastronomy and tourism. Collen, J. ve Richards, G. içinde *Gastronomy and tourism* (s. 54–74), Gravenwezel/Schilde, Belgium: Academie Voor de Streekgebonden Gastronomie.

Hoch, S. J. & Deighton, J. (1989). Managing what consumers learn from experience. *The Journal of Marketing*, 53(2), 1–20.

Horng, J. S. & Tsai, C. T. (2010). Government websites for promoting East Asian culinary tourism: A cross-national analysis. *Tourism Management,* 31(1), 74–85.

Ignatov, E. & Smith, S. (2006), Segmenting Canadian culinary tourists. *Current Issues in Tourism*, 9(3), 235–255.

Jamie, A., Levitt, P. Z., DiPetro, R. B. & Fang, M. (2017). Food tourist segmentation: Attitude, behavioral intentions and travel planning behavior based on food involvement and motivation. *International Journal of Hospitality & Tourism Administration*, 20(2), 129–155.

Jang, S. & Namkung, Y. (2009). Perceived quality, emotions, and behavioral intentions: Application of an extended Mehrabian–Russell model to restaurants. *Journal of Business Research*, 62(4), 451–460.

Karim, A. S. & Chi, C. G. Q. (2010). Culinary tourism as a destination attraction: An empirical examination of destinations' food image. *Journal of Hospitality Marketing & Management,*19(6), 531–555.

Kim, J. H., Ritchie, J. R. B. & McCormick, B. (2010). Development of a scale to measure memorable tourism experiences. *Journal of Travel Research,* 51(1), 12–25.

Kim, S., Kim, M., Agrusa, J. & Lee, A. (2012). Does a food-themed TV drama affect perceptions of national image and intention to visit a country? An empirical study of Korea TV drama. *Journal of Travel & Tourism Marketing,* 29(4), 313–326.

Kivela, J. & Crotts, J. C. (2006). Tourism and gastronomy: Gastronomy's influence on how tourists experience a destination. *Journal of Hospitality & Tourism Research*, 30(3), 354–377.

Kozak, M. (2001). Repeaters' behaviour at two distinct destinations. *Annals of Tourism Research*, 28(3), 784–807.

Lai, M. Y., Khoo-Lattimore, C. & Wang, Y. (2017). Food and cuisine image in destination branding: Toward a conceptual model, *Toursim and Hospitality Research,* 19(2), 1–14.

Lehto, X. Y., O'Leary, J. T. & Morrison, A. M. (2004). The effect of prior experience on vacation behaviour. *Annals of Tourism Research,* 31(4), 801–818.

Long, L. M. (2004). *Culinary Tourism*. Lexington: The University Press of Kentucky.

Long, L. M. (2006). Food pilgrimages: Seeking the sacred and the authentic in food. *Appetite,* 47(3), 393.

López-Guzmán, T. & Sánchesz-Canizares, S. (2012). Culinary tourism in Córdoba (Spain). *British Food Journal*, 114(2), 168–179.

Marschall, S. (2012). Personal memory tourism and a wider exploration of the tourism–memory nexus. *Journal of Tourism and Cultural Change*, 10(4), 321–335.

Mazursky, D. (1989). Past experience and future tourism decisions. *Annals of Tourism Research,* 16(3), 333–344.

Miao, L., Lehto, X. & Wei, W. (2014). The Hedonic value of hospitality consumption: Evidence from Spring break experiences, *Journal of Hospitality Marketing & Management*, 23(2), 99–121,

Morgan, M. (2006). Making space for experiences. *Journal of Retail and Leisure Property*, 5(4), 305–313.

Morgan, M. & Xu, F. (2009). Student travel experiences: Memories and dreams. *Journal of Hospitality Marketing & Management*, 18, 216–236.

Noguchi, P. (1992). Savor slowly: Ekiben the fast food of high-speed Japan. *Ethnology,* 33, 317–330.

Oliver, R. L. (1980). A cognitive model of the antecedents and consequences of satisfaction decisions. *Journal of Marketing Research,* 17(4), 460–469.

Ottenbacher, M. & Harrington, R. (2010). Culinary tourism: A case study of the gastronomic capital. *Journal of Culinary Science & Technology,* 8(1), 14–32.

Piqueras-Fiszman, B. & Jaeger, S. R. (2015). What makes meals 'memorable'? A consumer-centric exploration. *Food Research International*, 76, 233 –242.

Qing-Chi, C. G., Chua, B. L., Othman, M. & Ab Karim, S. (2013). Investigating the structural relationships between food image, food satisfaction, culinary quality and behavioral intentions: the case of Malaysia. *International Journal of Hospitality & Tourism Administration*, 14(2), 99–120.

Quan, S. & Wang, N. (2004). Towards a structural model of the tourist experience: An illustration from food experience in tourism. *Tourism Management*, 25(3), 297–305.

Sajna, S. S. (2005). *Food tourism and the culinary tourist (Unpublished doctoral dissertation)*, Clemson University, Clemson, SC.

Seyitoğlu, F. & Ivanov, S. (2020). A conceptual study of the strategic role of gastronomy in tourism destinations. *International Journal of Gastronomy and Food Science,* 100230.

Solunoglu, A. & Yayla, Ö. (2020). Deneyim Kalitesinin Memnuniyet ve Yaşam Kalitesine Etkisi: Kırıkkale Büyük Şehir Parkında Bir Uygulama, *Seyahat ve Otel İşletmeciliği Dergisi*, 17(1), 20–31.

Sparks, B., Bowen, J. & Klag, S. (2003). Restaurants and the tourist market. *International Journal of Contemporary Hospitality Management,* 15(1), 6–13.

Sthapit, E. (2018). A netnographic examination of tourists memorable hotel experiences. *Anatolia*, 29(1), 108–128.

Stone, M. J., Migacz, S. & Wolf, E. (2018). Beyond the journey: the lasting impact of culinary tourism activities. *Current Issues in Tourism*, 22(2), 147–152.

Stone, M. J., Soulard, J., Mihacz, S. & Wolf, E. (2017). Elements of memorable food, drink and culinary tourism experiences. *Journal of Travel Research*, 57(8), 1121–1132.

Telfer, D. & Wall, G. (1996). Linkages between tourism and food production. *Annals of Tourism Research*, 23(3), 635–653

Tikkanen, I. (2007). Maslow's hierarchy and food tourism in Finland: Five cases. *British Food Journal*, 109(3), 635–653.

Tsai, C. T. S. (2016). Memorable tourist experiences and place attachment when consuming local food. *International Journal of Tourism Research*, 18(6), 536–548.

Tsaur, S. H. & Lo, P. C. (2020). Measuring memorable dining experiences and related emotions in fine dining restaurants, *Journal of Hospitality Marketing & Management*, 1–24.

Tung, V. W. S. & Ritchie, J. R. B. (2011a). Exploring the essence of memorable tourism experiences. *Annals of Tourism Research*, 38(4), 1367–1386.

Tung, V. W. S. & Ritchie, J. R. B. (2011b). Investigating the memorable experiences of the senior travel market: An examination of the reminiscence bump. *Journal of Travel & Tourism Marketing*, 28, 331–343.

Wolf, E. (2002). *Culinary Tourism: A tasty economic proposition.* International Culinary Tourism Task Force. Portland, OR.

Zhong, Y. Y. S. Busser, J. & Baloglu, S. (2017). A model of memorable tourism experience: The effects on satisfaction, affective commitment, and sstorytelling. *Tourism Analysis*, 22(2), 201–217.

Zhu, J., Jiang, L., Dou, W. & Liang, L. (2019). Post, eat, change: The effects of posting food photos on consumers' dining experiences and brand evaluation. *Journal of Interactive Marketing*, 46, 101–112.

Tufan SÜREN

18 Traditional Street Foods in Turkey

1 Introduction

Even though the history of people's fulfilling their food needs outside of their homes by paying a fee goes back to ancient times; the fact that modern eating outside home first appeared in the west of Europe, with modernization which is the result of urbanization and industrialization (Scholliers, 2008, p. 334). With urbanization and industrialization, people's obligation to fulfill their nutritional needs away from home and the frequency of traveling have caused eating out-of-home to become a commercial activity (Beardsworth and Keil, 2012, p. 35).

In Turkey, out-of-home food sector, is a growing field in the economic dimension, diversity and quality factors. It is possible to identify this development when studies on out-of-home food consumption are examined. As a result of the semi-structured interview of Akarcay and Suğur (2015, p. 24–25) with 23 interviewers in Eskisehir; it was observed that the participants consumed fast food a few times a year at most, and mostly preferred local fast food restaurants. Onurlubas, Dogan and Gurler (2015, p. 921) examined the out-of-home food consumption of consumers in their research conducted in Istanbul, Ankara, İzmir, Samsun, Antalya, Gaziantep, and Erzurum with the participation of 1282 people. According to the research; the average monthly food expenditure of the participants is 907.10 TL. Monthly total out-of-home food consumption expenditure is 362.20 TL. When the preferred food consumption places of the participants were examined, it was determined that they preferred mostly the restaurants and the least the bars. The percentage of those who definitely prefer luxury restaurants is 11.3 %, the percentage of those who definitely prefer kebab restaurants is 14.3 %, the percentage of those who definitely prefer restaurants is 18.5 %, the percentage of those who definitely prefer fast food restaurants is 15.6 %, the percentage of those who definitely prefer pizza restaurants is 8.3 % and the percentage of those who definitely prefer the wrap restaurants is 7.9 %.

One of the developing elements of out-of-home food consumption all over the world is the street foods. Even though street food has become an area where demand is increasing today, it is a sector open to development with the elimination of the concerns of consumers such as food safety and hygiene and healthy nutrition. Turkey has a rich variety street foods and culture as a result of food culture and taste differences between regions. The fact that street foods, which have become an important part of people's food consumption, as a gastronomic trend, increased their importance in terms of frequency of consumption (Karsavuran, 2018, p. 255). It is also predicted that it will play an important role in the development of gastronomic tourism. This study was carried out to examine the developing place of street foods in the gastronomy and gastronomy tourism in Turkish cuisine, the most demanded and known types of traditional street foods in Turkish cuisine, which regions, and cities these foods belong to and examine their ingredients.

Out-of-Home Food Consumption

Out-of-home food consumption is defined as the consumption of food and beverages prepared and served by a food and beverage establishment for a fee (Jaafar, Lumbers and Eves, 2009 p. 1; Petterson and Fjellström, 2007 p. 207).

Out-of-home food consumption has become commonplace, especially in developed countries. The increase in income, acceleration of urbanization, technological developments and time constraints cause increases in food quality rather than increasing the amount of food (Blisard, Biin-Hwan, Cromartie and Ballenger, 2002 p. 7). It is also possible to say that development of the situation in Turkey is presenting a similar way: in the study that Yazırlı (2015, p. 54) investigated the factors affecting 230 families out-of-home food consumption in Nazilli; it was determined that 83.9 % of the participants consume food outside of home and 76.7 % of these participants consume food outside for dinner. It was observed that the participants consumed out-of-home food at a rate of 41.5 % at least once a week, at least once in fifteen days at a rate of 35.3 %, at a rate of 23.2 % at least once a month. In another study conducted with 392 people; according to the frequency of eating out; 13.5 % of consumers daily, 17.1 % 4–5 times a week, 33.7 % 1–3 times a week, and 21.2 % eat out once or twice a month (Bekar and Gümüş Dönmez, 2016, p. 6).

The most prominent reasons of eating out are socialization, luxury and pretentious consumption, imperativeness, and pleasure (Bugge and Lavik, 2010 p. 217). As a result of these reasons caused through time, out-of-home food sector, as parallel to its development all over the world, has also gained importance economically, it has diversified and developed in Turkey, as well. According to a study carried out in Turkey, the size of gastronomy economy was determined as $24 billion (Yıldız, 2018). Additionally, according to the statistics of the Ministry of Commerce; there are 42379 restaurants, 13 320 pieces of snack food businesses, 13288 cafés and breakfast places, 12926 patisseries, 7997 pita and lahmacun restaurants, 3362 pieces of pizza, toast and baked potato vendors, and 3038 units simit-poğaça vendors in Turkey (T.C. Ticaret Bakanlığı, 2016).

According to the data of the Household Consumption Expenditure Data Base presented by Turkey Statistical Institute (TÜİK, 2018); food and non-alcoholic beverages constitute 20.3 %, restaurant and hotel spending constitutes 6.2 % of the total expenditure in Turkey. When these data are considered, the share of food and restaurant expenditures in Turkey, reveals the importance of the sector.

Street Foods

One of the elements that make up the out-of-home food sector is street food, which has come from ancient times until today and can be considered as the beginning of the culture of eating out. Street food is the name given to foods that are sold by street vendors in the streets or in public places and consumed by the consumer at that moment or in a short time without any other process (Long-Solis, 2007, p. 213; WHO, 1996, p. 2).

Street foods are usually located in the city center and crowded places. Street foods are also defined as mobile food services. Their sales can be made with motorized/non-motorized vehicles, trays, baskets, portable small benches, or temporary walls surrounded by simple and portable poles. The benches are mostly located under a simple roof in the street. These businesses can also have low cost seats. Equipment such as disposable utensils, plates, glasses, forks and spoons are used in their service as well as they can be served between bread or lavash according to the characteristics of the product offered (Choi, Lee and Ok, 2013, p. 223; Honey, 2016, p. 9; Njaya, 2014, p. 53).

Street food is an important representative of the culinary and eating culture of countries. Consumption in fast and haphazard environments provides more opportunities for people's communication and social sharing compared to businesses that sit at tables and eat. These characteristics of street food can provide tourists with the opportunity to get to know the local flavors and food culture unique to the region, as well as the opportunity to communicate with the local people, to observe, to live and to be a part of the regional culture. For this reason, street food is becoming more and more prominent as a new tool in diversifying and developing gastronomic tourism in a region (Choi, Lee, and Ok, 2013, p. 233). Street food is also an important element shaping the common areas where street and city culture can be observed and open to the use of city residents (Karsavuran, 2018, p. 247).

Street foods are considered as a low-priced food group that is consumed quickly and practically, but it differs from fast food restaurants in terms of ownership structure and regulations. In addition, it is predominantly prepared with local foods and traditional cooking techniques, it contains ethnic elements; it also has slow food features (Calloni, 2013, p. 3411; Karsavuran, 2018, p. 247).

Due to the reasons such as being cheap and easily accessible and being considered problematic in terms of food safety, street foods were generally preferred by workers, students, and low-income groups in the past. The situation is changing in countries such as Europe and the USA and they are also preferred by high-income local residents and tourists (Calloni, 2013, p. 3407; Karsavuran, 2018, p. 247; Pirivitera and Nesci, 2015, p. 721).

The Classification of Street Foods

When the classifications made for street foods are examined; in a study, a classification has been made on the products offered for sale and it has been stated that there are four classes of street foods in the form of *meals, snacks, drinks and fruits* (Steyn and Labadarios, 2011, p. 462). Another classification is in the form of *dishes, drinks, snacks, desserts,* and *fruits* (Ballı, 2016, p. 10). Another classification was compiled under 7 different headings: *seafood, foods of animal origin, pastries, beverages, desserts, fruits, and other gastronomic values* (Demir, Akdağ, Sormaz and Özata, 2017, p. 596). In the classification made by the World Health Organization, they are divided into four classes as *food prepared in small-scale enterprises or traditional workshops, at home, on the marketplace, and in the street* (WHO, 2010, p. 1).

The Consumption Status of Street Foods

Studies on the frequency of consumption of street foods and the reasons for the preference of consumers have provided important results. A study in Ghana found that men with low education levels, young individuals under the age of 35 prefer street foods more than singles, on average, 6.76 times a week (Hiamey, Amuquandoh and Boison, 2013, p. 9–10).

In a study conducted in South Africa, consumers of low-income black Africans consumed street foods more than other ethnic groups, and consumers often bought cold drinks, fruit, snacks, desserts and biscuits, and cooked meals, and by decreasing of the age range (16–24) they consume more of street food rather than fast food (Steyn and Labadorios, 2011, p. 463).

In a study conducted in Mexico City, the attitudes of tourists in six different dimensions, including interest in street food, food quality, service quality, hygiene, satisfaction, and value perception, were examined. It was concluded that tourists are interested in street foods and satisfied with the quality of service and food, accordingly, they tend to consume street foods again and recommend them to others (Torres-Chavarria and Phakdee-Auksorn, 2017, p. 72–73).

In the study of Bektarım, Çakıcı and Ballı (2019, p. 680–682) in Mersin; it can be observed that an average of 3.3 of 409 participants prefer tantuni, 2.9 of them prefer chicken döner kebab and these two foods are the most preferred foods. In conclusion of the analysis made in the study; it has been revealed that individuals with a monthly income of more than 5000 TL, holding graduate degree and aged 50–69 stay away from street flavors. Another finding is those who consume street delicacies frequently are single, 20–29 years old, university graduates with less than minimum wage. The others who rarely consume street flavors are married, elementary and high school graduates between the ages of 30–49, with a monthly income of 2000–5000 TL.

According to the results of a study conducted with 424 participants in Gaziantep; 25.0 % of the participants consumed street flavors at a frequency of 6–10 per month, and 16.9 % at a frequency of 11–15 per month (Solunoğlu and Nazik, 2018, p. 48).

In a study conducted with 308 members of gastronomy and food communities on social media; It was determined that the street food most consumed by the participants was the simit and the products such as döner, ice cream and kokoreç followed the simit. According to other results obtained from the study; it has been determined that men and young people have a more positive attitude toward street foods. (Sünnetçioğlu and Yıldırım, 2019, p. 2814).

The Diversity of Traditional Street Foods Sold in Turkey

Turkey has a rich history in terms of street food. Döner, fish bread, simit, kokoreç, toast, meatball sandwich, kumpir, çiğ köfte, börek, pita bread, chickpeas rice pilaf, rolls, chestnuts, corn, stuffed mussels, etc. represents some of the traditional street food (Aşık, 2019, p. 49, Sokak Yemekleri, 2020). The richness of Turkish street food culture also results from the food culture between the geographical regions and each region has its own street food. (Dönmez Polat and Gezen, 2017, p. 119).

When the limited studies in the field are examined; It is seen that the most popular street flavors depending on different cities are sandwiches (meatballs, fish, döner, sausage), simit, toast, poğaça, börek, tantuni, gözleme, corn, çiğ köfte, chestnut and mussel (Aşık, 2019, s.53; Bektarım, Çakıcı and Ballı, 2019, p. 680; Sünnetçioğlu and Yıldırım 2019, p. 2810; Yıldırım and Albayrak, 2019, p. 1084). In the other studies which complied the street flavors of various cities, it is stated that flavors such as kebabs and offal, simit, gözleme, kumpir and boza come to the fore (Ballı, 2016, p. 10; Demir, Akdağ, Sormaz and Özata, 2017, p. 596).

With a wide geography, well-established and diverse culture, and rich cuisine in Turkey, it does not seem possible to classify and list all of the street foods. As a result of the regional differences, it is known that various street foods in different cities and regions are consumed in region-specific rituals. For example, cartlak (liver) kebab (Sabbag, 2015, p. 206), which is consumed for breakfast in Gaziantep, is often consumed for lunch and dinner in the western regions. It is understood that the regional differences in Turkish cuisine and the richness created by these differences are also valid for street foods. However, even though regional differences, it is possible to talk about a large variety of street foods that have place at the national level, produced, sold, and consumed in almost every town with minor changes in Turkey (Dönmez Polat and Gezen, 2017).

In conclusion of these data, the most commonly consumed street foods in Turkey is likely to compile in a classification as follows:

1. Bakery Products

Simit: It is a bakery product obtained by giving ring shape to the dough prepared with flour, water, salt, yeast, dipping molasses and sesame seeds and baking in the oven (Şentürk and Ötleş, 2017, p. 432). There are 6 different simits registered by obtaining a geographical indication document from the Turkish Patent and Trademark Office. These are Samsun, Ankara, Manisa, Rize, İzmit, Kastamonu simits (Özbay, 2020, p. 672).

Poğaca (Pastry): It is a kind of soft bun obtained by cooking yeast dough (Süren, 2020, p. 419). It has varieties such as plain, cheese, cheddar cheese, olives and potatoes. **Açma** is also a kind of fatty, soft bun and has a variety of pastry.

Kol Böreği: It is a type of patty that is placed in a wafer, rolled with ground meat, cheese, potatoes, spinach, etc., wrapped in an arm shape, laid on a tray and baked in the oven (TDK, 2020).

Gözleme: It is a pastry that is prepared from wafer by putting cheese, ground meat, potatoes, etc., fried on sheet or in oil (TDK, 2020).

Toast: It is made by frying cheese, sausage, etc., placed between two slices of bread, with a special machine. (TDK, 2020).

Ayvalık Toast: It is a sandwich toast, cheese, sausage, salami, sausage, tomato, and pickles placed between its unique chickpea yeast bread (Ayvalık Esnaf Odası, 2020). It is a local street food which took its name from Ayvalık district of Balikesir City.

İzmir Boyoz: It is a round, crispy bun with caramelized color prepared with flour, water and salt baked after the dough has been given the shape of a round belly. (Türk Patent ve Marka Kurumu, 2017a).

Kumru: It is a type of sandwich prepared by putting brynza, tomatoes, peppers in its own bread, served cold or served hot with the addition of sausage, salami and cheddar cheese and tomatoes. (Türk Patent ve Marka Kurumu, 2017b).

Lahmacun: Mortar prepared with ground meat, parsley, onion, garlic, and various spices is applied on thin and rounded dough, and it is prepared by cooking in a stone oven. There are varieties specific to the provinces of Şanlıurfa and Gaziantep (Türk Patent ve Marka Kurumu, 2017c).

Kır Pidesi: It is a kind of pita obtained by placing the inner mortar inside the pocket yeast dough and cooking it in the oven (Kır Pidesi, 2018). There are ground meat, cheese, and potato varieties, it is smaller in weight than known pita. It is a street food that belongs to Kastamonu.

Halka Tatlısı: It is a sherbet dumpling deep fried (Tarınç, 2019, p. 33) and takes its name from its shape.

Lokma Tatlısı: It is a street dessert that is shaped by squeezing the soft yeast dough with the help of a spoon and cooked in hot deep oil (Tarınç, 2019, p. 36).

Kağıt Helva: It is a type of halva obtained by pouring the liquid yeast dough thinly into round molds. It has two layers and two pieces are glued with caramel. It is mostly consumed together with ice cream.

2. Sandwiches

Meatball Sandwich: It is a food that contains small meatballs cooked on the grill, onions, tomatoes, and green peppers and consumed especially in sports, arts, and events.

Fried Fish Sandwich: It is a food that contains fried fish such as anchovy, whiting, sea bass, and garnishes such as onion and parsley.

Döner Sandwich: It is a street food that contains döner kebab (beef or chicken), onion, parsley, tomato, lettuce, and vegetables.

Kokoreç Sandwich: Kokoreç is a product made by wrapping small intestines around mesenterial oils, cooked in charcoal grills after a slight heat treatment, consumed spicy or without seasoning (Temelli, Saltan Evrensel, Anar ve Tayar, 2002, p. 468), mostly served as sandwich.

Tantuni: Tantuni is prepared with finely chopped beef is roasted in its original pan with water and oil, (Kadıoğlu, 2019, p. 46) it is served in lavash or as sandwich with tomatoes, onions, parsley. It is a local street food of Mersin City.

Dürüm: It is a wafer or thin pita wrapped with various additives (TDK, 2020). Dürüm; döner, chicken döner, Adana Kebab or Urfa Kebab can be prepared with products such as, lamb liver, boiled chickpeas. In addition to these products, suitable side dishes, spices and sauces are also wrapped.

3. Fruits and Vegetables

Kumpir: It is a street food that is consumed simply by adding cheddar cheese and butter into the baked oversized potato or enriched with plain various vegetables (peas, corn) and meat products (grilled steak, grilled chicken meat, sausage, salami).

Corn: Light yellow colored, soft and juicy sweet corn (TDK, 2020) is cooked by frying on charcoal fire or boiled in water.

Chestnuts Kebab: It is the food prepared by roasting drawn chestnuts when raw, over grill or charcoal fire.

Taffy Apple: It is prepared by putting the whole red apples on a stick and dipping them into colored syrup.

4. Cereals and Pulses

Chickpea and Rice Pilaf: It is a street food prepared by adding boiled chickpeas to rice pilaf. It is sold in unique mobile cars.

Çiğ Köfte: It is a food prepared by kneading; either all of cracked wheat, onion, garlic, red pepper, tomato paste, tomato sauce, pepper paste, various spices, sugar, vinegar, pomegranate syrup, walnut, almond, hazelnut, water, and vegetable oil, or some of them (Turkish Standards Institute, 2018). Although the original çiğ köfte contains raw ground beef, it is forbidden to have raw beef in çiğ köfte sold as take-home food. It is the traditional food of Şanlıurfa and Adıyaman Cities.

5. Sea Food

Fried Mussels: It is prepared by laying the mussels on sticks and making flour coating then frying in hot oil (Güngörür and Mol, 2019, p. 5). It can be served in portions with various sauces or as sandwich.

Stuffed Mussels: It is prepared by filling the cleaned shelled mussels with rice and cooking in a pot with a little water. It is served with the addition of lemon juice.

6. Dairy Products

Maras Ice Cream: It is a white, hard-flexible consistency ice cream made from goat milk, salep and sugar (Yilmazoğlu, 2014, p. 290). It is a street food originating in Kahramanmaraş City, sold with its own rituals.

2 Conclusion

Street foods; are generally considered as a category preferred by young, single people, students, and workers in the low-income group, due to concerns on issues such as food safety, heavy atmospheres, informality and being uncontrolled. With the developments such as ensuring food security, regular inspection, increasing trust in sellers, and improving the environment of the enterprises in the near term; it has turned into a safe and modern structure reflecting the atmosphere of the city and street, and consumed by tourists and local people from all walks of life. Street foods encourage people to know more about the places they go and local foods, and also lead them to discover the cultures that these foods represent (Polat and Gezen, 2017, p. 119).

Registration of street foods by the administrations not only prevents tax losses but ensures that they are transformed into foods served by reliable businesses as a result of audits. Especially the supply of street foods consisted of local and ethnic foods, is an intersection point for gastronomy tourism with the demand of local food and authentic atmosphere by tourists; it also contributes to the promotion and production of local products, local economy, and the employment of local people (Njaya, 2014, p. 56). These changes increase the share of street foods in the out-of-home food sector and gastronomic tourism and contribute to the development of touristic destinations.

Turkey's wealth of culinary culture is seen at the traditional street foods as well. Traditional Turkish street foods with traditional cooking techniques, tools, local raw materials, and rituals have an important place in the out-of-home food sector. Simit, poğaça, sandwiches, dürüm, çiğ köfte, kumpir, chickpeas with rice pilaf, corn, chestnuts are popular street foods that are widely consumed in Turkey. Turkish street foods with more local varieties and richness are a reflection of Turkey's cultural diversity; they are increasing their importance and share in Turkish cuisine day by day.

Additionally, it is seen that in academic studies on gastronomic tourism products, while focusing on activities such as restaurants, farms, or festivals, have not given place to street foods (Karsavuran, 2018, p. 256). As long as the share of street food in gastronomic tourism increases, especially with the development of local tourism, it can be predicted that it will find more place in academic studies.

It is concluded that the promotion of street foods, supervision of the businesses that are selling them, supporting them with academic studies, and the introduction of local street food to the consumption of the locals and tourists at national and international levels will change the negative perceptions of the consumers and contribute to all the stakeholders of the sector.

References

Ağlamaz, E., & Öncüer Çivici, M. E. (2018). Yerel Halkın Gözünden İzmir'in Simge Yemeklerinin Belirlenmesi Üzerine Bir Araştırma. *Journal of Travel & Tourism Research*, 13, 50–69.

Akarçay, E. & Suğur, N. (2015). Dışarıda Yemek: Eskişehir'de Yeni Orta Sınıfın Fast-Food Yeme-İçme Örüntüleri. *Sosyoloji Araştırmaları Dergisi*, 18(1), 1–29.

Aşık, N. A. (2019). Yerli Turistlerin Sokak Yiyeceklerine Yönelik Tutumlarının Belirlenmesi. *Uluslararası Türk Dünyası Turizm Araştırmaları Dergisi*, 4(1), 47–61.

Ayvalık Esnaf Odası (n.d.). Ayvalık Tostunun Farkı Ne Diye Soruyorsanız. http://ayvalikesnafodasi.org/ayvalik-tostunun-farki-ne-diye-soruyorsaniz/ (Accessed: 07.01.2020).

Ballı, E. (2016). Gastronomi Turizmi Açısından Adana Sokak Lezzetleri. *Journal of Tourism and Gastronomy Studies*, 4(1), 3–17.

Beardsworth, A. & Keil, T. (2012). *Yemek sosyolojisi: yemek ve toplum çalışmasına davet.* (çev. A. Dede). Ankara: Phoenix.

Bekar, A. & Dönmez Gümüş, F. (2016). Tüketicilerin Dışarıda Yemek Yeme Nedenlerine İlişkin Bir Değerlendirme. *Social Sciences (NWSASOS)*, 11(1), 1–15. http://dx.doi.org/10.12739/NWSA.2016.11.1.3C0134

Bektarım, N., Çakıcı, A. & Ballı, E., (2019, Eylül). *Sokak Lezzetleri Tüketim Sıklığı.* Tam metin basılı bildiri, 3.Uluslararası Turizmin Geleceği Kongresi: İnovasyon, Girişimcilik Ve Sürdürebilirlik (678–684). Mersin, Türkiye.

Blisard, N., Biing-Hwan, L., Cromartie, J. & Ballenger, N. (2002). American's Changing Appetite: Food Consumption and Spending to 2020. *Food Review*, 25(1), 2–9.

Bugge, A. B. & Lavik, R. (2010). Eating Out: Multifaceted Activity on Contemporary Norway. *Food, Culture and Society*, 13(2), 215–240.

Calloni, M. (2013). Street Food on Move: A Socio-Philosophical Approach. *Journal of Food Science and Agriculture*, 93, 3406–3413.

Choi J., Lee, A. & Ok, C. (2013). The Effects of Consumers Perceived Risk and Benefit on Attitude and Behavioral Intention: A Study of Street Food. *Journal of Travel & Tourism Marketing*, 30, 222–237.

Demir, Ş., Akdağ, G., Sormaz, Ü. & Özata, E. (2018). Sokak Lezzetlerinin Gastronomik Değeri: İstanbul Sokak Lezzetleri. *Güncel Turizm Araştırmaları Dergisi*, 2(Ek.1), 589–601.

Güngörür, M. N. & Sühendan, M. O. L. (2019). Bir Gida Olarak Midye. *Aydın Gastronomy*, 3(2), 119–127.

Hiamey, S. E., Amuquandoh, F. E. & Boison, G. A. (2013). Are We Indeed What We Eat? Street Food Consumption in the Market Circle Area of Takoradi, Ghana. *Nutrition and Health*, 22, 215–235.

Jaafar, S. N., Lumbers, M. & Eves, A. (2009). Does Food Really Matters In The Eating Out Experience In Restaurants. *International Journal of Contemporary Hospitality Management*, 8(2), 1–11.

Kadıoğlu, C. T. (2019, Eylül). *Bir Tantuniciye Müdavim Olmak! Algılanan Değerin Sadakate Etkisi,* Tam metin basılı bildiri, 3.Uluslararası Turizmin Geleceği Kongresi: İnovasyon, Girişimcilik Ve Sürdürebilirlik (46–56). Mersin, Türkiye.

Karsavuran, Z. (2018). Sokak Yemekleri: Farklı Disiplinlerin Yaklaşımı Ve Gastronomi Turizmi Alanında Sokak Yemeklerinin Değerlendirilmesi. *Journal of Tourism and GastronomyStudies*, 6(1), 246–265.

Kır Pidesi. (n.d.). https://www.pide.gen.tr/kir-pidesi.html (Accessed: 12.04.2018).

Long-Solis, J. (2007). A Survey of Street Foods in Mexıco City. *Food and Foodways: Explorations in the History and Culture of Human Nourishment*, 15(3–4), 213–236.

Njaya, T. (2014). Nature, Operations and Socio-Economic Features of Street Food Entrepreneurs of Harare, Zimbabwe. *IOSR Journal of Humanities and Social Science (IOSR-JHSS)*, 19(4), 49–58.

Onurlubaş, E., Doğan, H. G. & Gürler, A. Z. (2015). Türkiye'de Ev Dışı Gıda Tüketimin Durumu Ve Tüketici Eğilimleri. *Uluslararası Sosyal Araştırmalar Dergisi*, 8(38), 917–924.

Özbay, G. (2020). Ulusal ve Uluslararası Platformda Gastronomik Kimlik Unsuru Olarak Simit. *Journal of Tourism and Gastronomy Studies*, 8(1), 670–683.

Petterson, A. & Fjellström, C. (2007). Restaurants As Friends Of The Family: Fuctions Of Restaurant Visits In Everyday Life. *Journal of Foodservice*,18, 207–217.

Polat, D. D. & Gezen, A. (2017). Gastronomide Sokak Yiyecekleri Ve Satıcıları: Teorik Bir Çalışma. *Journal of Tourism and Gastronomy Studies*, 5(2), 117–124.

Privitera, D. & Nesci, F. S. (2015).Globalization Vs. Local: The Role of Street Food in the Urban Food System. *Procedia Economics and Finance*, 22, 716–722.

Sabbağ, Ç. (2015). *Gaziantep yeme içme kültürü.* Ramilya Yarullina Yıldırım (Ed.), *Fırat'tan Volga'ya medeniyetler köprüsü* içinde (197–297). Adıyaman, Adıyaman Üniversitesi Yayınları (13).

Scholliers, P. (2008). *Yenilik ve gelenek: gastronominin önündeki yeni ufuk. Yemek: damak tadının tarihi* içinde (333–357). N. Elhüseyni (Çev.). İstanbul: Oğlak.

Sokak Yemekleri. (n.d.). *Türkiye Sokak Yemekleri.* http://www.sokakyemekleri.com/turkiye-sokak-yemekleri]/ (Accessed: 07.01.2020).

Solunoğlu, A. & Nazik, M. H. (2018). Sokak Lezzetlerine İlişkin Tüketici Tercihleri: Gaziantep Örneği. *Journal of Tourism and Gastronomy Studies,* 6(3), 40–59.

Steyn, N. P. & Labadarios, D. (2011). Street Foods and Fast Foods: How Much Do South Africans of Different Ethnic Groups Consume? *Ethnicity & Disease,* 21, 462–466.

Sünnetçioğlu, S. & Yıldırım, H. M. (2019). Sokak Yiyeceklerinin Tüketiminde Satıcıya Olan Güven, Algılanan Risk, Algılanan Fayda, Tutum ve Satın Alma Niyeti Arasındaki İlişkilerin İncelenmesi. *MANAS Sosyal Araştırmalar Dergisi,* 8(3), 2799–2818.

Süren, T. (2020). *Ekmek ve unlu mamuller.* Semra Akar Şahingöz & Tufan Süren (Ed.), *Mutfak uygulamaları* içinde (403–432). Ankara: Detay Yayıncılık.

Şentürk, A. & Ötleş, S. (2017). Farklı Düzeylerde Kefir Kullanımının Simidin Bazı Fiziksel, Kimyasal ve Duyusal Özellikleri Üzerine Etkisi. *Harran Tarım ve Gıda Bilimleri Dergisi,* 21(4), 431–443.

Tarınç, A. (2019). *Avrupa ve Türkiye'deki sokak tatlılarının karşılaştırılması.* Kadriye Alev Akmeşe (Ed.), *Turizmin temelleri* içinde (27–59). Ankara: İksad.

T.C. Ticaret Bakanlığı, (2016, Şubat 17). *Esnaf ve Sanatkar Verileri, 491 Meslek Koluna Göre İşyeri Sayıları,* https://ticaret.gov.tr/data/5d774a8313b876bdf cd7c334/8-491 %20Meslek%20Koluna%20Gore%20Isyeri%20Sayilari.pdf (Accessed: 06.29.2020).

TDK. (n.d.). *Güncel Türkçe Sözlük,* https://sozluk.gov.tr/ (Accessed: 07.01.2020).

Temelli, S., Saltan Evrensel, S., Anar, Ş. & Tayar, M. (2002). Bursa'da Tüketilen Kokoreçlerin Mikrobiyolojik Kalitesinin Belirlenmesi. *İstanbul Üniversitesi Veteriner Fakültesi Dergisi,* 28(2), 467–473.

TÜİK. (2019, Temmuz 26). *Hanehalkı Tüketim Harcaması,* http://www.tuik.gov.tr/PreHaberBultenleri.do?id=30584 (Accessed: 07.01.2020).

Türk Patent ve Marka Kurumu. (2017a, Aralık 8). *İzmir Boyozu,* Coğrafi İşaretler Portalı: https://www.ci.gov.tr/Files/GeographicalSigns/268.pdf (Accessed: 07.01.2020).

Türk Patent ve Marka Kurumu. (2017b, Aralık 7). *İzmir Kumrusu,* Coğrafi İşaretler Portalı: https://www.ci.gov.tr/Files/GeographicalSigns/262.pdf (Accessed: 07.01.2020).

Türk Patent ve Marka Kurumu. (2017c, Kasım 20). *Gaziantep Lahmacunu.* Coğrafi İşaretler Portalı: https://www.ci.gov.tr/Files/GeographicalSigns/236.pdf (Accessed: 07.01.2020).

Türk Standartları Enstitüsü. (2018, Ocak 23). *Etsiz Çiğ Köftenin Standardı Hazır.* Türk Standartları Enstitüsü: https://tse.org.tr/Icerik/HaberDetay?HaberID=14531 (Accessed: 07.01.2020).

WHO (1996). Essential Safety Requirements for Street-Vended Foods. Food Safety Unit Division of Food and Nutrition, World Health Organization.

WHO (2010). Basic Steps to İmprove Safety of Street-Vended Food. INFOSAN Information Note, No: 3.

Yazırlı, N. (2015). *Ev Dışı Gıda Tüketimini Etkileyen Faktörler: Nazilli Örneği.* Yüksek Lisans Tezi, Niğde Üniversitesi Sosyal Bilimler Enstitüsü, Niğde.

Yıldırım, Ö. & Albayrak, A. (2019). Yabancı Turistlerin İstanbul Sokak Yemeklerini Değerlendirmeleri Üzerine Bir Çalışma. *Journal of Tourism and Gastronomy Studies,* 7(2), 1077–1092.

Yıldız, D. (2018, Mart 16). *Türkiye'de Yeme-İçme Sektörünün Boyutları Gastronomi Ekonomisi Ara Rapor*, http://www.brandday.net/ekonomi/turkiyede-yeme-icme-sektorunun-boyutlari-gastronomi-ekonomisi-ara-rapor-2018-h8213.html (Accessed: 07.01.2020).

Yılmaz, G. & Akman, S. (2018). Sürdürülebilir Gastronomi Çerçevesinde Edremit Körfezi Yöresel Mutfakları. *Journal of Tourism and Gastronomy Studies,* 6(4), 852–872.

Yılmazoğlu, H. İ. (2014). *Kahramanmaraş'ın mutfak ve beslenme kültürü.* Orhan Doğan, Ramazan Avcı & Serdar Yakar (Ed.) *Akdeniz'in altın kenti Kahramanmaraş* içinde (268–298). Kahramanmaraş: Kahramanmaraş İl Kültür ve Turizm Müdürlüğü.

Fuat BAYRAM

19 Kitchen Planning in Food and Beverage Business

1 Introduction

In the second half of the 20th century, with the changes that occurred in economic, social, cultural, environmental, and psychological fields, people's expectations, habits, and needs have changed, with the effect of these changes, an international tourism movement has occurred (Usta, 1996). Tourism, is the combination of activities and services such as transportation, accommodation, eating and entertainment that offer many experiences (Goeldner and Ritchie, 2003); Which can be defined as Set of events and relationships that arise from people's temporary accommodation for traveling outside the places where they permanently reside, work and meet their usual needs, by demanding goods and services usually produced by tourism enterprises (Kozak, Kozak and Kozak, 2017).

Although tourism activities are an economic power that enables countries to earn foreign currency and tax income, they are also especially important in terms of providing employment for the citizens of the country (Hudson, 2008). As for all countries in tourism activities poses is also of great importance for Turkey, Turkey is among the countries that having large income from tourism activities (Akin and Gultekin, 2015). In addition to accommodation revenues, revenues from food and beverage services have a large share in these revenues. Especially in hotel businesses, the revenue generated from food and beverage services ranks second (25–50 %) after room revenues (Kozak et al., 2008).

In addition to its importance in generating income, food and beverage services complement the travel, tourism, accommodation, entertainment, and leisure sector and play a major role in the formation of customer satisfaction and customer loyalty (Turksoy, 2015; Bayram, 2018). So much so that Gundersen, Heide, and Olson (1996) stated in their study on individuals traveling for business purposes that the services provided in the reception, housekeeping and catering departments affect the general satisfaction of the guests. Once more, Kandampully and Suhartanto (2000) emphasized the positive effects of quality service for housekeeping, reception and food and beverage in the formation of customer loyalty in their study.

In addition to its positive effects on employment and the added value it provides to countries, food and beverage services make a great contribution to the promotion of the country and the branding of destinations, together with the regional food culture. So much so that in the destination branding process, meals have a great role in promoting the region and conducting gastronomy tourism activities (Aydogdu and Duman, 2017; Alacam, 2019; Boztoprak, 2019; Karamahmutoglu, 2019; Kizilgeci, 2019). In their study Sahin and Unver (2015) in which they investigated the effects of gastronomy in destination marketing and 93 Group A travel agency, emphasized that gastronomy and foods have great importance in the creation of brand value by introduction and marketing for tourism of Turkey and Istanbul.

The success of food and beverage businesses, which are the center of attention with their tourism activities and place in the economy, depends on the satisfaction of the customer base they serve. Kitchens, which are expressed as the heart of food and beverage businesses, have great importance in this success. In this context, food and beverage businesses that want to increase their competitiveness in the market by satisfying their customers should have kitchens that will facilitate the work flow during production, provide a comfortable and healthy working environment for the working personnel and where a hygienic production can be made (Aktas, 2011; Bayram, 2020).

The planning of kitchens, which is so important for food and beverage businesses, is a job that requires great care. Mistakes to be made during the planning of the kitchens can cause forward-looking business, time and customer losses, as well as cause serious difficulties for the personnel working during the production. This study was carried out to examine the points to be considered during the planning of kitchens, which have great importance in food and beverage businesses.

Kitchen Concept in Food and Beverage Business

People have traveled to different places from their places of residence for various reasons from past to present (Yilmaz, Yilmaz and Yilmaz, 2013), and during these travels, they needed places where they could accommodate and feed themselves. In the past, while the accommodation and food and beverage needs of the people during these travels were met by the inns and caravanserais built on the travel routes, today the same service is provided by the hotels and food and beverage establishments. With the increase in the disposable income of people and their orientation toward tourism activities, many food and beverage businesses were opened in order to meet their eating and drinking needs, while out-of-home food consumption expenditure in developed countries was 20 % of the total amount spent for food in the 1960s, this rate reached 26 % in the 1980s, and in the 2000's, it reached around 50 %, and food and beverage services became a major industry (Denizer, 2005; Turksoy, 2015). The form and content of nutrition began to change, with the fact that food products started to be seen as the basic element of a healthy and long life as well as they consumed for satiety as a physiological need, and people came to the point of enjoying what they eat as well as the satiation. In this process, cooking in food and beverage businesses has become an art, and this art has been realized in the kitchens where meals are produced (Turksoy, 2007; Mavis, 2008; Kucukaslan, 2011; Bayram, 2020).

Kitchens are, "the areas where food is produced in the targeted quantities, quality and standards in food and beverage enterprises" (Sokmen, 2009, p. 31). It has been derived into Turkish from the Arabian word "matbah", which means a place where food is cooked, and pronounced as "mutfak (kitchen)" in Turkish. With the existence of humanity, the kitchen has taken its place in the dwellings they live in, and parallel to the life styles and economic opportunities of the societies, it has been in constant change in terms of both physical and equipment used (Aktaş ve Ozdemir, 2007; Kucukaslan, 2011).

Kitchens, which are of great importance in the dwellings where people live, have started to gain importance in the industrial sense after the Industrial Revolution. Due to the increase in the production volume in the workplaces and the necessity of meeting the nutritional needs of the employees by the workplaces for the reason of longer stay in the workplaces have led to the formation of industrial kitchens. After the Industrial Revolution, people's adoption of eating habits outside for various reasons was added to these developments and requirements in the kitchen field, and the formation and planning of industrial kitchens became widespread (Kucukaslan, 2011; Sokmen, 2014). With the technological developments in the 20th century and afterwards, the preparation and storage equipment used in kitchens have diversified, kitchen ventilation and lighting equipment have developed, and modern kitchens have begun to be created after architects, engineers, dieticians and kitchen chefs began to scientifically investigate the production processes in kitchens (Powell, 2005; Guldemir, 2015).

Kitchen Planning in Food and Beverage Businesses

Regardless of the theme, the success of a restaurant or catering business depends primarily on the popularity of the food served in that business and the quality of the service. Therefore, the kitchen in which the food is prepared plays a big role in the success story of the catering service. Despite this fact, it seems that having a perfect kitchen facility is not the top priority of business owners. In order to give the guests a comfortable atmosphere and a comfortable feeling, a large amount of expenditure is spent on décor and furniture, and as the spending in this area increases, the expenditure on the kitchen space decreases. Therefore, the necessary importance is not given to an effective kitchen planning and design (Dareker ve Peshave, 2016). However, kitchen planning requires great care, just like planning a hospital or training center. In a hospital, planning according to the location and equipment of each department, the experience and areas of the staff to work is important for the life of the patient, such an importance kitchens are at the same level for the continuation of health in terms of their functions and services. With the planning that will be made, the work flow during production should be facilitated and a comfortable working environment should be provided for the personnel. So much so that the quality, taste and nutritional value of the meals produced for the establishment should be as important as the comfort and safety of the employees. For this reason, a wrong planning in the kitchens will adversely affect the health of both customers and employees (Bryant and McKay, 2005; Aktas, 2011).

In poorly planned kitchens, there will be problems in the production process, work efficiency will decrease, staff motivation will decrease, and work accidents will occur, and therefore, a loss of image in the business will occur. For this reason, when planning the kitchen, the type of menu which will be applied, the amount of production, the type of service and the number of meals, the equipment that will be needed for production and the way they are placed in the kitchen, the raw material purchase policies of the enterprise that will be used in production, the storage facilities and the number of personnel that will be employed should be taken into consideration (Kucukaslan, 2011).

Kitchen Architecture

The location of the food and beverage business kitchens inside the building is a matter of the architectural structure. Kitchens can be positioned on the upper floors of the building or on the ground floors of the building. Both of these positioning styles have advantages and disadvantages. When kitchens are located on the ground floors of the buildings, it will be easier to transfer food materials received for production to the storage areas and from the storage areas to the production line. However, when the kitchens are planned on the ground and underground floors, there will be problems in lighting and ventilation, so natural lighting and ventilation will not be used sufficiently, artificial lighting and ventilation methods will be used (Gokdemir, 2003; Aktas, 2011).

When kitchens are located on the upper floors of the building, natural daylight can be used for kitchen lighting and the energy consumption for lighting will be reduced. However, separate elevators will need to be included in the planning for the delivery of food items received for production to the kitchen and for the personnel to reach the kitchen. Likewise, the discharge of wastes that will occur after production in the kitchens on the upper floors will also cause additional costs.

One of the problems that can be experienced when the kitchens located on the upper floors is the water cuts that will be experienced when the water does not have sufficient pressure (Gokdemir, 2003; Aktas, 2011). The floor of the building where the kitchen will be located should be decided by the planning team, considering the above-mentioned advantages and disadvantages.

The layout of the kitchens is also an issue that needs to be carefully considered. Since the northern parts of the buildings are generally cooler than the south facing parts, the emplacement direction of the kitchens should be the northern part of the buildings (Turkan, 2010). In this way, during the conditioning of kitchens energy usage and storage areas will be saved and it will be helped to control the temperature increases in kitchens, especially in cooking areas.

Lighting in Kitchens

In order to ensure the production process in the kitchens without any problems, prevent any occupational accidents and food poisoning by cleaning the kitchen and equipment as desired, and helping to personnel to feel themselves good in psychological terms, the lighting in the kitchens of the food and beverage establishments must be perfect (Aktas, 2011).

Lighting in kitchens is provided in two ways: The first of these is natural lighting. In cases where the buildings are located above the ground floor, natural lighting can be used. Natural lighting should be preferred in kitchens because it does not change the natural colors of foodstuffs and its lighting power is higher than artificial lighting. In order to benefit from natural lighting at a sufficient rate, the ratio of kitchen windows to the kitchen floor should be ¼. However, since the kitchens of food and beverage business are generally located on the ground and underground floors, kitchen lighting should be supported by artificial lighting (Dizdar and Kurt, 2001; Turkan, 2010; Yigit, 2013). In fact, Bayram (2020) has determined that the kitchens of food and beverage business cannot get the benefit of natural lighting with a sufficient amount and found that kitchen window areas were less than ¼ of the kitchen floor in his study that he conducted in order to reveal planning errors by examining the physical condition within the kitchens of food and beverage business, on 23 food and beverage businesses in three touristic cities, Ankara, Antalya, and Istanbul of Turkey.

In cases where the kitchens are on the ground floors or there are not enough windows to benefit from natural lighting, artificial lighting is used. In this case, lighting should be provided by a light bulb or fluorescent that will be placed in the appropriate parts of the kitchen (Guldemir, 2015). Care should be taken that the light bulbs and fluorescents used for illumination emit a color that does not change the natural color of foodstuffs, and care should be taken that the light used does not cast a shadow on the work areas and the light does not reflect directly on the employees' eyes (Turkan, 2010). Since the lack of lighting in the kitchens may cause the risk factors that may occur in the kitchen to be overlooked, work accidents and food poisoning, the kitchen light level should be measured with luxmeters and supports should be made in areas with insufficient lighting. The lighting of the general areas in the kitchens should be 200 lux, and the lighting of the work areas such as stoves and countertops should be 300–400 lux (Aktas and Ozdemir, 2007). Lighting inside the kitchen should not be reduced to less than 100 lux in general areas other than work areas and 200 lux in work areas (Mullins, 1996; Arman, 2019). However, in the study conducted by Bayram in order to reveal the planning errors by examining the physical conditions within the kitchens of food and beverage business, although it is known that the lighting level in the kitchens is not at the desired level in more than half of the enterprises where the research has been carried out and the problems that may arise due to the insufficient lighting in the kitchen, the food and beverage business has not shown sufficient sensitivity for lighting in the kitchens.

Kitchen Ventilation

The ventilation of the kitchens refers to giving clean air to the working environment instead of the hot air, odor and moisture arising due to the cooking processes in the production in order for the personnel to work in a comfortable environment (Aktas ve Ozdemir, 2012). This process in kitchens is carried out in two ways: The first is natural ventilation. Natural ventilation is generally preferred in small kitchens, and in kitchens located on the upper floors, the gratings and air holes placed at certain points of the kitchen allow fresh air to enter. Evacuation of dirty air is carried out by using hoods placed on the stove tops. In this method, hoods should be placed on the stoves and the distance of the hoods from the floor should be 190–210 cm (Tumer, 2008; Turkan, 2010; Budak, 2015).

The second and most preferred method of ventilation of kitchens is mechanical ventilation. In the mechanical ventilation method, the dirty air in the kitchen is drawn from the mechanical air ducts and clean air is given instead. In this method, both the ambient air can be cleaned, and the ambient temperature can be adjusted. The temperature of the kitchen environment should be between 20–22 ° C in winter and 18–20 ° C in summer (Türkan, 2010).

Kitchen Floor

One of the most common occupational accidents in food and beverage businesses is caused by slipping or stumbling and falls. One of the major causes of falling injuries is slippery floors. For this reason, kitchen floors should be covered with non-slip floor coverings (Merdol and Birer, 1997; Aktas and Ozdemir, 2012). In addition, the coating material for the floor should be chosen from materials which are easy to clean, disinfect, resistant to impacts, acid and moisture, and light colored (Topal, 1996). In addition, a sufficient amount of limber holes should be considered at certain points of the floor in order to drain the liquids spilled on the kitchen floors or to wash the floors when necessary.

The flooring on the kitchen floors can wear out over time, and cracks or fractures may occur on them. Microorganisms can reproduce in these cracks and fractures and cause food poisoning by contaminating the foodstuffs produced, or the working personnel can get caught in this damaged floor covering and cause work accidents. For this reason, damaged and broken floors should be repaired immediately and any risks that may arise should be prevented (Sahingöz and Şık, 2015). Since the fact that the rate of accidents by sliding or stumbling cannot be underestimated (Fisek, 2009) necessitates taking precautions regarding the issue in the kitchens of catering businesses and it is seen that the employees are conscious on kitchen floor coverings.

So much so that, in the study conducted by Ozel and Comert (2015) for kitchen workers, it was observed that kitchen workers were conscious of the need to prefer non-slip flooring material in their kitchen floor covering material.

Ceiling

Kitchen ceilings should be planned higher than other areas in order to be less affected by the heat, humidity and bad odor that occur during the production process in the kitchen. If the kitchen ceiling is planned low, the ambient temperature will not only increase, but the working personnel will not feel well psychologically (Aktas, 2011). Kucukyaman and Demir (2019), in their study on the effects of hotel kitchen plans on the work performance of employees, stated that most of the employees were overwhelmed and restless during their work if the kitchen ceiling was low. Despite the importance of the floor-ceiling height for employees to work in an ergonomic environment, it is seen in the studies that sufficient attention is not paid to this issue during the planning phase of the catering business kitchens. In the study conducted by Bayram in the kitchens of 23 catering establishments in 2020, in which examined the physical conditions and planning of the kitchens, it was observed that the ceiling floor heights of 22 food and beverage kitchens were not within the desired range.

If the ceiling height of the kitchens is not taken into consideration during the projecting phase, problems that cannot be compensated will arise. For this reason, although ceiling heights of kitchens are planned between 4–5 meters at the planning stage, the height can be planned a little lower in the storage areas where the temperature is lower and the slaughterhouse section (Aktaş ve Özdemir, 2012).

Kitchen Walls

During the production in kitchens, kitchen walls may get dirty. In order for the walls to be easily cleaned, it is recommended that the walls to be covered with tiles completely, if not possible, at least 2 meters (Kucukaslan, 2011). Because tiles are the best wall covering material for kitchens due to their easy-to-clean structure and long life. (Ozdemir, 2001). Since the preferred color of the tile is light, it will contribute to better lighting as it will reflect the light better, white tiles will reflect the light by 80–90 % and make the environment brighter (Gokdemir, 2003).

Electricity and Sanitary Systems in Kitchens

The gas and electrical installations used for the operation of the equipment used for production in the kitchens should be designed before the kitchen is installed and should be prepared in accordance with the laws according to the type, location and the amount of energy to be consumed. Electrical installations should never be exposed and should be well insulated (Saritas and Yasar, 2017). In addition, the water systems to be used in the kitchen should be prepared in accordance with the rules, clean water should be delivered to the usage areas without contamination and it should be ensured that the waste water is collected and transported outside the building in a healthy way (Turkan, 2010; Ozoral and Yildirim, 2014).

Since almost all of the equipment used in kitchens is powered by electricity and gas, the unit with the highest probability of fire in food and beverage establishments is the kitchens. For this reason, when planning kitchens against possible fires, equipment such as fire and smoke warning system, smoke detector, heat detector, manual fire button, audible alarm system, gas detector, emergency lighting, smoke evacuation system and sprinkler system should be included in the planning (Ozoral and Yildirim, 2014).

Other Details to Consider in Kitchen Planning

Individuals who want to come to food and beverage businesses want to eat in a decent environment, but also want their health to be protected (Sokmen, 2008). Because if the production areas in the kitchens are not hygienic or the personnel does not pay attention to the hygiene rules, the individuals who come to the enterprise and consume the food produced may get poisoned and face with deterioration of their health (Sokmen, 2014).

The cleanliness and personal hygiene of the personnel is as important as the cleanliness and hygiene of the environment in food and beverage businesses. For this reason, areas where the personal hygiene of the personnel will be provided should be considered during the planning of the kitchen. Therefore, planning should be made to have a toilet and shower area for every 10–12 people working in the business, provided that it is at least 8 meters away from the kitchens (Denizer, 2005; Unuvar, 2007). There should be hand washing sinks in both washrooms and kitchen entrances and care should be taken that these sinks have sensors. Hand washing sinks in the kitchen should be separate from the sinks where food is washed (Merdol and Birer, 1997).

Another important point to be considered for a healthy and hygienic production in the kitchen is preventing the contact of raw and cooked foodstuffs and other foods that pose a potential risk to each other. For this reason, production in kitchens should be carried out in separate units according to food types and there should be compartments between the working sections (Turkan, 2009; Ozel and Comert, 2015).

In order to ensure a healthy production in kitchens, wastes in the kitchen should be disposed of regularly and thrown into the garbage rooms. The floors of the garbage rooms should be furnished with easy-to-clean floor covering material, and water taps and water grates should be placed at a point close to the garbage room so that the rooms can be washed after they are emptied. The temperature should be at most 10 0C so that the garbage kept in these rooms does not smell. (Koçak, 2007; Unuvar, 2007; Turkan, 2009).

2 Conclusion

The food and beverage service sector has developed since the Middle Ages and has become a part of our lives by progressing in parallel with urbanization and industrialization. With the increase of its participation in tourism activities, the intensity of business life and the participation of women in the working life have increased the frequency of people eating outside, which has led to the opening of many food and beverage businesses (Akin and Gultekin, 2015; Turksoy, 2015). With the increase in food and beverage businesses, businesses sought various ways to maintain their profitability, survive and become successful.

The success of a food and beverage business depends on the satisfaction of the customer base it will serve. Since this satisfaction can be achieved by meeting the needs of the guests by providing a fast, clean and desired service, there is a need for a well-planned and segmented kitchen in enterprises (Chan and Lam, 2009).

When it comes to kitchen planning, only the planning of an area where only meals are cooked and production process completed, should not come to mind. The planning should provide an environment suitable for the principles of occupational health and safety for production, facilitate the work flow, and create ergonomic conditions for the kitchen staff who will produce in kitchens. In an unplanned kitchen of the desired quality, production will not only be interrupted, businesses will suffer material and moral losses due to food poisoning or occupational accidents that employees will be exposed to, and worst of all, loss of life will be encountered.

While the planning of the kitchens where the production will be carried out in the food and beverage enterprises has great importance, unfortunately, the necessary attention is not paid to the planning of the kitchens. The biggest mistake that made during planning is investors empower a single person for kitchen planning, rather than a project team specialized in different fields (Arman, 2019). However, kitchen planning is a complex task that takes many issues into account such as: Positioning the kitchen, determining the direction of the settlement, planning the kitchen production line and sections, kitchen lighting, ventilation, making electricity and other installations, determining the food safety risks that may arise from planning during production, the necessity of planning taking into account the principles of occupational health safety and ergonomics. For this reason, the planning process should be carried out with a planning team consisting of experts in the field from beginning to end. The opinion of the industrial kitchen technician, architect, kitchen chef and other experts should be taken while planning the kitchen (Dogdubay and Karan, 2017). In the planning phase, in addition to the above-mentioned project team, getting the ideas of experts such as consulting employees in the field of food and beverage management, interior architects, industrial kitchen equipment manufacturers, civil engineers, lighting consultants, acoustic technicians, graphic consultants, technology consultants and sustainable environmental consultants, will help to plan a more functional kitchen (Gumussoy,

2014). With the kitchens which created as a result of the planning that carried out together by the expert stakeholders, there will be no disruptions in the work flow, and the production of foodstuffs will be carried out at the desired quality and on time. Waste of raw materials will be minimized with the storage and production line planned in the desired quality; thus, a high profitability rate will be achieved. In addition, with the principles of occupational health, safety and ergonomics to be followed in planning, an efficient and safe working area will be created for the personnel, and more efficiency will be obtained with less personnel.

References

Akın, G. & Gültekin, T. (2015). Günümüz Restoran Tasarımında Kriterler. *Süleyman Demirel Üniversitesi Mühendislik Bilimleri ve Tasarım Dergisi*, 3(3), 251–258.

Aktaş, A. & Özdemir, B. (2007). *Otel İşletmelerinde Mutfak Yönetimi*. (2. baskı). Ankara: Detay Yayıncılık.

Aktaş, A. & Özdemir, B. (2012). *Otel İşletmelerinde Mutfak Yönetimi*. (3. baskı). Ankara: Detay Yayıncılık.

Aktaş, A. (2011). *Ağırlama Hizmet İşletmelerinde Yiyecek İçecek Yönetimi Food and Beverage Management*. (3. baskı). Ankara: Detay Yayıncılık.

Alaçam, D. O. (2019). *Gastronomi Turizmine Türk Mutfağının Etkileri Trabzon İli Örneği*. (Yayımlanmamış Yüksek Lisans Tezi). İstanbul Okan Üniversitesi Sosyal Bilimler Enstitüsü, İstanbul, Türkiye.

Arman, A. (2019). *Endüstriyel Mutfak Tasarım Ölçütlerinde İşlevselliğe Bağlı Parametrelerin Değerlendirilmesi Ve Tasarıma Yönelik Öneriler* (Yayımlanmamış Doktora Tezi). Necmettin Erbakan Üniversitesi Sosyal Bilimler Enstitüsü, Konya, Türkiye.

Aydoğdu, A. & Duman, S. (2017). Destinasyon Çekicilik Unsuru Olarak Gastronomi Turizmi: Kastamonu Örneği. *Turar Turizm & Araştırma Dergisi*, 6(1), 4–23.

Bayram, F. (2020). *Mutfak Yöneticilerinin Ticari Mutfak Planlama İlkeleri Konusunda Bilgi Düzeyleri Sektörel Hatalar ve Çözüm önerileri*. (Yayımlanmamış Doktora Tezi). Ankara Hacı Bayram Veli Üniversitesi Lisan Üstü Eğitim Enstitüsü Gastronomi ve Mutfak Sanatları Anabilim Dalı, Ankara, Türkiye.

Bayram, M. (2018). Otel Konuk Memnuniyetini Etkileyen Unsurlar. *Journal of Travel and Hospitality Management*. 15(2) 352–369.

Boztoprak, F. (2019). *Gastronomi Turizminin Kentin Tanıtımına ve Markalaşmasına Etkisi: Erzurum İli Örneği*. (Yayımlanmamış Yüksek Lisans

Tezi). Gazi Üniversitesi Eğitim Bilimleri Enstitüsü Aile Ekonomisi ve Beslenme Eğitimi Anabilim Dalı, Ankara, Türkiye.

Bryant, W. & McKay, E. (2005). What's cooking? Theory and practice in the kitchen. *British Journal of Occupational Therapy*, 68(2), 68–74.

Budak, İ. (2015). *Otel İşletmeleri Mutfak Planlamasına İlişkin Sistematik Bir Süreç Analizi* (Yayımlanmamış Yüksek Lisans Tezi). Balıkesir Üniversitesi Sosyal Bilimler Enstitüsü, Balıkesir, Türkiye.

Chan, E. S. & Lam, L. (2009). Understanding attributes affecting selection of private kitchens. *International Journal of Contemporary Hospitality Management*, 21(7), 854–875. doi:10.1108/09596110910985322

Dareker, S. & Peshave, M. (2016). A Study of İmportance of Kitchen Designing in Standalone Restaurants. *International Journal of Research in IT and Management*, 6(6), 110–119.

Denizer, D. (2005). *Konaklama İşletmelerinde Yiyecek ve İçecek Yönetimi.* (1. baskı). Ankara: Detay Yayıncılık.

Dizdar, N. E. & Kurt, M. (2001). *İş Güvenliği.* Ankara: Gazi Üniversitesi Yayınları.

Doğdubay, M. & Karan, İ. (2017). Otel mutfaklarında sistematik iş yeri düzenleme planı (sidp) modelinin uygulanması. *Aydın Gastronomy*, 1 (1): 9–23. https://dergipark.org.tr/en/download/article-file/356872 (Accessed: 05.06.2020).

Fişek, G. A. (2009). *Çalışma yaşamında sağlık güvenlik.* (1. bs.). Ankara: Fişek Enstitüsü Çalışan Çocuklar Bilim ve Eylem Vakfı Yayını No.3-2 ISBN:978-975-98736-8-4.

Geoldner, C. R. & Ritchie, J. B. (2003). *Tourism-principles, practices.* Hoboken, New Jersey: John Wiley & Sons, 9th Edition.

Gökdemir, A. (2003). *Mutfak Hizmetleri Yönetimi.* A. Sökmen. (Ed.). (1. baskı). Ankara: Detay Yayıncılık.

Gundersen, M. G., Heide, M. & Olsson, U. H. (1996). Hotel guest satisfaction among business travelers. *Cornell Hospitality Quarterly*, 37(2), 72–81.

Güldemir, O. (2015). *Yiyecek İçecek Bölümü Mutfak Planlaması.* Murat Doğdubay. (Ed) *Turizm İşletmelerinde Yiyecek İçecek Yönetimi* içinde (ss.101–123). Ankara: Grafiker Yayınları.

Hudson, S. (2008). *Tourism and hospitality marketing: a global perspective.* London: Sage publication Ltd.

Kandampully, J. & Suhartanto, D. (2000). Customer loyalty in the hotel industry: the role of customer satisfaction and image. *International Journal of Contemporary Hospitality Management*, 12(6), 346–351.

Karamahmutoğlu, E. (2019). *Türkiye Turizm Tanıtımında Gastronomik Değerlerin Yeri ve İhtiyaç Analizi* (Yayımlanmamış Yüksek Lisans Tezi). Eskişehir Osman Gazi Üniversitesi Sosyal Bilimler Enstitüsü. Eskişehir, Türkiye.

Kızılgeçi, Ç. (2019). *Gastronomi Turizminin Alternatif Turizm Ürünü Olarak Değerlendirilmesi: Mardin İlinde Bulunan İşletmelere Yönelik Bir Uygulama* (Yayımlanmamış Doktora Tezi). Hasan Kalyoncu Üniversitesi Sosyal Bilimler Enstitüsü, Gaziantep, Türkiye.

Koçak, N. (2007). *Yiyecek İçecek İşletmelerinde Gıda ve Personel Hijyeni.* (1. baskı). Ankara: Detay Yayıncılık.

Kozak, M. A., Azaltun, M., Sarıışık, M., Çakıcı, A.C., Sökmen, A. & Çetinsöz, B. C. (2008). *Otel İşletmeciliği.* N. Kozak. (Ed.). (2. baskı). Ankara: Detay Yayıncılık.

Kozak, N., Kozak, A. M. & Kozak, M. (2017). *Genel Turizm İlkeler- Kavramlar.* (19. baskı). Ankara: Detay yayıncılık.

Küçükaslan, N. (2011). *Yiyecek İçecek İşletmelerinde Mutfak Hizmetleri Yönetimi.* (2. baskı). Bursa: Alfa Akademi.

Küçükyaman, A. M. & Demir, Ş. Ş. (2019). Otel mutfak Tasarımlarının Çalışanlar Tarafından Değerlendirilmesi. *Uluslararası Global Turizm Araştırmaları Dergisi*, 3(2), 125–139.

Maviş, F. (2008). *Endüstriyel Yiyecek Üretimi.* (2. baskı). Ankara: Detay Yayıncılık.

Merdol, K. T. & Birer S. (1997). *Kurum Beslenmesi.* (5. Baskı). İstanbul: MEB Yayını.

Mullins Laurie, J. (1996). *Hospitality Management.* UK: London, 3rd edition.

Özdemir, B. (2001). *Otel İşletmelerinde Mutfak Yönetimi ve Her Şey Dahil (All Inclusive) Uygulamasının Mutfak Yönetimine Etkileri Üzerine Sektörel Bir Araştırma.* (Yayımlanmamış Yüksek Lisans Tezi). Akdeniz Üniversitesi Sosyal Bilimler Enstitüsü, Antalya, Türkiye.

Özel, K. & Cömert, M. (2015). Otel İşletmeleri Mutfak Çalışanlarının Mutfak Planlaması ile İlgili Bilgi Düzeylerinin Belirlenmesi. *Journal of Tourism and Gastronomy Studies*, 3(2), 48–55.

Özoral, A. & Yıldırım, Ö. (2014). *Mutfak ve Stewardıng Profesyonel Mutfak Rehberi.* İstanbul: Otelci Online Yayınları.

Powell, C. B. (2005). *Architects Pocket Book of Kitchen Design.* Oxford, UK: Architectural Press.

Sarıtaş, İ. & Yaşar A. (2017). *Elektrikle Çalışanlarda İş Sağlığı ve Güvenliği Tedbirleri.* M. Altın ve Ş. Taşdemir (Ed.) *İSG İş Sağlığı ve Güvenliği* içinde (ss.591–612). Konya: Eğitim Yayınevi.

Sökmen, A. (2008). *Yiyecek İçecek Hizmetleri Yönetimi ve İşletmeciliği*. (4. baskı). Ankara: Detay Yayıncılık.

Sökmen, A. (2009). *Konaklama ve Yiyecek İçecek İşletmelerinde Servis Tekniği ve Uygulamaları*. (2. baskı). Ankara: Detay Yayıncılık.

Sökmen, A. (2014). *Yiyecek İçecek Hizmetleri Yönetimi ve İşletmeciliği*. (7. baskı). Ankara: Detay Yayıncılık.

Şahin, G. G. & Ünver, G. (2015). Destinasyon pazarlama aracı olarak gastronomi turizmi: İstanbul'un gastronomi turizmi potansiyeli üzerine bir araştırma. *Journal of Tourism and Gastronomy Studies*, 3(2), 63–73.

Şahingöz, A. S. & Şık, A. (2015). *Konaklama ve Beslenme İşletmelerinde İş Sağlığı ve Güvenliği*. (1. baskı). Ankara: Detay Yayıncılık.

Topal, Ş. (1996). *Gıda Güvenliği ve Kalite Yönetim Sistemleri. Kocaeli*: TÜBİTAK Marmara Araştırma Merkezi Matbaası.

Tümer, H. (2008). *Toplu Beslenme Hizmeti Verilen Kurumlarda Mutfak Planının İncelenmesi ve Mutfak Planının İş Akışına Etkisi Üzerine Bir Araştırma*. (Yayımlanmamış Yüksek Lisans Tezi), Hacettepe Üniversitesi Sağlık Bilimleri Enstitüsü, Ankara, Türkiye.

Türkan, C. (2009). *Mutfak Hizmetleri Yönetimi*. (3. baskı). Bolu: Cemalturkan Yayınları.

Türkan, C. (2010). *Mutfak Teknolojisi*. (4. baskı). Ankara: Sistem Ofset.

Türksoy, A. (2007). *Yiyecek ve İçecek İşletmeleri Yönetimi*. (3. baskı). Ankara: Turhan Kitapevi.

Türksoy, A. (2015) *Yiyecek ve içecek hizmetleri yönetimi*. (1. baskı). Ankara: Detay Yayıncılık.

Usta, Ö. (1996). *Turizm*. İstanbul: Altın Kitaplar.

Ünüvar, Ş. (2007). *Gıda/Besin Teknolojisi*. (1. baskı). Ankara: Nobel Yayım Basım.

Yiğit, A. (2013). *İş güvenliği*. Ankara: Alfa Aktüel Yayınları.

Yılmaz, Ö., Yılmaz, Y. & Yılmaz, Ö. (2013). *Yiyecek İçecek İşletmeciliği*. (1. baskı). Ankara: Detay Yayıncılık.

Serkan AYLAN & Fatmanur Kübra AYLAN

20 The Place and Importance of Recreation Activities in Health Tourism

1 Introduction

According to the World Health Organization, health is not a complete state of not having illness or weakness, but a state of mental, physical and social well-being (WHO, 1986). Health tourism is generally defined as individuals traveling from one place to another in order to get preventive, therapeutic and rehabilitative services (Ova, Biçer and Yurcu, 2019: 141). Health tourism, as can be understood from the definition, aims both protecting health and treating diseases. Activities aimed at protecting health are as important as treating diseases. Because the inactive and stressful life, which is a routine of modern life, causes many negativities such as adiposeness and obesity, smoking and alcohol over time. The realization of the effects of difficult living conditions on human health causing to change people's demands and form a new health consciousness (Hacıoğlu, Gökdeniz and Dinç, 2009: 64). At the same time, the problems people have experienced in their family, school or business life can disrupt the neural system, and cause fatigue and decrease in endurance, and may cause great physiological and psychological discomfort for the person (Karaküçük, 1997). In this context, recreation has many benefits for individuals and society. Especially, when the increasing stress and difficult living conditions are considered, the effects of recreation on the body and mental health can be understood better (Değer, 2020: 310). In conclusion, the participation of people in recreational activities within the scope of health tourism contributes to the increase in the quality of life and a healthy life, and plays a supportive role in the treatment process of the patients.

In this chapter, health tourism and its types, the concept of recreation, health tourism and recreation, participation in recreational activities by age within the scope of health tourism are examined under separate headings to reveal the place and importance of recreational activities in health tourism. In the conclusion part, suggestions are made for recreational activities which are thought may contribute to the development of health tourism.

2 Health Tourism and Types of Health Tourism

According to the World Health Organization, health is the level of fulfilling the aspirations of an individual or group, meeting needs and changing or coping with the environment, health is not the purpose of life, but a resource for everyday life and is a positive concept that affects social and personal situations as well as physical condition (WHO, 1984). In addition, health is not a complete state of not having illness or weakness, but a state of mental, physical and social well-being (WHO, 1986). Health tourism refers to travel to a country (or region) different from the place of normal residence to get medical (surgeries, tooth-eye treatments, organ transplantation, diagnostic services, etc.) and non-medical (aesthetic operations, healthy life and wellness, third age care-treatment) services. Although health tourists benefit from different holiday opportunities in parallel with their travels, their main purpose is getting a certain health service (Bookman and Bookman, 2007). It is seen that different services are offered to patients or health tourists within the scope of health tourism. Health tourism services are divided into three groups according to Gonzales et al (2001) (Gonzales, Brenzel and Sancho, 2001): Health Promotion Services (hot spring, herbal therapy, massage, etc.), Treatment Services (Aesthetic surgery, heart surgery, eye surgery, cancer treatment, etc.), Rehabilitation Services (Dialysis, addiction program, elderly care program).

Medical Tourism: Medical tourism can be defined as traveling for the purpose of receiving medical care to prevent various medical interventions that require expertise or in order to prevent potential diseases. Along with the developments in the medical industry and technology, significant developments are observed in the demand for medical tourism due to the increasing demand for health services. Due to the high levels of economic, education and welfare conditions in developed and developing countries, healthcare costs can also be quite high. In addition, due to the changing demographic structure, the healthcare demands of the aging population may increase, and health expenditures can increase. Countries that have difficulties in meeting these expenses with social security institutions try to get health services at low costs by making agreements with other countries (Daştan, 2014: 147). From this point of view, medical tourism can be defined as the travel made from high-income countries to low and middle-income countries to get cheaper health services (Noree, Hanefeld and Smith, 2015).

Thermal Tourism: Thermal tourism is the use of thermal waters for various treatment methods and recreational purposes (SB, 2012: 112). In general, it is in the form of utilizing thermal spring waters in the form of physical therapy and exercise. It is the application of accommodation and physical therapy in hotel-style places where underground hot water sources come out (Özer and Songur, 2013; Kördeve, 2016: 53).

Wellness–Spa Tourism: The word "wellness" is a combination of the words well-being and fitness. While "well-being" means health and prosperity, the concept of "fitness" means being fit and vitality. On the other hand, wellness tourism is a health tourism service offered to help a person to be in good physical, spiritual and mental health (Ağaoğlu, Karagöz and Zabun, 2019: 13). Body treatments performed with nature and natural products such as skincare, mud and algae baths, bathtub treatments, thalassotherapy and healthy living methods are included in the scope of wellness (Topuz, 2012: 11). SPA "is an abbreviation of the Latin phrase" Salus Per Aquam "and means" goodness coming with water" or "health coming with water ". SPA service removes fatigue and pain from the body by adjusting the water and salt balance in the body. SPA treatment has a history dating back to the Roman Empire. Throughout history, baths were used to heal soldiers who returned from wars weak and tired. Today, SPA has various application methods such as hot water pools, massage therapies, fragrance therapies, beauty and care services, cure treatments, as well as water and mud therapies (Ağaoğlu et al., 2019: 13).

Elderly Tourism: Another type of tourism for health purposes is elderly tourism, which is also called "geriatric tourism" or "third age tourism". Elderly tourism refers to travels carried out for the purpose of protecting the health of elderly people, prevention of possible diseases, diagnosis and treatment. The issue of caring for elderly individuals in another country has also combined elderly care and tourism in a sense (Bülbül, 2015: 38).

Accessible Tourism: This type of tourism is a type of tourism that developed for the traveling, traveling, entertainment and treatment needs of the disabled (Topuz, 2012: 10). "Disabled tourism", which is accepted as a tourism activity created by disabled individuals, is defined as the whole of tourism products, services and environments designed independently and equally universally that allow people to access mobility, vision, hearing and mental aspects (UNWTO, 2013: 4).

3 The Concept of Recreation

The concept of recreation, which has derived from the English word "recreation" to Turkish, etymologically means re-creation, animation, re-created thing, entertainment, resting, relaxation, playtime (at school). Recreation is the activities in which people participate voluntarily and provide satisfaction during the period of people's limited life other than mandatory works (in spare time). Recreation is aimed at the use of touristic leisure, sportive leisure, social leisure, creative leisure (Hazar, 2014: 35). Recreation is the involvement of human beings in an activity which is suitable for their nature and enjoys after participating, after leaving aside the life that has become ordinary and living between work and home as if repeating the same routine, performing spiritual and mental rest apart from these compulsory situations, and socializing with other people (Gül, 2014: 10). Recreation includes passive and active, group or individual participation in sports, cultural activities, appreciation of nature or human history, excursions, entertainment, pleasurable events and informal education (Hacıoğlu et al., 2017: 27). It is possible to list the personal benefits of recreational activities as having a positive effect on the development of mental and physical health, integrating the individual into social life, the individual's awareness of his/her own skills and abilities, bringing the creativity in thoughts into the forefront, increasing the working efficiency by relieving the stress and fatigue of the person, and success as the reflection of this situation in the working life of happy individuals (Karaküçük, 1999: 72).

4 Recreation and Health Tourism

Health tourism is generally a type of tourism formed by people traveling to protect or regain their health (Topuz, 2012; Aslan and Güzel, 2018: 95). Recreational activities have a protective and supportive role in health tourism travels. In terms of health, recreation is basically evaluated in two parts. The first of these is recreation services provided to healthy individuals within the scope of preventive health services, and the second is provided to disabled and incompetent people (those with health problems) within the scope of rehabilitation services (Tütüncü, 2020: 86). Recreation services that provided to healthy individuals in the world are considered within the scope of "Recreation", and recreation services provided to people with disabilities and health problems are evaluated within the scope of "Recreation Therapy". In this context, based on recreation therapy, rehabilitation-oriented well-being, and recreation services are applied to individuals with disease or disability in order to support treatment in their spare time (Tütüncü, 2012: 249). Recreational therapy is carried out in order to strengthen the morale motivation of patients who have been treated in hospital for a long time, who have chronic diseases and who require constant follow-up, and in order to support treatment. First of all, the main purpose of all activities is adapting the disabled individual to an independent lifestyle. Recreation therapy has an important place in nursing homes, clinics, rehabilitation centers, community mental health centers and tourism enterprises as well as in hospitals. In this context, people who will specialize in the field of recreation therapy should have a strong information infrastructure for health (Tütüncü, 2012: 251). Recreation therapy is a therapeutic training branch that helps treatment (Özdemir, 2020);

Therapeutic recreation activities can be divided into two main groups (Özdemir, 2020); (1) Recreational activities for treatment, education and rehabilitation, which are often compulsory, applied in addition to the education of disabled individuals; (2) Recreational activities in free time with complete freedom.

Therapeutic recreation has applications for different groups as well as patients, elderly and disabled groups. These groups can be listed as follows (Özdemir, 2020); (1) Activities carried out to reintegrate prisoners in prisons or who have been released from prison; (2) Activities for preventive purposes for individuals with criminal potential or drug addiction; (3) Special recreational activities are applied for amputees (when the person does not have a part of his/her arm, leg, or foot).

Some studies revealing the positive contributions of recreational activities to human health show that physical activities increase the health, physical fitness and well-being levels of individuals with disabilities (Kaya and Sarı, 2018: 21). At the same time, atypical leisure behavior activities (alcohol, smoking, etc.) that individuals will choose without being aware of or not will have negative consequences in their physiological, sociological and psychological well-being. When recreative activities are selected and carried out consciously, it is possible to improve individuals physiologically, ensure their intellectual development, improve their social relations, improve their mood and increase their happiness (Yurcu, 2017: 27). Recreative activities play an important role in increasing the well-being of individuals. The recreational activities that individuals will actively participate in line with their own desires and abilities will keep them away from negative affections; increase their physiological, psychological, sociological well-being and increase their life satisfaction (Yurcu, 2017: 38). In conclusion, recreational activities such as music, handicrafts, dance, nature trips, sports, games, etc. which have physiological, psychological and sociological benefits to people are used as curative or therapeutic in medicine (Hazar, 2014: 49).

5 Recreation by Age Within the Scope of Health Tourism

5.1 Children's Health and Recreation

Children's recreation includes recreational activities for individuals in the 0–18 age group. The recreational activities that will be carried out in this period provide important benefits such as the physical development of the child, as well as the development of personality and self-confidence, the development of social relations and mental health. Additionally, well-organized recreation programs provide healthy individual development and prevent negative consequences that may arise in the future of the child. By promoting skill development, it allows the child to discover himself/herself (Gültekin, 2018: 53).

In order to protect the health of children and support their treatment, recreational activities can be organized within the scope of thermal spring tourism. The recreational activities organized within the scope of thermal tourism will contribute to the children getting away from the psychology of disease by being in a social environment, having fun and having a good time. Supportive applications for healthy life such as thermal therapies for children, play baths, cure application races can be presented as an animation activity. Taplak et al (2017) stated in her study that she thinks that hot springs have relaxing and healing effects in terms of health for the children of mothers who constitute the sample of the study, and it is good for joint disorders and positively affects the bone development and growth of children (Taplak et al, 2017: 204). In another study, it was determined that thermal therapies were recommended for children with skin diseases and musculoskeletal disorders in thermal springs (Polat et al., 2017). In order for children to lead a healthy life, recreational activities organized in nature also contribute to the mental and physical development of children. Outdoor recreational activities, which are organized by taking advantage of the richness of nature, have an important place in reducing the negative effects of today's generation Z and Alpha generation children living with technology. For example, forest bathing activities can be organized for children. Ergüven (2019) stated in his study that forest bathing (Shinrin Yoku) is a method of relaxation and stress reduction recognized in Japan. Fresh air, exercise, and conscious perception of the environment relax the body and mind. Forest bathing, which is integrated with Japanese culture, is not only a walk of people in the forest, but includes perceiving and experiencing the environment with a calm, comfortable and awake consciousness during the walk (Ergüven, 2019: 68). Recreational activities organized in nature will allow children to be more confident, protected from various diseases that may arise from stress, and get rid of technology addiction.

For children with recreational restrictions, recreational therapy activities can also be a supportive element to get rid of their troubles. For example, the recreational activities organized in the process of protecting or improving the mental and physical health of children who are taken under protection in children's homes and in the process of supporting their personal and social development can positively contribute. Because it is important to spend leisure time positively in ensuring the mental health of the individual. Social relationships provide people with a sense of trust at different levels, raise their dignity and creativity to higher levels, and enable them to realize themselves (Birol, Temel and Aydın, 2018: 471).

5.2 Youth Health and Recreation

Youth recreation includes activities that will generally benefit youth development. Youth recreation is an important element in the development of personality in an individual's life and in the development of communication with individuals of the same age group. In addition, it enables the person to strengthen his/her skills and has psychological and behavioral benefits such as building self-confidence, problem solving, creativity and vitality (Gültekin, 2018: 54). Because, the individuals who are in a high period in terms of energy want to spend their energy in one way or another. Spending this energy by participating in recreational activities will enable them to be protected from bad habits, acquire knowledge and skills and use them beneficially for the society (Güçlü, 2013: 160). Lack of areas and opportunities for recreational activities in the youth years when people are most physically active may cause this physical energy to appear negatively in another way. In fact, the negative accumulations caused by the insufficient recreational possibilities sow the seeds of neuroses that may emerge in various forms for the individuals of the future (Tütüncü, 2012). It is stated that leisure activities for socialization reduce anti-social behaviors in young people. Activities where young people meet with their friends help their social development. For this reason, determining activities for the need of socialization is important in terms of completing the psycho-social development process (Birol et al, 2018: 471). Therefore, recreation activities play an important role in the good use of their leisure time.

Within the scope of preventive health tourism, forest bathing activities can be organized for young people to spend their time in nature like children. It has been demonstrated by researchers that young people can take a conscious walk in the forest, release their energy and stay away from stress, as well as its help to fight with cancer at a young age. While taking a forest bath, sounds, smells and colors in the forest are perceived consciously. For a forest bath, it is not much needed except smell, sound and colors. For example, it is enough for young people to walk slowly in the forest with bare feet and breathe deeply in order to release their excess energy and calm down. Although it is scientifically proven that the forest air reduces blood pressure, it has been proven that staying in the forest for a long time increases oxygen intake and the number of some immune cells (Ergüven, 2019: 67).

It has been supported by a number of scientific researches where recreational activities have an important place as a supportive factor in the protection of the health of young people and the treatment of their diseases. For example, Tercan (2015) found in his study that students with a high level of participation in recreational activities among university students had a higher level of life satisfaction. High life satisfaction of young people will contribute to a healthy life away from stress. At the same time, in a study conducted for refugee youth, sports recreation activities as a therapeutic recreation activity contributed positively to the physical development, psychological state and social skills of refugee students, helping them in order to gain positive emotions such as self-control and self-confidence, making it difficult for them to enter a negative mood. It has also been determined that learning the concepts such as solidarity, teamwork, and cooperation contributes significantly to the perception of life satisfaction and life quality of the individual (Kök, 2019: 51).

5.3 Adult Health and Recreation

Adults are people who have reached a certain mental and physical maturity. Adulthood consists of 3 stages. These processes are divided into three as first adulthood, middle adulthood, and advanced adulthood. First adulthood refers to the period from the end of adolescence until the age of thirty-five; middle adulthood refers to the period from the age of thirty-five to the age of sixty; advanced adulthood refers to the period of age sixty and after. Within the scope of these processes, all recreation activities that they participate actively or passively are considered within the scope of adult recreation (Gültekin, 2018: 54).

As an area that can be evaluated within the scope of health tourism, healing gardens are designed to make people feel better psychologically and physically. The causes of the most important diseases in our age are seen as stress, fatigue and mental depressions. Healing gardens are places that eliminate such factors and allow patients to be completely alone with nature. In the Far East, healing gardens are used to support the treatment processes and healing processes of cancer patients. Recreative activities such as yoga, meditation, water and color therapy, which will contribute to the healthy aging of adults in healing gardens, also allow patients to be treated spiritually (Çalışkan and Çanga, 2018: 545).

Another area that can be considered within the scope of health protection and supportive treatment in touch with nature is Kazdağı National Park. The air values and air flora in Kaz Dağları have therapeutic characteristics there.

The airborne flora can cure asthma, COPD and shortness of breath. As a matter of fact, the oxygen here is recommended by doctors for asthma, COPD, and shortness of breath diseases. A scientist stated that they can measure the active ingredients of endemic plants in Kaz Dağları and increase the effects of the active ingredients. He also stated that he managed to defeat kidney cancer by living in Kaz Dağları (Yılmaz, 2018: 31). The recreational activities carried out in Kaz Dağları, which can be evaluated within the scope of health tourism, are also very important. Some of these recreational activities that can be done are hiking and jeep safari tours.

In Spa & Wellness tourism for adults, recreation therapies such as; aesthetic and beauty treatments, massages, phytotherapy, aromatherapy, fangotherapy, snow and ice rooms, salt rooms, sauna, stone therapy and color therapy, etc. can be arranged. In addition, the Spa provides supplementary recreational therapy support in which the feelings of rest and relaxation gained through the applications of hot, cold and different forms of pouring, drip, spraying and shower, etc. (Değer, 2020: 310).

There are also scientific studies showing that the participation of adults in recreational activities contributes to a healthy life. For example, Sevin and Şen (2019) revealed a positive relationship between the level of teachers' participation in recreational activities and their life satisfaction in their study. It is expected that the individual will lead a happier and healthier life due to the increase in life satisfaction and the low probability of suffering from stress-related diseases.

5.4 Elderly Health and Recreation

The term aging means the prolongation of life expectancy and the differentiation of physical, mental, social and environmental health problems and becoming specific to advanced age (Turaman, 2001: 23). Considering that the health problems of the advanced age group will tend to increase, it is necessary to plan preventive, early diagnosis, treatment and rehabilitative health and social support services for elderly people and managing the effective and efficient use of resource management (Akın and Ersoy, 2012: 23). Traveling to other countries for the purpose of meeting the care and other needs of the elderly is called elderly tourism. Elderly care and tourism gains great importance depending on the welfare levels of the countries, technological and economic developments and changes in their demographic structures (Daştan, 2014: 146–147). Elderly tourism creates an increasing customer potential for the tourism industry by offering customers the opportunity to travel over the country where they will be treated, in order to make the service attractive, as well as health tourism. A type of tourism that fits the lifestyle of the elderly, responds to their needs, effective, improves people and colors their life has emerged. Clinical hotels, recreation areas, holiday villages and nursing homes, sightseeing tours, occupational therapy are examples of this situation (Bektaş and Şimşek, 2016: 179–180).

For older people, recreational activities are important to participate in the community and stay active after retirement. Recreative activities provide long-term advantages for the elderly. It provides a therapeutic regeneration both mentally and physically. The most common exercise that elderly people do is walking (Başarangil, 2016: 3174–3175). For example, in the Healthy Life Villages and Long Life Villages located in Kaz Dağları recreational activities are offered for the purposes of both protecting their health and supporting their treatments. Among the recreational services offered to residents in health villages; there are excursions with scooter, bowling, spa pool, gym, beauty care centers, cinema, theater, indoor bowling, swimming pool, library, bar, internet cafe, shopping areas, billiards, hobby gardens, art workshops and courtyards for walking, recreational activities and venues (Başarangil, 2016: 3174–3175). What's more, the scientist who could measure the active ingredients of endemic plants in the Healthy Life Village in Kaz Dağları, has measured those active ingredients in the treatments that would be applied to cancer patients and stated that these substances played an important role in the healing of the diseases (Yılmaz, 2018: 31). Long Life Villages are considered as an important element of health tourism. These health villages are centers that can offer health services such as wellness, rehabilitation, recreation, etc. to elderly individuals in a holistic manner. In this way, some treatments, such as preventation of the occurrence of some diseases that may arise in advanced ages, delaying the aging, treat for emerged diseases, providing rejuvenation with regeneration and relieving pain with rehabilitation are carried out by various professionals in Long Life Villages (Başarangil, 2016: 3174–3175). Another recreational activity that can be carried out within the scope of health tourism for the elderly is forest walk. Pulse, blood pressure and cortisol levels decrease as well as parasympathetic nerve with forest walk. In addition to the positive effects on the cardiovascular system, the production of anti-aging proteins and cancer-killing cells is also effectively increased. Briefly, being in nature increases resistance to stress. Trees also have a positive effect on mental state (Ergüven, 2019: 69). Şener, Terzioğlu and Karabulut (2007) revealed that participation in recreational activities is the most important dimension that explains life satisfaction as a result of the research conducted to determine the relationship between the frequency of retirees' participation in recreational activities and their life satisfaction.

6 Conclusion

Factors such as the compelling effect of living conditions on human beings, the increase of the responsibilities placed on people, the change in the economy, the increase in the population, the proliferation of diseases, the rapid development of technology have caused life to be more complex and stressful. These reasons cause physiological and psychological deterioration and increase the need for recreation. Recreation is born out of a need that creates a sense of renewal by relieving the fatigue and stress of working life, alleviating the burden of living conditions to some extent (Gültekin, 2018: 49–50). Recreational activities that individuals do in their spare time contribute greatly to personal and social development. Recreation activities provide personal physical health development, mental health, socialization of the individual, development of creativity, individual skills and abilities, work success and contribution to work efficiency. There are benefits such as improving physical health, making new friendships, learning new information, eliminating daily monotony, increasing self-confidence, improving mental health, contributing to intelligence development and gaining leadership skills in eliminating negative emotions that will affect productivity with recreational activities of people (Birol et al, 2018: 462). At the same time, recreational activities help disadvantaged groups to overcome the obstacles they encounter in their struggle for life, cope with psychological problems, socialize and adapt to other individuals. Within the scope of health tourism, recreational activities contribute to the increase of people's life quality and a positive course of treatment. Below, some recreational activities have been suggested to be implemented for both the protection of health and the support of the treatment process during the travels within the scope of the health holiday.

(1) In order to protect health, anti-aging, anti-stress and empowering cures and recreational therapy applications can be included in a wider scope; (2) The holiday process can be enriched with sports recreation activities for a healthy life for all ages. For example, since today's children have a high rate of obesity, recreational activities can be organized that will allow them to lose energy during the holiday process; (3) Fun and instructive recreational activities can be organized for young people to spend their energy in a positive way; (4) Recreational activities such as nature walks, jeep safari tours, fishing, which will contribute to the healthy aging of adults and their physical and mental well-being, can be organized more; (4) Within the scope of health tourism, recreation areas should be created that will make individuals forget that they are in the hospital environment during the treatment process; (5) In

order to contribute positively to the treatment process of the patients, tours and excursions can be organized within the scope of outdoor recreation activities to increase the morale level and in order to benefit from the beneficial effect of clean air.

References

Ağaoğlu, F. O., Karagöz, N., and Zabun, S. (2019). Sağlık Turizmi Uygulamaları ve Türkiye'ye Getirilerine İlişkin Bir Değerlendirme, S.C.Ü. İktisadi ve İdari Bilimler Dergisi, 20 (2): 1–23.

Akın, A., and Ersoy, K. (2012). 2050'ye Doğru Nüfusbilim ve Yönetim: Sağlık Sistemine Bakış, TUSİAD- T/2012-11/533.

Aslan, Ş., and Güzel, Ş. (2018). Dünyada ve Türkiye'de Yaşlanan Nüfus ve Sağlık Turizminin Bir Boyutu Olarak Yaşlı Turizminin Artan Önemi, 1.Uluslararası Turizmde Yeni Jenerasyonlar ve Yeni Trendler Konferansı: 92–102, 01–03 Kasım Sapanca, Türkiye.

Aydın, G., and Aydın, B. K. (2015). Dünyada ve Türkiye'de Sağlık Turizmi Pazarlama Uygulamaları ve Karşılaştırmalı Durum Analizi, Pazarlama ve Pazarlama Araştırmaları Dergisi, 16: 1–21.

Başarangil, İ. (2016). Sağlık Turizminde Uzun Yaşam Köyleri: Rekreatif Aktiviteler Açısından Bir Değerlendirme, ICOMEP 2016 | International Congress of Management Economy and Policy, İstanbul, Turkey.

Bektaş, G., and Şimşek, F. (2016). İleri Yaş Sağlık Turizminde Mobil Sağlık Hizmetlerinin Önemi, Sağlık Akademisyenleri Dergisi, 3 (4): 179–185.

Birol, S. Ş., Temel, V., and Aydın, E. (2018). Karaman İli Bünyesindeki Çocuk Evlerinde Koruma Altına Alınan Çocukların Rekreatif Faaliyetlere Katılım Düzeylerinin ve Psiko-Sosyal Durumlarının Belirlenmesi. Journal of Human Sciences, 15 (1): 459–477.

Bookman, Z. M., and Bookman, R. K. (2007). Medical Tourism in Developing Countries, Palgrave Macmillan: New York.

Bülbül, F. (2015). Sağlık Turizminin Türkiye'deki Gelişimi. Yayımlanmamış Yüksek Lisans Tezi, Beykent Üniversitesi Sosyal Bilimler Enstitüsü, İstanbul.

Çalışkan, H., and Çelik Çanga, A. (2018). Şifa Bahçeleri ve Turizmdeki Yeri, Uluslararası Turizm, İşletme, Ekonomi Dergisi, 2 (2): 541–548.

Daştan, İ. (2014). Türkiye'de Sağlık Turizmi: Türkiye ve Özelinde İzmir'de Sağlık Turizminin Mevcut Durum Analizi ve Strateji Önerileri, Mehmet Akif Ersoy Üniversitesi Sosyal Bilimler Enstitüsü Dergisi, 6 (10): 143–163.

Değer, M. (2020). Rekreasyonda Yeni Trendler: Spa & Wellness, Journal of Recreation and Tourism Research, 7 (2): 308–324.

Ergüven, M. H. (2019). Sağlıklı Yaşamda Doğal Bir Wellness Trendi: Orman Banyosu, Jass Studies- The Journal of Academic Social Science Studies, 76, 65–73.

Gonzales, A., Brenzel, L., and Sancho, J.(2001). Health Tourism and Related Services: Caribbean Development and International Trade. Final Report. https://caricom.org/documents/10177-health_study.pdf Erişim Tarihi: 07.07.2020.

Güçlü, M. (2013). Gençlik Döneminde Boş Zaman Faaliyetlerinin Yeri ve Önemi, Gençlik Araştırmaları Dergisi, 1 (1): 158–169.

Gül, T. (2014). Rekreasyon Olgusuna Genel Yaklaşım. (Ed.) A. Yaylı içinde, Rekreasyona Giriş (s. 2–62). Ankara: Detay Yayıncılık.

Gültekin Cıbıt, Ö. (2018). Medikal Turizmde Rekreasyon Planlaması ve Müşteri Tatmini İlişkisi: İstanbul Örneği, Balıkesir Üniversitesi Sosyal Bilimler Enstitüsü (Yayınlanmamış Yüksek Lisans Tezi), Balıkesir.

Hacıoğlu, N., Gökdeniz, A., and Dinç, Y. (2009). Boş Zaman ve Rekreasyon Yönetimi, Ankara: Detay Yayıncılık.

Hazar, A. (2014). Rekreasyon ve Animasyon. Ankara: Detay Yayıncılık.

İçöz, O. (2009). Sağlık Turizmi Kapsamında Medikal (Tıbbi) Turizm ve Türkiye'nin Olanakları, Journal of Yasar University, 4 (14): 2257–2279.

Karaküçük, S. (1997). Rekreasyon Serbest zamanları Değerlendirme, Seren Ofset, (2. Baskı), Ankara.

Karaküçük, S. (1999). Rekreasyon: Boş Zamanları Değerlendirme, Bağırgan Yayınevi, (3. Baskı), Ankara.

Kaya, E., and Sarı, İ. (2018). Bedensel ve Görme Engelli Bireylerin Çeşitli Değişkenler Açısından Boş Zaman Tutumlarının İncelenmesi, Gaziantep Üniversitesi Spor Bilimleri Dergisi, 3 (4): 19–31.

Kök, A. (2019). Düzenli Olarak Yapılan Rekreatif Faaliyetlerin Barınma Merkezinde Kalan Mülteci Ortaöğretim Öğrencilerinin Üzerine Etkisi, Necmettin Erbakan Üniversitesi Sosyal Bilimler Enstitüsü Rekreasyon Yönetimi Anabilim Dalı, Konya.

Kördeve, M. K. (2016). Sağlık Turizmine Genel Bir Bakış ve Türkiye'nin Sağlık Turizmindeki Yeri. Uluslararası Sağlık Yönetimi ve Stratejileri Araştırma Dergisi, 2 (1): 51–61.

Noree, T., Hanefeld, J.v, and Smith, R. (2016). Medical tourism in Thailand: A cross-sectional study. Bull World Health Organ, 94: 30–36

Ova, P. Ö., Biçer, E. B., and Yurcu, G. (2019). Yaşlılarda Termal Sağlık Turizmi Hizmet Algısı: Sivas İli Örneği, Gümüşhane Üniversitesi Sosyal Bilimler Enstitüsü Elektronik Dergisi (GUSBEED), 10 (1): 140–153.

Özdemir, M. (2020). Terapatik Rekreasyon Nedir? https://mesutozdemir.org/terapatik-rekreasyon-nedir/ Erişim Tarihi: 07.07.2020.

Özer, Ö., and Sonğur, C. (2013). Türkiye'nin Dünya Sağlık Turizmindeki Yeri ve Ekonomik Boyutu, Mehmet Akif Ersoy Üniversitesi Sosyal Bilimler Enstitüsü Dergisi, 4 (7): 69–81.

Polat, S., Akturan, E., Parlak Gürol, A., and Şener Taplak, A. (2017). Termal Kaplıca Terapisinin Çocuklarda Kullanımına İlişkin Hemşirelerin Görüşleri, Yozgat Örneği, II. Uluslararası Bozok Sempozyumu Yozgat'ın Turizm Potansiyelleri ve Sorunları, 191–203.

Sevin, H. D., and Şen, K. (2019). Öğretmenlerin Rekreasyon Aktivitelerine Katılım Düzeyleri ile Yaşam Mutluluğu ve İş Performansları Arasındaki İlişki, Dini Araştırmalar, 22 (55): 213–232.

SB (2012). T.C. Sağlık Bakanlığı. Sağlık Turizmi El Kitabı.

Şener, A., Terzioğlu, R. G., and Karabulut, E. (2007). Life Satisfaction and Leisure Activites During Men's Retirement: A Turkish Sample, Aging and Mental Health, 11(1), 30–36.

Taplak, Ş. A., Parlak Gürol, A., Polat, S., Tubaş, F., and Polat, M. F. (2017). Termal Kaplıca Kullanımının Çocuk Sağlığına Etkilerine İlişkin Anne Görüşleri, Yozgat Örneği, II. Uluslararası Bozok Sempozyumu Yozgat'ın Turizm Potansiyelleri ve Sorunları, 204–2017.

Tercan, E. (2015). An Examination of Leisure Participation, Family Assessment and Life Satisfaction in University Students, Procedia-Social and Behavioral Sciences, 186, 58–63.

Topuz, N. (2012). Türkiye Sağlık (Medikal) Turizmi Stratejisi 2023. Ankara: Uzmanlık Tezi.

Turaman, C. (2001). Yaşlı Sağlığı Hizmetlerinin Birinci Basamakta Planlanması, Turkish Journal of Geriatrics, 4 (1): 22–27.

Tütüncü, Ö., and Kuşluvan, Z. (1997). Çevre Sorun1arının Doğada Rekreasyon Faaliyetlerine Duyulan Gereksinimi Artırıcı Etkisi, Anatolia: Turizm Araştırmaları Dergisi, 8 (1–2): 9–11.

Tütüncü, Ö. (2012). Rekreasyon ve Rekreasyon Terapisinin Yaşam Kalitesindeki Rolü, Anatolia: Turizm Araştırmaları Dergisi, 23 (2): 248–252.

Tütüncü, Ö. (2020). Şiddet, Suç ve Na-Hoş Zaman İlişkisi, Anatolia: Turizm Araştırmaları Dergisi, 31 (1): 85–91.

UNWTO. (2013). Recommendations on Accesible Tourism. https://www.e-unwto.org/doi/epdf/10.18111/9789284415984, Erişim Tarihi: 10.07.2020.

WHO (1984). Health promotion: a discussion document on the concept and principles: summary report of the Working Group on Concept and Principles of Health Promotion, Copenhagen, 9–13 July 1984.

WHO (1986). Ottawa Charter for Health Promotion, First International Conference on Health Promotion, November.

Yılmaz, G. (2018). İda Dağı (Kazdağları)nın Sağlık Turizmi Çerçevesinde İncelenmesi, International Anatolia Academic Online Journal/Social Science Journal, 15–35.

Yurcu, G. (2017). Rekreatif Faaliyet Katılım Alanlarına Göre Bireylerin Öfke İfade Tarzlarının Negatif-Pozitif Duygulanım ve Yaşam Doyumları Üzerindeki Etkisi, Süleyman Demirel Üniversitesi Vizyoner Dergisi, 8 (18): 26–40.

Üzeyir KEMENT

21 Tourism and Recreation

1 Tourism and Recreation Conceptual Review

Recreation are gradually gaining importance within the organizational structure of tourism sector. In this section, the phenomena of recreation and tourism are examined from a conceptual perspective and the relationship between recreation and tourism is emphasized.

Recreational activities are carried out by individuals in leisure time. In this context, primarily the concept of leisure time is examined. According to Hazar (2009: 5) time is defined *"as the past and passing time in a job or event"*. Kemp and Pearson (1997: 2) express leisure time as the period of time that increases after people's time for working, sleeping, eating, and other compulsory tasks. In other words, leisure time is the time when an individual is freed from all obligations or connections for herself and others and engages in an activity of her/his own will (Parker, 1979). According to the definitions made in the literature, leisure time is not compulsory and it is done with free will. It is emphasized that in this period of time, which is completely belonging to people, they act freely in line with their own wishes without feeling any obligation. Another situation that stands out in the definitions is activity. Free time is not about people doing nothing, but rather doing activities that make leisure time valuable and self-fulfilling.

Most of the leisure activities can be shown as recreational activities. Recreation is defined as *"passive or active activities that people do, individually or in groups, to take physical and mental enjoyment of sports, cultural, nature-related, historical activities and activities without any personal coercion or pressure"* (Lundberg, 1980: 4). Carlson et al. (1972: 7) defines recreation as *"recreational activities that individuals participate voluntarily in their leisure time and as a result reach satisfaction"*. Broadhurst (2001: 2) explains recreation as *"the activities people choose to do in their spare time"*. In the definitions of recreation, activity/action discourse comes to the fore. Because when leisure is evaluated with an activity, it causes the phenomenon of recreation. Recreational activities include traveling, having fun, relaxing, self-improvement, etc. It contains many activities.

When the activities/actions performed by people are examined in terms of time with the help of the scheme, they are grouped as work time, free time, leisure time, a-typical leisure time and recreation (Fig. 1). People work to survive economically. The process in which people work is grouped as "work time". The

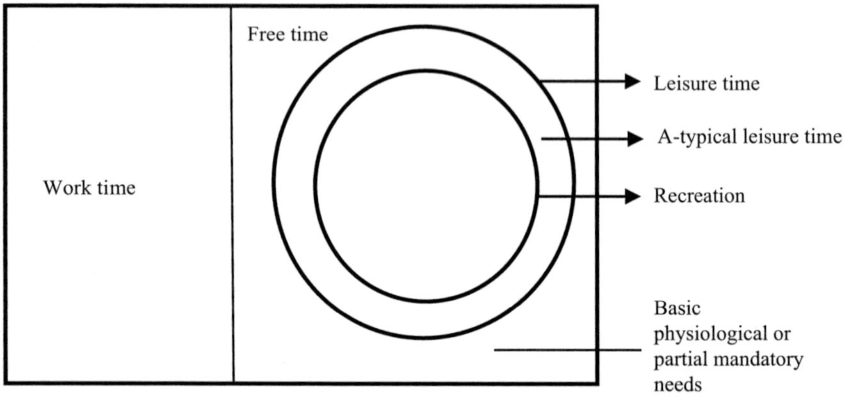

Fig. 1: Leisure Time and Recreation

actions/activities performed by people in time outside of work time are called "free time". People sleep, eat, etc. It has basic physiological needs. They also perform partial compulsory actions such as shopping to fulfill their basic physiological needs. People perform all these actions in free time. The period of time other than work time and activities covering basic physiological needs is called "leisure time". Leisure time is divided into recreational activities and a-typical leisure behaviors. Actions that are health or socially damaging, that do not allow people to be physically, mentally or socially regenerated, take place in a-typical leisure time. Activities in which people renew themselves physically, mentally or socially without pursuing a commercial purpose with their free will and that do not violate universal ethical principles are called "recreation". Most of the tourism activities are recreational activities that people do in their leisure time.

Tourism activities are one of the leisure activities. "*Tourism is all the events and relationships that occur due to people traveling and staying in a way that does not turn into permanent settlement and is not associated with any profitable activity*" (Burkart and Medlik, 1980: 40). More comprehensively, tourism is activities carried out in order not to settle outside of their permanent residence, to earn money or to pursue a political or military purpose. In addition, tourism is carried out in a free environment for purposes such as business, curiosity, religion, health, sports, recreation, culture and snobbism or for reasons such as family visits, congresses and seminars. Finally, tourism is a consumption event, an industry and a social phenomenon that emerges from personal or collective travels and accommodation for more than 24 hours (Olalı and Timur, 1988). It is seen that the concepts of recreation and tourism are interrelated.

2 Relationship Between Tourism and Recreation

Tourism and recreation have the feature of increasing people's quality of life and thus satisfying the person. This situation creates a common relationship between recreation and tourism. This relationship between both areas creates an increasing interest and curiosity (Sönmez et al., 1994). Tourism is a phenomenon that occurs in people's leisure time and creates an opportunity for people in the selection and testing of recreational activities (Mathison and Will, 1982). Recreation and tourism can use the same types of activities and activity areas. In particular, supply and demand of tourism and outdoor recreation activities share common economic and social characteristics. Thus, recreationists and tourists perform the same/similar activities (Brotherton, 1991). Many of the tourism definitions made fall within the concepts of recreation and leisure. This shows that tourism shares strong characteristics, theories and foundations with the concepts of leisure and recreation (Swarbrooke et al., 2003: 5). Another relationship between recreation and tourism is the effect of recreational activities on tourism. Recreational activities have an impact on the image of the tourist destination. Especially in the context of event tourism, recreational activities provide a serious image to the touristic region (Tassiopoulos, 2005). Another effect of recreational activities on tourism is to reduce the seasonality effect of tourism and to use off-season natural attractions (McDonnell et al., 1998). In the concepts of tourism and recreation, there are conditions such as not aiming for economic gain, the condition of having free time and the individual to realize it on her/his own. However, both phenomena have different characteristics due to their nature. The most important difference between tourism and recreation is place. Because there is an obligation to leave the place of residence in tourism. Tourism has the obligation (relocation) of people to arrive somewhere other than their place of residence. Recreational activities can be carried out even in people's own homes. The second important difference is time. Tourism activity has to stay at least twenty-four hours. Recreational activity can take an hour or less. In addition, it is necessary to stay one night at the destination reached in tourism activities. There is no accommodation requirement for recreation.

The fact that both cases have their own principles has led to the necessity of classifying touristic and recreational activities differently. If it is necessary to explain this situation with an example, while a person's cycling to spend time in her/his leisure time is a recreational activity, it is not a tourist activity since

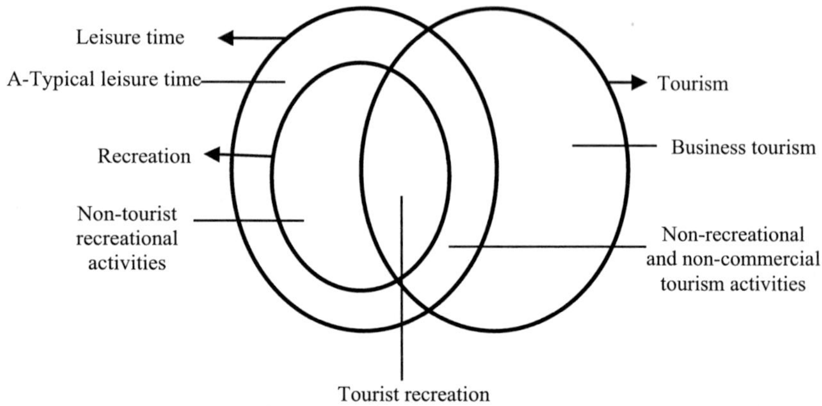

Fig. 2: Tourism, Leisure Time and Recreation

the same action does not take place outside of the place of residence. However, cycling in a different destination where people go for holiday purposes is both a recreational and touristic activity. In this case, it shows recreation as a wider phenomenon that also includes tourism. As a result, it can be said that there is a close relationship between both phenomena (tourism-recreation). However, it can be expressed as a general belief that there are different disciplines (Fig. 2).

The common point in the relationship between tourism, recreation and leisure is tourist recreation activities. Tourist recreation activities include both tourism and recreational features. These common features; It is to be carried out in leisure time, to be carried out with free will, to operate in a destination other than the destination of residence and to have no commercial character. In summer, tourists can benefit from the sea-sand-sun, cultural tours, nature and historical tourism, ecotourism, etc. many activities can be cited as examples of touristic recreation. Activities that are not touristic and recreational but are carried out in leisure time are considered within the scope of a-typical leisure behaviors. These types of behaviors are actions that harm people rather than physical, psychological or social regeneration. Activities/activities such as smoking, hunting, etc. are examples of a-typical leisure behavior. Non-tourist recreational activities; It includes the activities carried out by individuals in the fields of health, sports, education, cultural, artistic, etc. without a touristic purpose. Non-touristic recreational activities; It covers the situations where at least one of the conditions of the accommodation is not fulfilled, or staying at the place of residence for at least 24 hours. Hiking, daily tours, fishing, etc. activities can be given as an example. Recreational and non-commercial tourism activities include touristic activities that do not fulfill at least one of the conditions of free will, non-commercial quality, physical, mental or social renewal. Hunting tourism is the best example of these activities. It includes tourist activities that include working time, such as business tourism activities, incentive tourism and congress tourism. An example of this activity is that people go to business holiday destinations and take a vacation while interviewing.

3 Recreation in Tourism Businesses

Intensive work and city stress that have occurred over time since the industrial revolution cause people to experience psychological wear. Therefore, recreational activities emerge as a need in individuals. The touristic activities that people carry out in order to relax, have fun and make use of their leisure time are also a recreational activity (Stetic, 1992; Yayla and Çetiner, 2019). Recreation activities are under the umbrella of tourism, especially in developed countries. With the regulations made in these countries, recreational opportunities are included in touristic organizations. In addition, in many tourism development projects, touristic opportunities are combined with recreational activities in harmony and succeed when planned (Topuzlu, 1990). People's choice of destination for vacation does not only include rest and travel needs, but also entertainment and animation activities.

When the structure of tourism enterprises is examined, it is seen that it consists of four different groups. These; accommodation, travel, food and beverage and recreation businesses. Recreation establishments; *"It can be defined as businesses that offer people the opportunity to have fun, rest and leisure time"* (Kozak et al., 2007: 61). In the tourism sector, where the competitive environment is extremely important, the recreation and animation activities offered by the enterprises give the enterprises a significant competitive power. For this reason, the importance of the animation department in tourism enterprises is increasing day by day (Fig. 3).

In hotels, the animation department is managed by the animation chief. The Animation chief reports directly to the hotel manager and assistant. Animation chief is subordinated to the director, mini club, sports and game animators. The director's animator is responsible for organizing and planning stage events. Below it is the sound and lighting staff, choreographer and dance team. Mini club animators organize activities related to children's entertainment. Sports animators organize sports activities on water and land resources in different

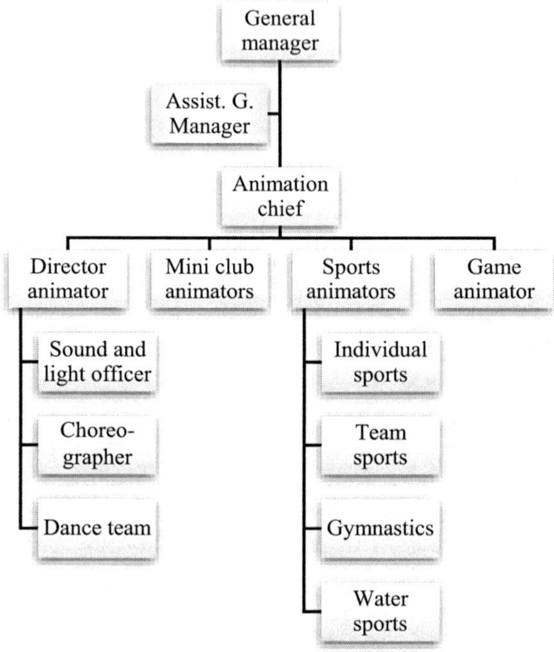

Fig. 3: Animation Department Organization Chart

ways, individually and as a team. Game animators offer short-term games and entertainment activities to hotel guests.

In tourism enterprises, the animation department is the department that is responsible for increasing the leisure quality of the guests coming to the hotel and offering activities with health, sports and cultural entertainment in order to appeal to different age groups (Costa et al., 2004). Animation department is responsible for planning, implementing, managing animation activities and getting feedback. All these stages in animation activities are a process that requires attention and care. Because the quality of the shows and the performance of the shows by talented animators increases customer satisfaction (Mikulic and Prebezac, 2011). The recreational activities at the location as well as the recreational activities offered within the tourism business also add an element of attraction to the business. Disneyland in the Florida and in Paris can be shown as an example (Akat, 2000).

4 Tourism and Recreation in Healthcare

Health tourism is carried out in order to protect and improve the psychological, physiological and sociological health of individuals. In addition, individuals need to travel from where they live for a certain period of time (Ross, 2001; Bennett et al., 2004; Mueller and Kaufmann, 2011). Health and therapeutic recreation is when individuals go to thermal, health, cure and SPA centers in or outside of the city where they live in order to be treated or live healthy, on the advice of doctors and therapists or in line with their individual preferences (Gül, 2014). However, if the individual is sick and goes to the relevant places under the obligation of treatment, this is neither tourism nor recreation. In general, recreational activities that individuals carry out to protect their health or to regenerate; thermalism, climatism, vaulism and SPA activities (Hazar, 2003). Health recreation should be evaluated from two different angles. The first of these is the services provided to healthy individuals within the scope of preventive health services, and the second is the recreation services provided to the disabled and inadequate people within the scope of rehabilitation services (Tütüncü, 2012). Another dimension of the relationship between health and recreation is therapeutic recreation. Therapeutic recreation is a type of recreation that aims to make special groups (patients, elderly, disabled) happy through recreational activities and to help them relax physically and mentally. Therapeutic recreation activities; can be listed as sports, art, music, handicrafts, drama and dances (Karaküçük, 2012).

5 Sports Based Tourism and Recreation Activities

"The type of recreation, which is based on physical exercise or the application of various sports branches for recreational purposes, is called sportive recreation" (Birol and Karaküçük, 2014: 80). Sports tourism is the set of events and relationships resulting from the participation of people who are interested in sports (Batman and Eraslan, 2007).

The most defining difference with regard to sports tourism and sports recreation is the element of displacement. Because in sports recreation, there is no obligation to go anywhere other than the place where you live. Because in sports recreation, there is no obligation to go anywhere other than the place where you live. Sports recreation activities allow individuals to maintain a healthy life in terms of spiritual and mental aspects as well as providing body development naturally (Kement, 2014). Sports recreation activities are basically divided into two as active and passive participation. Active participation is defined as the actual participation of a person in any sportive activity of her/his own will, while passive participation is defined as watching any sports branch competition on-site or in one's own environment.

The sociological dimension of sports recreation is also very important. Because it can be said that sportive recreation is beneficial for individuals' physical and mental development and relaxation as well as socio-culturally. This is because it is accepted that society members improve their healthy communication, sense of integration and acting together (Türkay and Çeviker, 2016). Hazar (2003) defined sports activities that require active and passive participation in indoor and outdoor place as physical recreation and discussed in four sections.

- **Pool, Stream, Sea and Lake Activities**
  - Swimming, underwater diving,
  - Marine and fresh water hunting,
  - Rafting, sailing, windsurfing, water skiing activities,
  - Boat, yacht, pedalo, submarine trips,
  - Pool games (water ballet, water volleyball)

- **Mountain Sports**
  - Climbing, trekking, hunting, snow skiing,
  - Snowmobile and sled trips,
  - Various competitions, etc. Events

- **Activities at Sports Recreation Stations**
  - Dangerous sports (jumping, paragliding)
  - Basketball, football, handball, volleyball,
  - Golf, mini golf, table tennis, boxing, etc. events,
  - Defense sports, wrestling, gymnastics,
  - Riding, archery, etc.

- **Physical Skill Games and Competitions**
  - Children's games and competitions,
  - Games and competitions for young people,
  - Games and competitions for adults,
  - Games and competitions for the third age group

6 Nature Based Tourism and Recreation Activities

People's desire to be intertwined with nature has brought the concept of recreation ecology, which has been discussed by many scientists for years (Kement and Çavuşoğlu, 2017; Çavuşoğlu and Durmaz, 2019). It is an area that examines the ecological impact of recreationists visiting natural areas (Liddle, 1997; Hammit and Cole, 1998). In a broader sense, it is the study of touristic and recreational activities in order to ensure the sustainability of ecologically protected destinations and natural areas (Leung et al., 2008). To be intertwined with flora, fauna and existing cultural texture, it can be defined as a type of tourism that involves traveling to natural areas that are relatively undisturbed and non-polluted (Luzar et al., 1995: 544). Nature-based tourism activities include safari and wildlife watching, nature photography, botanical study or plant watching, camping, fishing, caving, butterfly watching, horseback riding, archeology research (Ingram and Durst, 1989). In addition to these activities, the life to be seen in the places to be visited, the plateau or village houses to be chosen for accommodation, the choice of local food in meeting the food needs are also the characteristics of ecotourism activities (Yeşilyurt et al., 2012). Ecotourism activities are also nature-based recreational activities. Recreational activities carried out for nature are also called eco-recreation (Kement, 2019). Ecocreative activities can be classified as active, passive and causative. For example, ecocreative camping activities; while the activities that individuals will do in any camp area are active, it is passive to participate in the games that take place during a camp activity as a spectator. In addition, receiving training from a person who is an expert in setting up a tent in the camp as a group can be shown as an example of causative participation recreation (Kement, 2015). The difference between ecocreative activity and ecotourism is the duration of the activity and the obligation to go to another destination than the destination. In touristic activities, there is an obligation to exceed at least 24 hours and to go to a destination other than the destination. However, these two requirements are not required in recreational activities.

7 Tourism and Recreation in Cultural and Artistic Events

Leisure time refers to the basic physiological needs of people and the period of time outside of working time. The activities that people perform actively or passively within this time period refer to recreation. Among these activities, those with cultural and artistic content can be defined as cultural and artistic recreational activities.

Cultural and artistic events are often prime examples of recreational activities. Artistic recreation and cultural recreation, which are functional recreation types, should be evaluated differently. Artistic recreation is the use of leisure time by dealing with one or more of the many branches of art. Cultural recreation includes old art and historical artifacts, museums, and activities for other abstract-concrete cultural elements (Karaküçük, 2012). Artistic recreational activities that individuals apply to make use of their leisure time can be exemplified by separating them within their own branches of art. In this context, activities such as weaving, jewelry, shoemaking can be given as examples of recreational activities evaluated in the context of industrial arts. In addition, art branches such as music, painting, sculpture, theater, and ballet, which include the auditory, plastic and rhythmic areas, can be shown as examples of recreational activities. In mixed arts, it can be shown as the most preferred artistic recreational activities such as cinema, opera and folk dances. Carpenter (2006) describes artistic recreation activities by diversifying them under the headings of garden arts, home economy arts, athletic arts, folk arts, cultural arts, fine arts, performing arts, communication arts, perceptual arts and children's arts. While artistic recreational activities rest the soul of individuals, they also contribute to the development of social harmony and community spirit (Fletcher, 1988).

Culture is another area that takes place in all kinds of leisure activities aimed at increasing the knowledge and skills of the participants (Hazar, 2003). Cultural recreation activities include visiting old art and historical artifacts, museums and other cultural activities (Karaküçük, 2012). The most important feature that distinguishes cultural tourism and cultural recreation activities from each other is that the necessity of going outside the place where you live is not a condition in recreational activity. While people reading books at home in their spare time, doing any handicraft can be considered a recreational activity, it is not considered a cultural tourist activity. In participating in cultural and artistic recreational activities, the individual's direct action or only in the audience position determines the passive/active participation in the event. The piano, in which a person plays in his spare time to satisfy his soul at home is an active recreational activity. In addition, his/her piano piece, which he/she listens to in a concert hall or home through a music player, is an artistic/cultural recreational activity with passive participation.

8 Recreational View on Halal Tourism

Halal tourism activities indicating businesses have begun to emerge since the end of 1996 in Turkey, and especially since 2002, and an increase in the number of these businesses and activities diversity has occurred (Bayhan and Keskin, 2020). Since the 2000s, both tourism and recreation enterprises have been involved in various planning studies in terms of physical and active type in line with the demands of the market in order to meet the demands and needs of the market formed within the scope of halal tourism. In this direction, the features that tourism and recreation enterprises should have in terms of entertainment and recreational activities within the concept of halal tourism are determined by the Turkish Standards Institute (TSE) Halal Management System 13683 and the Organization of Islamic Cooperation (OIC) as follows (Soybalı et al., 2017: 266);

- Providing entertainment facilities suitable for the guest profile and not contrary to Islamic morality,
- Having separate recreational activity areas and opportunities for men and women,
- Presentation of separate animation shows for women,
- Lack of casino, gambling and alcoholic nightclub activities,
- Planning pools and beaches separate for men and women,
- Offering free time activities with religious content,
- Offering optional religious education and activity opportunities specific to children.

Similarly, Tekin and Yılmaz (2016) in their study to determine the profile of the halal tourism potential coming to Antalya destination; they determined that the participant group wanted to have entertainment and activities in accordance with Islamic principles. In addition, they determined that the group wanted to participate in different venues for men and women in the organization of religious and Sufi recreational activities during Ramadan.

Activities organized in tourism enterprises or recreation enterprises should coincide with elements such as the country, community, community, custom, value judgment that are carried out within the framework of individual preferences (Gül, 2014). For this reason, organizing recreational activities in line with the expectations of halal tourists who want to spend their leisure time in accordance with Islamic rules and conditions will be of great importance in terms of the satisfaction of customers\tourists\participants.

The acts of worship performed by people based on religion are basically based on free will. Because people have the opportunity to freely choose their religious beliefs. In addition, meetings are held without a commercial purpose. People have the opportunity to renew themselves physically, mentally and socially while praying. All of these show that many of people's activities of worship are recreational. For example, the pilgrimage visits of Muslims are both touristic and touristic.

References

Akat, Ö. (2000). *Pazarlama Ağırlıklı Turizm İşletmeciliği*. Motif Press, 2nd Edition, İstanbul.

Batman, O., and Eraslan, I. H., (2007). Spor Turizmi, Melih Bulu ve Hakkı Eraslan (Ed.), Sürdürülebilir Rekabet Avantajı Elde Etmede Turizm Sektörü: Sektörel Stratejiler ve Uygulamalar, *International Competition Research Association (URAK) Publications,* Publication No: 2007/1, İstanbul, pp.197–220.

Bayhan, I., and Keskin, I. (2020). Türkiye'de Helal Turizm Gelişiminin Helal Endüstrisi Profesyonelleri ile. *Journal of Tourism and Gastronomy Studies, 8*(2), 1413–1438.

Bennett, M., King, B., and Milner, L. (2004). The health resort sector in australia: a positioning study. *Journal of Vacation Marketing, 10*(2), 122–137.

Birol, S. Ş., and Karaküçük, S. (2014). Sportif Rekreasyon Uygulama Politikaları Üzerine Bir Araştırma. *KMÜ Sosyal ve Ekonomik Araştırma Dergisi, 16*(27), 80–86.

Boyacı, C., Aktaş, A., Ehtiyar, R., Aksu. A., Kurcan, F., Başer, G. G., and Köksal, C. D. (1996). Antalya yöresine gelen turistlerin profilini belirleme araştırması. *Anatolia: Turizm Araştırmaları Dergisi, 7*(1), 36–51.

Broadhurst, R. (2001). *Managing Environments for Leisure and Recreation*. Routledge, Taylor & Francis Group, London.

Brotherton, B. (1991). Konaklama, Turizm ve Eğlence Tanımları; Görüş Açıları, Problemler ve Çıkan Sonuçlar. *Turizmde Seçme Makaleler*; 11. TUGEV Publication, İstanbul.

Burkart, J. A., and Medlik, S. (1980). *Tourism Past, Present and Future*. William Heinmann Ltd, London.

Carlson, R. E., Deppe, T. R., and MacLean, J. R. (1972). *Recreation in American Life*. The National Agricultural Library USA.

Carpenter, G. (2006). *Art and Culture,* In, Kinetics, H. (Ed.), Introduction to Recreation and Leisure, Second Edition, pp. 333–352, Leeds, UK.

Costa, G., Tsitskari, E., Tzetzis, G., and Goudas, M. (2004). The factor for evaluating service quality in athletic camps: A case study, *Europen Sport Management Quarterly, 4*, 22–35.

Çavuşoğlu, S., and Durmaz, Y. (2019). Yeşil Davranışlara Karşı Tutumun Ziyaret Niyetine Etkisinde Yeşil İmajın Düzenleyicilik Rolü: Yeşil Oteller Örneği. *Fırat Üniversitesi Sosyal Bilimler Dergisi, 29*(2), 303–315.

Fletcher, J. (1988). *A Look at Leisure. Alberto, Recreation and Parks Recretion on Development Division*. Edmonton, Alberta, May, No: 23.

Gül, T. (2014). *Rekreasyon Olgusuna Genel Yaklaşım*. In, Yaylı A. (Ed.), Rekreasyona Giriş, First edition, Detay publishing, Ankara.

Hazar, A. (2003). *Rekreasyon ve Animasyon*. Detay publishing, Ankara.

Hazar, A. (2009). *Rekreasyon ve Animasyon*. Detay publishing, Ankara.

Hammitt, W. E., and Cole, D. N. (1998). *Wildland Recreation: Ecology and Management*. New York: John Wiley & Sons.

Ingram, C. D., and Durst, P. B. (1989). Nature-oriented tour operators: Travel to developing countries, *Journal of Travel Research, 28*(2), 10–15.

Karaküçük, S. (2012). *Terapatik Rekreasyon*. Gazi Publishing, Ankara.

Kement, Ü. (2014). *Rekreasyonun Diğer Alanlar ile İlişkisi*. In, Yaylı A. (Ed.), Rekreasyona Giriş, First edition, Detay publishing, Ankara.

Kement, Ü. (2015). *Doğaya Yönelik Gerçekleştirilen Rekreatif Etkinliklerde Tüketicilerin Çevre Dostu Rekreasyon Davranışlarının İncelenmesi: Kamping Örneği*. Unpublished Doctoral Thesis. Gazi University Institute of Social Sciences, Ankara.

Kement, Ü. (2019). *Ecorecreation*, In, Yazıcıoğlu, I., Yayla, Ö., and Solunoğlu, A. (Ed.), Current Issues in Tourism and Hospitality Management, SRA Strategic Researches Acdemy, Lithuanian.

Kement, Ü., and Çavuşoğlu, S. (2017). Hafızaya Yönelik Müşteri Deneyimlerinin Müşteri Sadakatine Etkisi: Yeşil Oteller Örneği. *Uluslararası Sosyal ve Eğitim Bilimleri Dergisi, 4*(8), 172–194.

Kozak, N., Kozak, M., and Kozak-Akoğlan, M. (2007). *Genel Turizm İlkeler Kavramlar*, Detay publishing, Ankara.

Kemp, K., and Pearson, S. (1997). *Leisure and Tourism,* Great Britain: Longman Press

Leung, Y. F., Marion, J. L., and Farrell, T. A. (2008). Recreation Ecology in Sustainable Tourism and Ecotourism: A Strengthening Role. In, Stephen F. McCool and R. Neil Moisey (Eds). *Tourism Recreation and Sustainability*, pp. 19–37.

Liddle, M. (1997). *Recreation Ecology: The Ecological Impact of Outdoor Recreation and Ecotourism*. Chapman & Hall Ltd, London, UK.

Lundberg, D. E. (1980). *The Tourist Business*, 4th Edition, Boston: CBI Publishing Inc.

Luzar, E. J., Diagne, A., Gan, C., and Henning, B. R. (1995). Evaluating nature-based tourism using the new environmental paradigm, *Journal of Agricultural and Applied Economics, 27*(2), 544–555.

Mathison, A., and Will, G. (1982). *Tourism Economic, Physical and Social Impact*, Lengman Group LTD, New York.

McDonnell, I., Johnny, A., and O'toole, W. (1998). *Festival and Special Event Management*, John Wiley & Sons Australia Ltd., Brisbane.

Mikulic, J., and Prebezac, D. (2011). Evaluating hotel animation programs at Mediterranean Sun and Sea Resorts: An impact-asymmetry analysis. *Tourism Management, 32*(3), 688–696.

Mueller, H., and Kaufmann, E. L. (2001). Wellness tourism: Market analysis of a special health tourism segment and implications for the hotel industry. *Journal of Vacation Marketing, 7*(1), 5–17.

Olalı, H., and Timur, A. (1988). *Turizm Ekonomisi.* İzmir Ofis Ticaret Matbaacılık.

Parker, S. (1979). Retirement-leisure or not? *Society and Leisure, 2*(2), 329–340.

Ross, K. (2001). *Health tourism: An overview,* HSMAI Marketing Review, (http://www.hospitalitynet.org/news/4010521. html), Accessed date: 06.08.2020.

Soybalı, H. H., Boyraz, M., and Dikmen, H. (2017). Helal Konseptli Konaklama İşletmeleri Kriter ve Uygulamalarının Karşılaştırılması, *1. International Halal Tourism Congress Procedings Book*, pp. 259–275. Alanya.

Sönmez, S. F., Backman, S. J., and Allen, L. (1994). *Managing Tourism Crises: A Guidebook*. Department of Parks, Recreation and Tourism Management, Clemson University.

Stetic, S. (1992). *Health Tourism and Recreation Reality and Perspectives. Proocedings of International Congress on Health and Recreation Management.* Boğaziçi University Press, Istanbul.

Swarbrooke, J., Beard, C., Leckie, S., and Pomfret, G. (2003). *Adventure Tourism: The New Frontier*. Routledge, London, U.K.

Tassiopoulos, D. (2005). *Event Management: A Professional and Developmental Approach*, 2nd Editon, Juta Academic, South Africa.

Tekin, Ö. A., and Yılmaz, E. (2016). İslami Turizm Konseptinde Hizmet Veren Konaklama İşletmeleri Üzerine Bir İnceleme. *Journal of International Social Research, 9*(42), 2046–2058.

Topuzlu, F. K. (1990). Turizm ve rekreasyon olgusunun geleceği. *Anatolia: Turizm Araştırmaları Dergisi, 1*(6), 45–46.

Türkay, H., and Çeviker, A. (2016). Sportif Rekreasyon Kavramına Sosyolojik Bir Bakış, *Atatürk Üniversitesi Beden Eğitimi ve Spor Bilimleri Dergisi, 4*(18), 99–108.

Tütüncü, Ö. (2012). Rekreasyon ve rekreasyon terapisinin yaşam kalitesindeki rolü. *Anatolia: Turizm Araştırmaları Dergisi, 23*(2), 248–252.

Yayla, Ö., and Çetiner, H. (2019). Boş zamanların değerlendirilmesinde etkili olan faktörlerin boş zaman tatminine etkisi. *Turizm Akademik Dergisi, 6*(1), 219–228.

Yeşilyurt, C., Yücel, H., and Yılmaz, B. S. (2012). Yavaş Şehirlerin Rekreasyon Olanakları: Seferihisar'da Rekreasyonel Bir Faaliyet Olarak Eko-Turizm. *Rekreasyon Araştırmaları Kongresi*, pp. 866–877.

Muhammet Cenk BİRİNCİ & Yusuf KARAKUŞ

22 Halal Recreation and Reflections on the Hospitality Industry

1 Introduction

Leisure time and religion are two striking phenomena that recur from the existence of humanity to the present day. These two phenomena have witnessed every stage of human history. Despite the fact that these two concepts have been arranged according to the customs, menstruations and traditions in the social environment, regardless of the development levels of the society, their functions have never changed. While the concept of religion serves to stabilize the existing social order by embellishing generally accepted traditions with holiness, entertainment is seen as a means of expressing oneself.

Since the existence of humanity, the concept of religion has come to the fore in ways such as helping societies to organize their behavior, attitude, consumption, lifestyle in different situations and seeking help from God, whom they assigned supernatural powers when they cannot handle a problem. Religion guides each member of society. By means of religion, humanity is always directed toward what is right and what is considered beautiful. Throughout history, different societies have lived from the existence of humanity to the present day. Every society has its own traditions. These traditions varied according to the lifestyle, the nature of the location, the climate or the rules of the religion they believed in. Religion has an important effect on guiding society's behavior, attitudes and habits. For example, according to the Christianity, while it is acceptable to engage in combat sports in their leisure time, whereas it is not acceptable to engage in combat sports in their leisure time according to the Islam.

The concept of leisure time is as old as the existence of humanity. It is seen that the habits, attitudes and behaviors of societies have the same effect on leisure time as the religion factor is sufficient. From the use of leisure time as a method of punishment in the historical process to its use for entertainment purposes, it has come humanity's way in many different ways. In the Middle Ages, while working according to the Christian belief was considered good, spending time by hacking around was considered infernal. According to the Islamic belief, hacking around and wasting the time is not considered acceptable.

The inescapable link between religion and leisure time appears throughout history. Throughout history, both scientists and clergy have tried to explain both concepts in various ways. In the following parts of the study, we aim to mention the concepts of "Islam" and "Leisure Time".

2 Halal and Haram According to Islam

The concept of halal is a word of Arabic origin and its meaning is permitted. When the concept of halal is mentioned by non-Muslims, food and drinks forbidden by Islam usually come to mind. However, the concept of halal refers to a much broader concept than dietary rules. Battour and Ismail (2016) used the concept of halal as "That which is permitted, with respect to which no restriction exists, and the doing of which the law-giver, Allah, is allowed" in their study. As can be understood from this definition, halal refers to all kinds of actions and attitudes permitted by Islam in the lives of Muslims. From the opposite point of view, everything that Islam prohibits Muslims is defined as haram. The concepts of halal and haram can actually be seen as a set of rules that regulate people's behavior from an Islamic perspective. Within these rules, it regulates the relations of every person with Allah and the relations of people with other people. Humans are social beings and living together is a requirement of their nature. However, in a social community, people's behavior must be within specific rules. Otherwise, it existence of social life is not possible for many reasons such as personal interests and the uncertain boundaries of the concept of freedom. There is a set of rules such as legal rules, ethical rules, moral rules, customs and traditions in order to ensure this order in people's lives. Religious rules are also among the rules that function in this sense. From the point of view of Islam, these rules correspond to the fulfillment of the duty of servitude by obeying halal and harams (Mustafa, 1997: 152)

The source of halal and harams is the Quran, the holy book of Muslims. According to Islamic belief, the Quran is the words of Allah. Prophet Muhammad expresses us Allah's words. Therefore, what is forbidden and what is halal has been determined by Allah. Halal and haram are taught to people through the Quran and the Sunnah (what Allah taught and made Prophet Muhammad say). In the twentieth verse of the Surah Luqman, "Have not you seen that Allah has given all that is in the heavens and the earth to your service and has completed His blessings on you openly or secretly?" He says. In the twenty-ninth verse of the chapter of al-Baqarah, He tells that whatever he created on the earth is for human. On the forehead, the main thing is halal, and haram is an extrinsic situation. In other words, the number of those which are halal is much more than the number of those which are haram. With the Quran and the Sunnah, those which are not stated as haram are made halal.

The above criteria are also valid if the recreational activities of people are halal or haram. According to Islamic belief, the human body is a blessing entrusted to them. For this reason, taking care of themselves is a duty for Muslims and their actions to achieve this are halal. From this point of view, doing sports is halal for Muslims, even a duty if necessary. However, there are other criteria that should be considered at this point. The activity to be carried out should not prevent people from Allah or other duties, should not prevent the rights and freedoms of other people, should not be the basis for the occurrence of other bad things, etc. From the opposite point of view, any action that harms oneself is haram. For example, using drugs or alcohol are behaviors that should be avoided in this sense and defined as haram. From the same point of view, recreational activities that people will do to protect their mental health will be included in the halal category. However, when there is any doubt about whether it is halal or not, Muslims prefer to avoid. (Wilson and Liu, 2010).

3 Times

The concept of time is as old as human history. Scientists, religious figures, scholars, and ulama have sought the answer to this concept. Some tried to explain it abstractly, some tried to prove its concrete existence. Although it has many different definitions according to religions and scientific constructions, time is basically defined as "the period in which a task passed, passes or passing through" (Turkish Language Association). According to Islamic belief, while time is being accepted as its concrete existence, it has been tried to be explained as a spiritual phenomenon (Fig. 1). According to Islamic belief, time is a concept that expresses life, attitudes and behaviors, habits, movements, continuity, stabilities, interactions, processes and events (Okumus, 2011: 11).

Islam divides the concept of time into three dimensions. These are worship, control and contemplation of one's own life, and gaining the sustenance. The working individuals find the opportunity to contemplate with the peace of fulfilling their worldly obligations. (Canan, 1991, p. 79, Aktaran Karaküçük, 2017: 20). The following hadith is a good example of this. "On the Day of Judgment, people are held accountable for some things. These:

- Where and how they consumed their life,
- Where and on which work they spent their youth,
- Where they earned their wealth and where they spent it,
- And the last thing, what they did with their knowledge (Islam with Hadiths, v.7, p.461).

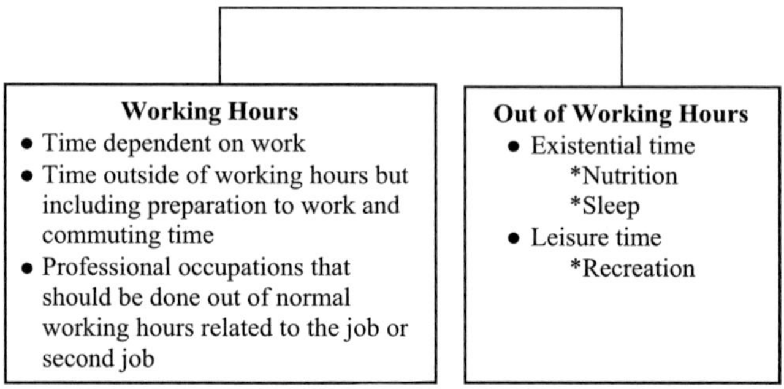

Fig. 1: Time Usage Sections.

Source: Meyer et. al. (1969). Community Recreation a Guide to Its Organization. p. 29

3.1 Islam and Leisure Time

Leisure time is an ancient view that comes from the earliest days of civilization. This concept was first derived from the concept of "schole" in Ancient Greek. It was seen as a cultural value, the opposite of the work (Torkildsen, 1999: 73). Over time, this concept has changed as a period of time left over from work and having fun in line with the individual's wishes by getting rid of the obligations (Encyclopedia of Leisure and Outdoor Recreation, 2004: 279). The concept of leisure time is mentioned in the Surah of Inshirah in the Holy Book of Islam, the Holy Qur'an. In this surah, it has been mentioned on the fifth "So there is an ease with difficulty", the sixth "yes, there is a convenience beside every difficulty", the seventh "so when you finish an important task, immediately put on the other", and the eighth "and only go to your Lord" verses. According to the Islamic belief, there is a prevalent belief that human life is very precious, that time will not come back, so every moment of every moment should be spent in a pleasant and full-bodied manner. The pursuit of worldly pleasures and wasting time is not considered right. It is useful to behave meticulously under the influence of religious culture in the evaluation of leisure time in Islamic belief. Since the individual will be responsible in the hereafter for the worthless time spent in this world, the time spent should be spent qualitatively, not wasted. It is thought that the time spent without any benefit for the individual or humanity is not correct. While doing activities in leisure time benefits the body, and also the time allocated to help other people is considered as not wasted. According to Islamic belief, leisure time is perceived positively and considered as an opportunity to do good and useful work. Since the individuals believe that they will be rewarded in the hereafter for their halal behavior during their worldly life, it is important to examine the concept of halal recreation.

4 Recreation

After the development of the industrial revolution, the increased leisure time gave people the opportunity to be refreshed by participating in activities such as resting, entertainment, sports, hobbies, and getting away from the intensity of daily life. Recreation is a concept expressing these activities which people do in their leisure time (Karakucuk, 2008: 58). When the concept of recreation is examined etymologically, its origin is derived from the Latin words "recreare and recreatio", which mean renewed or restored (Jensen and Naylor, 2000: 2, Torkildsen, 2005: 57, Mclean, Hurd and Brattain-Rogers, 2008: 40). When the literature is examined, it is seen that many definitions have been made on the concept of recreation (Stebbins, 2005, Madrigal, 2006). These are the activities that the individual participates in a non-profit way of free will in order to get away from the stress, fatigue and adverse weather conditions brought about by routine working life (Karakucuk, 1997: 54). Recreation is a whole of hobbies that include willing participation, satisfaction, adventure, creativity and socialization throughout lifetime. (Corbin, 1970: 9). Recreation are activities such as trekking (Kendir, Arslan and Asan, 2019), cultural heritage tour (Yayla, 2019) in which individuals participate actively or passively in their leisure time (How and Carpenter 1985: 15). These are activities that allow the individual to have fun, relaxation, physical, mental and spiritual renewal by means of the activity that enjoyed alone or in groups in the leisure time of the individual (Ardahan, 2016:11). The Turkish language institution defines the concept of recreation in two different ways: "activities in which people voluntarily participate in their leisure time for entertainment and sports purposes" and "bringing a region into a state where people can use for recreation and entertainment" (Turkish Language Institution).

In the light of this information, the concept of recreation can be defined as the whole of activities that an individual participates without feeling any pressure, completely with his/her free will, individually or as a group, actively or passively, mentally, physically or spiritually for the purpose of having fun, resting and finding peace within the leisure time left after the individual comes home after work and fulfills his/her external responsibilities such as his/her family, social environment, beliefs, sleep and food.

If we look at the Qur'an's approach to worldliness in the Surah of An'am: "World life is nothing than a game and entertainment. The abode of the hereafter is better for those who avoid it. Still can't understand?" (An'am 6/32).

4.1 Classification of Recreation

The concept of recreation is shaped according to the needs and demands of the individual and society. Due to the differences in the demands and needs of the society or the individual, it is not possible to collect recreation under a single heading. Therefore, scientists felt the need to classify recreation. Classification of the recreation depends on the type of action (active, passive), in terms of space (open, closed), aims (recreation, cultural, social, sports, tourism, artistic), various criteria (age, mode of participation, time, sociological content), special function (commercial, campus, hospital), local classification (rural, urban), structural classification (challenging, competitive, creative) and lifestyle (serious, indifferent) (Birinci, 2018). The way he or she evaluates leisure time varies according to the person's behaviors and attitudes. The way of participating in recreational activities gives information about the needs of the society's or the individual's, life style, culture, and economic status.

According to Getz (2008), recreational activities are grouped in four main groups. These:

- Cultural celebrations (festivals, carnivals, commemoration, religious events), political and official events (summits, official ceremonies, political events, VIP visits), arts and entertainment (concerts, award ceremonies)
- Business and trade (meetings, agreements, consumer and trade shows, fairs and markets, promotional events), education and science (conferences, seminars, congresses)
- Sports competitions (amateur/professional, audience/participant) and entertainment (recreational games and sports events)
- Special events (weddings, parties, communal [special entertainment, anniversaries, family gatherings]).

The phenomenon of religion has also a place of its own in recreational activities. Participation of individuals in recreational activities is significantly influenced by their economic, social and religious beliefs (Tel, 2007). It is seen that the change in the thoughts in the belief values supports the recreational activities. For example, in Turkey, which was once quite common "playing football is a sin" concept changed and left its place to an opinion that exercising in sports benefits people physically, mentally and socially. (Canan, 1994: 111). This is not the only way that religious beliefs affect leisure time. At the same time, it is seen that behaviors that tolerated in some religions are not considered appropriate or even prohibited in other religions. For example, sports and recreational activities that keep the body fit are generally accepted, while leisure-time habits of west such as alcohol and gambling are prohibited in the religion of Islam (William and Mason 2003). Because, according to the Islamic belief, it is common to believe that alcohol and gambling harm the individual and his/her environment. Therefore, the individual should be free when determining the recreational activity to participate in, but should not be allowed to participate in activities which are not suitable for religious beliefs (Koknel, 1979: 37, Hall and Page, 2006). In other words, it is seen that participation in recreational activities can be preventive if there is no activity in accordance with religious rules.

4.2 Benefits of Recreation

Technology, which is in constant development, has increased the number of desk hours while reducing the work requiring physical strength. (Karakucuk, 2008: 68, Kilbas, 2010: 32). It is seen that in the primitive times of human beings, health problems were not as they are today. It is seen that technological developments bring negativities as well as positive aspects. Especially developments after the 19th century have changed the tempo of life and reduced body and mental activities. (Ilhan, 2010). In other words, as a continuing result of the industrial revolution, mechanization enabled more production to take place per unit time, so people's working hours tended to decrease gradually. Every member of the society has regressed physically, mentally and socially due to the decrease in daily activity (Blair et al., 1984, Blair et al., 1989, Berlin and Goldizt, 1990, Manson et al., 1992, Lee, 1994 and Larson et al., 2004). However, the conscious evaluation of the leisure time of the individual ensures that these impairments are eliminated (Camliyer, 1992, Tezcan, 1994, Philipp, 1997, Durhan, and Karakucuk, 2017). It is known that recreation provides benefits in many similar ways. However, when the literature is examined, the benefits of recreation are gathered under three main headings as physical, spiritual and social (Bammel, 1996, Mensink et al., 1999, Heintzman, 2009, Hung, 2012, Yue, 2012 and Guzel et al., 2014).

Physical Benefit: In parallel with the developing technology, it is seen that the physical movement of people during the day is getting more restricted day by day. The tools used during the day ensure that less effort is spent while doing a job (Yalcinkaya, 2019). For example, the movements of an agricultural society that does not use tractors differ significantly. Restriction or reduction in physical activity adversely affects public health (Schwazer, 1996: 32, Kuter, 2006, Zorba, 2006: 36). Increasing physical activity contributes to the development of muscle strength and body flexibility, strengthening the cardiovascular system, increasing endurance, reaching the ideal body weight and decreasing fat tissue, increasing the well-feeling and well-being. (Ozer and Baltaci 2008: 13, Tunay 2008: 8).

Social Benefit: The individual, who is a social entity, has been in search of reasons such as reducing loneliness, being a component of the social environment, being a family (Human Kinetics, 2006: 51), receiving support, feeling sincerity, and belonging to a group and a group since the existence of humanity (Cordes and Ibrahim, 2003: 49). Recreational activities give the individual the opportunity to win all of these phenomena. A group activity helps the individual to meet his desire to socialize, to belong to a group. For example, an individual who complains of loneliness finds the opportunity to be at the same level with the same status by moving away from the attributes of society such as language, religion, economic status, position, race, nationality, education. According to another example, it enables individuals who cannot come together in a chess tournament organized between departments in a large factory to meet each other.

Spiritual Benefit: The mental health of the individual, which expresses the emotional status, is expressed as the individual's constant harmony with his environment. Being a social entity, the daily intense life rush such as work, home, school in daily life affects the nervous system of the person negatively. In this case, the individual becomes stressed at first and then causes a decrease in his resistance to adverse events in the face of daily events. The mental health of the individual who is exposed to stress decreases. Recreational activities have a great effect on the recovery of mental health by getting away from the stress of daily life. Psychiatrists argue that recreational activities that are attended at regular intervals during the day are good for mental and emotional illnesses. (Kraus, 1977: 75).

As a result, recreational activities help the individual to gain physical, social and mental health.

5 Halal Recreation

Individuals have unlimited alternatives to spend their own leisure time. However, there are unique criteria that affect the decision-making process of each individual in terms of which of these alternatives to choose. Therefore, the perspective of recreational activities may differ for each individual. One of the most important criteria affecting the decisions of individuals is their regional beliefs and the culture formed by the religion they believe. Halal recreation concept emerges as a concept that should be examined at this point. Considering that approximately 1.9 billion Muslims live in the world (worldpopulationreview.com, 2020), it is particularly important to consider the concept of halal recreation in the creation of recreational plans, policies and strategies.

In the light of all these explanations, it would not be wrong to define the concept of halal recreation as follows: whole of recreation activities that individual participate freely, alone or in groups, with pleasure, as active or passive participation, in order to evaluate the time of the individual outside of the activities related to working and existence (such as feeding, sleeping), and all activities aiming at, physical, mental and spiritual renewal, taking into account the requirements of the religion of Islam. These activities may be activities that are regarded as worship in terms of the religion of Islam (such as visiting holy shrines) or activities such as doing sports that individuals of other faiths prefer. At this point, the most basic element to be considered is whether participation in recreational activities causes a situation that can be considered haram.

While leisure activities are thought to be attended by the individual for entertainment purposes at first, it is seen that they always participate in order to find themselves over time. The individual, who first participates in various activities to find himself, then leaves himself in the lap of nature. Because the individual finds the opportunity to purify in nature, to contemplate, to reach the highest level of spiritual peace. In this way, the individual put aside worldly pleasures and tries to find his/her spiritual wealth. Individual finds this richness sometimes in sacred temples suitable for his/her religion and sometimes by staying alone with nature. Individuals who go to sacred shrines and seek collective or individual peace of mind participate in different places and beliefs according to their religion. Some of the shrines which are regarded as sacred are seen as churches for Christians, temples for Buddhists and mosques for Muslims. It is possible to evaluate all leisure activities are generally accepted according to the belief and moral framework, which are not considered haram according to the Islamic belief, within the understanding of halal recreation.

5.1 Halal Recreation and Hospitality Industry Relationship

When we think of recreational activities, a wide range comes to our mind. The products of the tourism and hospitality industry are a well-known category among them. Although the concept of halal has been extensively studied in tourism, this richness is relatively low in terms of hospitality. The number of Muslims worldwide, the size of the hospitality industry and future trends make the reflections of the concept of halal recreation in the hospitality industry important. Although the concepts of halal recreation may seem like a concept related to Muslim destinations at first glance, they come to the fore as they express the preferences of a very important market in terms of non-Muslim destinations (Samori et al., 2016). In other words, the size of the market, the amount of expenditures (Mohsin et al., 2016) and the tendency to grow more in the future (M. M. Battour et al., 2010) highlight the concepts of halal recreation and therefore halal hospitality for the whole world.

One of the factors that direct attention to halal recreation in terms of the hospitality industry is the increasing awareness of consumers in this sense. The increase in the supply for halal recreation also enables individuals with religious sensitivity to shape their preferences accordingly. When deciding on a recreational activity, the fact that there are more halal recreational activities among the alternatives can enable the consumer to be more sensitive in this sense.

6 Conclusion and Suggestions

In today's conditions, many developments have shortened people's daily working hours and increased their purchasing power (Guneren and Karakus, 2015). Therefore, individuals have a significant amount of time apart from working hours and times devoted to vital activities such as nutrition and sleep. Changing consumer preferences in the post-modern world has led to the growth of sectors producing recreational products. When the demand population and the expenditure trends of this population are considered, the concept of halal recreation comes to the fore. Considering the sensitivities of individuals with Islamic beliefs, the concept of halal recreation must first be understood in order to meet their needs. The concept of halal, which is thought to be limited to the allowed food and drink, actually refers to a much wider area. It is necessary to provide necessary hospitality facilities for Muslims, which is an especially important market segment in terms of the hospitality industry. The tourism industry, which is intertwined with the hospitality industry, has many opportunities in this sense. At the destination level, these issues should be designed and they should be presented in a way that complies with the halal concept from transportation to eating and drinking, and from accommodation to entertainment. Currently, there are institutions that issue halal certificates to businesses. Businesses which are considering targeting this market should be involved in this certification process. Employees in non-Muslim destinations should be educated about the concept and practices of halal recreation. It is important that local governments design infrastructure and superstructure facilities to support halal recreation facilities.

References

Ardahan, F. (2016). Her yönüyle rekreasyon. Ankara: Detay Yayıncılık.

Bammel, G., & Burrus Bammel, L. L. (1996). Leisure and human behavior. Dubuque, Iowa: W. C. Brown.

Battour, M., & Ismail, M. N. (2016). Halal tourism: Concepts, practises, challenges and future. In *Tourism Management Perspectives* (Vol. 19, pp. 150–154). Elsevier. https://doi.org/10.1016/j.tmp.2015.12.008

Battour, M. M., Ismail, M. N., & Battor, M. (2010). Toward a halal tourism market. *Tourism Analysis*, 15(4), 461–470. https://doi.org/10.3727/1083542 10X12864727453304

Berlin J. A., & Golditz A. G. (1990). "A Meta Analysis of Physical Activity in the Prevention of Coronary Heart Diseases", Am. J.Epidemiol. 132: 612–628.

Biçer, T. (2008). Sporda Toplam Kalite Yönetimi ve Futbol Uygulamaları. İstanbul: Beyaz Yayınları.

Blair, S. N., Kohl, H. W., Barlow, C. E. (1989). "Physical Activity, Physical Fitness, and All-Cause Mortality a Prospective Study of Helthy Men and Women", J. Am Med Assoc: 262: 2395–2401.

Blair, S. N., Goodyear, N. N., Gibbons, L. W., Cooper, K. N., (1984). "Physical Fitness and İncidence of Hypertansyon İn Healthy Normottensive Men and Women", J. Am. Med Assoc. 252: 487–490.

Canan, İ. (1991). İslam'da Zaman Tanzimi, Cihan Yayınları, İstanbul.

Canan, İ. (1994). Vakti En İyi Değerlendirme Esasları. İslam'da Zaman Tanzimi, İstanbul.

Corbin, H. D. (1970). Recreation leadership. (Third Edition). USA: Prentice Hall Inc, 8–9.

Cordes, K. A., & Ibrahim, H. M. (1999). Applications in recreation and leisure: for today and the future (No. Ed. 2). McGraw-Hill Book Company Europe.

Çamlıyer, H. (1992). "Spor ve Serbest Zaman Etkinliği", M.E.B. Basımevi, Ankara: 63

Durhan, T. A., & Karaküçük, S. (2017). Çocuklarıyla Baby Gym aktivitelerine katılan ebeveynlerin rekreasyon fayda düzeylerinin incelenmesi. Gaziantep Üniversitesi Spor Bilimleri Dergisi, 2(4), 43–53.

Getz, G. (2008). Event tourism: Definition, evolution, and research, Tourism Management, 29 (3) 406 (403–428)

Güneren, E., & Karakuş, Y. (2015). Turizmde Postmodern Pazarlama. In B. Kılıç & Z. Öter (Eds.), *Turizm Pazarlamasında Güncel Yaklaşımla*. Beta Yayım Dağıtım.

Güzel, Ö., Türker, G. Ö., & Türker, A. (2014). Bir Pazarlama Enstrümanı Olarak Rekreasyon Faaliyetlerinin Otellerin Web Sitelerinde Kullanımı: Antalya Destinasyonunda Bir Araştırma, İnternet Uygulamaları ve Yönetimi Dergisi, 5(2), 5–18.

Heintzman, P. (2009). The spiritual benefits of leisure. Leisure/Loisir, 33(1), 419–445.

Howe, C. Z. and Carpenter, G. M. (1985). Programming leisure experiences. New Jersey: Prentice Hall Inc.

Human Kinetics. (2013). Introduction to recreation and leisure. Human Kinetics, Champaing IL.

Hung, H. J. (2012). A study on leisure benefits breaking through leisure activities. *Journal of National Taiwan Normal University*, 3(4):77–92.

İlhan, L. (2010). Hareketsiz yaşamlar kültürü ve beraberinde getirdikleri. Verimlilik Dergisi, 2010(3), 195–210.

Jenkins, J., & Pigram, J. (Eds.). (2004). Encyclopedia of leisure and outdoor recreation. London: Routledge.

Jensen, C. R., & Naylor, J. (2000). Opportunities in recreation and leisure careers. USA.VGM Career Books.

Karaküçük, S. (1997). Rekreasyon Boş Zamanları Değerlendirme-Kavram Kapsam ve Bir Araştırma, Ankara: Seren Ofset, 54.

Karaküçük, S. (2008). Rekreasyon: Boş zamanları değerlendirme. Ankara: Gazi Kitabevi.

Karaküçük, S. (2014). Rekreasyon: Boş zamanları değerlendirme. Geliştirilmiş 7. Baskı Ankara: Gazi Kitabevi.

Kendir, H., Arslan, E., & Asan, H. (2019). Determination of winter trekking routes within the scope of winter tourism potential in Sivas province, *Journal of Recreation and Tourism Research*, 6(3), 294–305.

Köknel, Ö. (1979). Cumhuriyet Gençleri ve Sorunları. Cem Yayınevi. İstanbul.

Kraus, R. G. (1977). Recreation Today, Program, Planning and Leadership, Goodyear Publishing Company, inc., Santa Monica, California

Kuter, M. (2006)." İnsan Uygarlığını Bekleyen İki Tehlike: Hipokinetik Yaşam Şekli ve Küresel Isınma", 9. Uluslararası Spor Bilimleri Kongresi Bildiriler Kitabı, Muğla, Nobel Yayınları: 1086.

Larsson, I., Lissner, L., Naslund, I., Lindroos, A. K., (2004). "Leisure and Occupational Physical Activity İn Relation To Body Mass İndex İn Men And Women", Scandinavian Journal Of Nutrition; Dec, Vol. 48 Issue 4:165.

Lee, I. M. (1994). "Physical Activity, Fitness and Cancer", In: Bouchard, C., Shepard, R. J., Stephens, T., Eds. Physical Activity, Fitness and Health: International, Proceedings And Consensus Statement. Champaign Human Kinetics, 814–831.

Madrigal, R. (2006). Measuring the Multidimensional Nature of Sporting Event Performance Consumption. *Journal of Leisure Research*, 38(3): 267–292.

Manson, L. E., Nathan, D. M., Krolewski, A. S., Stampfer, M. J., Willett, W.C., & Hennekens, C. H. (1992). "A Prospective Study of Exercise and İncidence of Diabetes Among" US Male Physicians, JAMA 268: 63–67.

Martin, W. H., & Mason, S. (2003). Leisure in three Middle Eastern countries. *World Leisure Journal*, 45(1), 35–44.

Mclean, D., Hurd, A. R., & Brattain-Rogers, N. (2008). Kraus' recreation and leisure in modern society. (Eighth Edition). Jones and Barlett Publishers, Sudbury, Massachusetts.

Mensink, G. B. M., Ziese, T., & Kok, F. J. (1999). Benefits of Leisure-time physical activity on the cardiovascular risk profile at older age. *International Journal of Epidemiology*, 28: 659–666.

Meyer, D. H., Brightbill, K. C., & Sessoms, D. H. (1969). Community recreation a guide to its organization. (4th ed.). USA: Prentice Hall.

Mohsin, A., Ramli, N., & Alkhulayfi, B. A. (2016). Halal tourism: Emerging opportunities. In *Tourism Management Perspectives* (Vol. 19, pp. 137–143). Elsevier. https://doi.org/10.1016/j.tmp.2015.12.010

Mustafa, E. (1997). İlahi dinlerin kutsal kitaplarında helal ve haram anlayışı üzerine bir araştırma. *Ankara Üniversitesi İlahiyat Fakültesi Dergisi*, 37(1), 1. https://doi.org/10.1501/Ilhfak_0000000896

Okumuş, E. (2013). Odunpazarı Evleri'nde Ramazan Gecelerinde Boş Zaman Mekanları. Journal of Social Sciences Eskisehir Osmangazi University/ Eskisehir Osmangazi Üniversitesi Sosyal Bilimler Dergisi.

Özer, D., Baltacı, G., & Tedavi, F. (2008). İş yerinde fiziksel aktivite. Klasmat Matbaacılık. Ankara.

Philipp, S. F. (1997). Race, gender, and leisure benefits. Leisure sciences, 19(3), 191–207.

Samori, Z., Md Salleh, N. Z., & Khalid, M. M. (2016). Current trends on Halal tourism: Cases on selected Asian countries. In *Tourism Management Perspectives* (Vol. 19, pp. 131–136). Elsevier. https://doi.org/10.1016/j.tmp.2015.12.011

Schwazer, R. (1996). "Psychologie des Gesundheitsverhal tens." Göttingen, Hogrete:32.

Sevinç, Ö. S., & Aktan, O. (2017). Üniversite Öğrencilerinin Serbest Zaman Etkinliklerinin Belirlenmesi (Manisa Celal Bayar Üniversitesi Eğitim Fakültesi Örneği) Asya Öğretim Dergisi5 (1), 28–43

Stebbins, R. A. (2005). Choice and Experiential Definitions of Leisure.Leisure Sciences; 27(4), 348.

Tel. M. (2007). Öğretim Üyelerinin Boş Zaman Etkinlikleri Üzerine Sosyolojik Bir Araştırma: Doğu Anadolu Örneği Fırat Üniversitesi Sosyal Bilimler Enstitüsü (Yayımlanmamış Doktora Tezi). Elâzığ

Tezcan, M. (1994). "Sosyolojik Açıdan Boş Zamanların Değerlendirilmesi", Atilla Yayınları, Ankara:76

Torkildsen, G. (1999). Leisure and recreation management. London: Spon Press.

Torkildsen, R. (2005). Leisure and recreation management. Fifth Edition. New York: Routledge.

Tunay V. B. (2008). Yetişkinlerde Fiziksel Aktivite. Klasmat Matbaacılık. Ankara.

Wilson, J. A. J., & Liu, J. (2010). Shaping the Halal into a brand? In *Journal of Islamic Marketing* (Vol. 1, Issue 2, pp. 107–123). Emerald Group Publishing Limited. https://doi.org/10.1108/17590831011055851

worldpopulationreview.com. (2020). *Muslim Population By Country 2020*. https://worldpopulationreview.com/countries/muslim-population-by-country/

Yalçınkaya. N. (2019). Üniversite Öğrencilerinin Rekreasyon Fayda Algıları ile Mutluluk Düzeyleri Arasındaki İlişkinin Belirlenmesi Sakarya Uygulamalı Bilimler Üniversitesi Lisansüstü Eğitim Enstitüsü (Yayımlanmamış Yüksek Lisans Tezi). Sakarya.

Yayla, Ö. (2019). Use of culture heritage elements in tourism, in *Current ıssues in tourism and hospitality management* (eds: Yazıcıoğlu, Yayla and Solunoğlu), SRA Publishing, 17–28.

Yue, G. J. (2012). The relationship between leisure involvement of remote control plane activities and leisure benefits. University of the Great Leaf. School of Management. Master Thesis.

Zorba, E. (2006). "Yaşam Boyu Spor", 2. Baskı, Nobel Yayınları, Ankara: 36.

Ebru GÖZEN

23 Intercultural Comparison of Attitudes Toward Leisure Time Activity: Antalya-Kuwait Sample

1 Introduction

Many personnel and environmental factors are effective in forming and shaping attitudes toward leisure time. This situation can be observed better especially in people in grow up in different cultures. The state of Kuwait located on the coast of Persian Gulf in the northeast of the Arabian Peninsula, is ranked 57th in the highest human development category according to the Human Development Index 2019 Report, and is among high income countries. Turkey ranks 59th in the same index (https://tr.wikipedia.org/wiki/%C4 %B0nsani_Geli%C5 %9Fme_Endeksi#%C3 %87ok_y%C3 %BCksek_geli%C5 %9Fme).

Culture which is an integrated behavioral pattern that seperates human beings from non-human creatures and is passed down from generation to generation, is one of the important concepts that shape the behavior of the individual. People living in the same country may have different cultural tendencies. For example, from a variety of countries hosting a large number of immigrants in Kuwait and can be observed within the antinomy as traditional and modernism which has thousands of years of unique and rich culture in Turkey.

When the studies in the literature are examined it is seen that although there are various studies examining people's leisure attitudes particularly a few was seen a comperative study of the attitudes leisure time in Turkey (Akgül, 2011, Baştuğ et al.).

It's aimed to examine the attitudes toward leisure time activities of individuals from the perspective of Turkey/Antalya and State of Kuwait/Kuwait.

1.1 The Concept of Attitude

A psychological definition of attitude identifies a verbal expression as behavior. A sociological definition of attitude looks at verbal expression as an intention to act (Chaiklin, 2011, p. 32). In the early days of attitude research, most investigators accepted as a given that human behavior is guided by social attitudes (Ajzen & Fishbein, 2005, p. 174). Scientific investigation of the attitude which started in the 19th century, means "ready for action" in its Latin origin

(Çöllü ve Öztürk, 2014, p. 376). Attitude is defined as a response to an antecedent stimulus or attitude object by Breckler (1984). Attitudes are not innate, they are acquired through experience. This acquisition is cultural and shapes lives (Mıdık, 2018, p. 20). It has been accepted by many researchers that examining the result of the existing attitude through the triple structure that includes emotion, cognition and behavior (Eagly & Chaiken, 2007). These three attitude elements are mutually interactive. When there is a change in one, the other items also change. According to this assumption if what an individual knows about a subject requires a positive view of that subject (cognitive element) the individual is positive about that subject (affective element). He/she also shows this with words or behaviors (behavioral element) (Ekici, 2012, p. 561). A core assumption underlying the attitude concept is that the three attitude components vary on a common evaluative continuum (Breckler, 1984, p. 1191). However, these components may not always be very interrelated and all these elements have specific verbal or non-verbal responses to the attitude object (Akgül, 2011, p. 7). The cognitive component of an attitude is the individual's thoughts and beliefs about the subject of attitude. Affective component is whether the individual likes or dislikes the subject of attitude. On the other hand, behavioral component is the behaviors of the individual regarding the subject of attitude (Demirci, 2006, p. 17). Measuring attitudes and beliefs in the history of psychology has been limited to self-report questionnaries, semantic scales and emotion thermometers. The use of such measures necessarily assumes that individuals have both the ability and the motivation to report attitudes and beliefs accurately, assumptions that do not stand up to scrutiny (Cunningham, Preacher & Banaji, 2001, p. 163). If we know the attitude of a person on a subject, it is thought that his/her behavior on that subject can also be predicted (Ajzen, 2001). It is important to know whether the attitudes toward the subject are positive or negative as a result of measuring the attitudes for future planning. Likert Scale, Thurstone Scale, Osgood's Differential Meaning Scale, Bogardus' Social Distance Scale and Guttman Scale are used in the literature to measure attitudes.

1.2 The Relationship Between Attitudes and Leisure Time

Karaküçük and Başaran (1996) defined leisure time as a part of time that is out of the working hours and the time that a person has to spare for him/herself and that he/she can freely use and have the choice. The components of the attitude toward leisure time are included in Tab. 1.

As can be seen Tab. 1 people should be aware of some things related to leisure time activities first of all. Also, feelings about leisure time activities are among

Tab. 1: Components of the Attitude toward Leisure. Source: Adapted from Akgül (2011)

Cognitive element of leisure time	Affective element of leisure time	Behavioral element of leisure time
- General knowledge and beliefs individuals acquire about leisure time activities, - Recognizing the relationship between leisure time and other concepts related to this field such as health, happiness and business life, - Being aware of the benefits of leisure time.	- Evaluating leisure time activities and experiences of individuals, - Whether or not they like the leisure time activities they experience, - Emotions felt during leisure time activities.	- Behavioral tendencies of individuals toward leisure time choices, - Frequency of leisure time activities he/she participated in the pas tor today.

the issues that guide the attitude. Finally, behavioral reflections of the attitude toward leisure time activities should be taken into account.

1.3 An Overview of Leisure Time Activities in Turkey and Kuwait

Recreation was used in conversations that Atatürk made in 1923 and 1937 in Turkey (Baştuğ, Zorba, Duyan, ve Çakır, 2017, p. 3896). It is possible to see the regulations and policies relating to recreation in Turkey development plans and government programs since 1963 (Karaküçük, Akgül, Karaman and Ayyıldız, 2016).

Climbing, skiing, hunting, hiking and fishing, outdoor recreational and sport activities are the best example of function that enables communication with nature. Another dimension as recreational activities are neighbor visits, picnic activity, time spend at the coffee house, events, cinema, theatre, park trips, activities in line with traditions and customs in Turkey (Baştuğ et al., 2017). According to the results of a research regarding the use of time in Turkey; the most time is spent to sleeping. It was observed that eating and personal care activities come second after sleep. According to the result of the same research, as the education level increases, the time allocated to social life and entertainment decreases, people between the age of 10–14 spend the most time on hobby, game, sports and nature sports activities, most popular social activity is watching TV, women spends less time on sports activities than men and the most popular sports activities are running and walking (TÜİK, 2015).

When the recreational life in Kuwait is examined it is seen that the wealth of Kuwait originating from oil is also reflected in the life of Kuwaitis. However, one of the biggest handicaps of the country is extreme heat. Therefore people do not spend much time outdoors. They spend most of their time indoors, in large shopping malls (https://ticaret.gov.tr/yurtdisi-teskilati/orta-dogu-ve-korfez/kuveyt/ulke-profili/genel-bilgiler). Kuwait lies at the northwest part of the Arabian Gulf. It is located at the northeast tip of the Arabian Desert (Mahgoub, 2007).

Kuwait has experienced deep and widespread societal changes, including massive urbanization and a decrease in the amount of physical activity associated with work and daily living, in just a few decades. The decrease in physical activity at work and in transportation has not been compensated by leisure time physical activity and sports (Ramadan, Vuori, Lankenau, Schmid and Pratt, 2010, p. 53). Engagement in sports activities has always been one of the priorities according to Kuwaiti culture, wherein, traditional sport activities such as horse racing and falconry are important features of the sport landscape (UNDP, 2017).

Recreation and leisure have become important issues in the small but rich desert country. Despite excellent facilities for recreation the people have still to be convinced of the quality of services offered in their own country (Al Moosa, 1989). Kuwait also holds both horse and camel racing events. Weekend desert camping is also popular. (Mahgoub, 2007). Many Kuwaiti families construct short-term tents in large camping communities in the desert during the winter months. Many of which are furnished with all the comforts of home. Additionally, there is large number of beach chalets, where Kuwaiti's spend time pursuing leisure activities (Paris, 2012, p. 4)

As a result of the study of Al-Baho, Al-Naar, Al-Shuaib, Panicker and Gaber (2016), 19.13 % of the participants of Kuwaiti adults did not perform any physical activity and 38.1 % had low physical activity level, with no significant difference between males and females. The most common perceived barriers were hot weather (75.9 %), work duties, laziness, lack of time, family responsibilities and chronic diseases. In the other study conducted by Ramadan et al. (2010), physical activity levels are low in all population groups. Approximately 70 % of adult men and approximately 80 % of women do not participate in any recreational activities.

2 Method

A descriptive and screening method was applied in this study and the research was carried out in two stages. The attitudes of people who live in Antalya (Turkey) and Kuwait (Kuwait) are examined toward to leisure time. In order to create a theoretical framework on the subject the current informations were presented purposefully by scanning the relevant literature in the first phase of the study. The questionnaire developed by Akgül (2011) was applied to the sample group in the second phase of the study.

2.1 Participants

The universe of this study consists of people residing in Antalya and Kuwaiti between May and October 2019. The sample of the study consists of 298 people living in Antalya and volunteering to participate in the study and 203 people living in Kuwaiti and volunteering to participate in the study. The total number of participants is 501 people. Tabachnick and Fidell's method was used to understand the adequacy of the appropriate sample size. According to Tabachnick and Fidell (2001), the sample size should be at least 5 times the expressions used in the research. When calculated from 66 expressions of the measurement tool used in the study, the sample size should be at least 330. According to this statement, the number of participants is considered sufficient.

2.2 Data Collection Tools

The data collection tool used in the study consists of two parts. There are expressions to determine the demographic information of the participants in the research group at the first part. The second part consists of "Leisure Atitude Scale (LAS)" developed in 1982 by Ragheb and Beard and adapted to Turkish by Akgül and Gürbüz (2010) to measure the attitudes toward leisure time of people living in Antalya and Kuwait. The original English version was used in Kuwait and the Turkish adaptation by Akgül and Gürbüz (2010) was used in Antalya.

LAS, developed by Ragheb and Beard (1982) consists of 3 evenly distributed sub-dimensions (cognitive, affective and behavioral) and a total of 36 statements. The scale is a 5-point Likert type (strongly disagree – strongly agree). The lowest score that can be obtained from the scale is 36 and the highest score is 180. The internal consistency coefficient of the scale was .97 for the whole scale, .92 for the affective dimension, .81 for the cognitive dimension and .91 for the behavioral dimension in the study by Akgül and Gürbüz (2010). The reliability coefficient of the scale was .94, .89 for the affective dimension, .92 for the cognitive dimension and .90 for the behavioral dimension. Scales with a reliability coefficient of .70 and above are considered reliable (Büyüköztürk, 2010). In this case, it can be said that the whole scale and sub-dimensions used in the study are reliable enough. EFA was used to determine the construct validity of the scale. The results of KMO and Barlett tests performed to determine the appropriateness of the scale for factor analysis were checked. KMO value should be greater than .80 in a very good factor analysis, but values higher than .50 are also acceptable (Turanlı, Taşpınar Cengiz and Bozkır, 2012, p. 49). Barlett sphericity test shows whether there is a sufficient relationship between variables such as KMO. When the p value of Barlett test is less than 0.05 significance level, there is a sufficient level of relationship between variables for factor analysis. If the result of the test is not meaningful, it is not suitable for factor analysis (Alp, 2019). KMO values of LAS are as follows: .873 for cognitive dimension, .836 for affective dimension and .878 for behavioral dimension ($p<.05$). This result shows that the relationship between variables is perfect (Durmuş, Yurtkoru and Çinko, 2013). After deciding the suitability of the data for factor analysis, factor load values and explained variance rates were calculated (Tab. 2). The factor load value is expected to be at least .30. Items with a load value of .60 and above are defined as high, and between .30-.59 regardless of the sign of the factor load as medium size (Alp, 2019).

The coefficient of .40 was taken as the cut-off point for the items in the standardization study of the scale. Analysis results support the 3-factor structure (cognitive, affective and behavioral) for the LAS as in the original scale and explain 52 % of the scale for 492 participants. Since it is not possible to reach very high variance rates for social sciences, variance rates between 40 % and 60 % are considered sufficient. This ratio can be considered high.

Tab. 2: Factor Loads of the LAS

Factors	**Factor Loadings**	**Variance Explained**
Cognitive Dimension		**33.942**
1: Engaging in leisure activities is a wise use of time.	.623	
2: Leisure activities are beneficial to individuals and society.	.673	
3: People often develop friendships in their leisure.	.621	
4: Leisure activities contribute to one's health.	.686	
5: Leisure activities increase one's happiness.	.558	
6: Leisure increases one's work productivity.	.643	
7: Leisure activities help to renew one's energy.	.752	
8: Leisure activities can be a tool for self-improvement.	.692	
9: Leisure activities help individuals to relax.	.753	
10: People need leisure activities.	.678	
11: Leisure activities are good opportunities for social contacts.	.671	
12: Leisure activities are important.	.741	
Affective Dimension		**10.370**
13: When I am engaged in leisure activities, the time flies.	.871	
14: My leisure activities give me pleasure.	.803	
15: I value my leisure activities.	.515	
16: I can be myself during my leisure.	.531	
17: My leisure activities provide me with delightful experiences.	.613	
18: I feel that leisure is good for me.	.883	
19: I like to take my time while I am engaged in leisure activities.	.518	
20: My leisure activities are refreshing.	.572	
21: I consider it appropriate to engage in leisure activities frequently	.514	
22: I feel that the time I spend on leisure activities is not wasted.	.530	
23: I like my leisure activities.	.604	
24: My leisure activities absorb or get my full attention.	.876	
Behavioral Dimension		**8.118**
25: I do leisure activities frequently	.450	
26: Given a choice I would increase the amount of time I spend in leisure activities	.433	
27: I buy goods and equipment to use in my leisure activities as my income allows.	.659	
28: I would do more new leisure activities if I could afford the time and money.	.715	

(*continued on next page*)

Tab. 2: Continued

Factors	Factor Loadings	Variance Explained
29: I spend considerable time and effort to be more competent in my leisure activities.	.525	
30: Given a choice I would live in environment or city which provides for leisure.	.455	
31: I do some leisure activities even when they have not been planned.	.631	
32: I would attend a seminar or a class to be able to do leisure activities better.	.776	
33: I support the idea of increasing my free time to engage in leisure activities	.479	
34: I engage in leisure activities even when I am busy.	.744	
35: I would spend time in education and preparation for leisure activities.	.744	
36: I give my leisure high priority among other activities.	.617	
Total Variance Explanation Rate (%): 52.430		

2.3 Analysis of Data

The data of the participates with distorting the homogeneity of covariances (Box's Test of Homogenity of Covariance Matrices), normality of distributions and equality of error variances (Levene's Test of Equality of Error Variances), missing items and each item filling in 1 or 5 were excluded from the calculation in the statistical method of Multivariate ANOVA (MANOVA) which was made to determine the differences in attitudes toward leisure time before analyzing the data. As a result, data from 492 of the 501 participants were included in the analysis. Descriptive statistical methods were used in the analysis of demographic data. Multivariate ANOVA (MANOVA) statistical method was used to find differences in attitudes toward leisure time according to the variables of country/city (Antalya-Kuwait), gender, age and belief effect. The Chi-Square statistical method was used to determine the cross table distributions according to country/city and country/city according to gender variables and the significance of these distributions according to country/city and gender variables.

3. Results

In this part of the study, the data related to the demographic characteristics of the participants in the sample, the validity and reliability results of the scale, findings and comments obtained from the comparison analysis of the attitudes toward leisure time activity of people living in Antalya and Kuwait are included. The frequency distributions and percentage values of the demographic data of the participants are included in Tab. 3.

It is seen that 16,7 % of the participants living in Antalya are female and 42,9 % are male while 10,9 % of the participants living in Kuwait are female and 29,5 % are male. 59,5 % of the total participants live in Antalya and 40,5 % live in Kuwait. According to the monthly income variable, the people living in Antalya (21,1 %) and Kuwait mostly (21,1 %) have a monthly income between 2001–4000 TL. The majority of the participants living in Antalya (22,7 %) and Kuwait (16 %) answered "no" to the situation that their belief and values affect leisure activities.

The results of the scores of the participants from the sub-dimensions of the LAS according to demographic variables are given in Tab. 4. The significance value fort his table is obtained by the number of dependent variables in the analysis.

Tab. 3: Frequency Distributions and Percentage (%) Values of Demographic Data

Variables			N	%
Gender	Antalya	Female	82	16.7
		Male	211	42.9
	Kuwait	Female	54	10.9
		Male	145	29.5
Country/City	Antalya		293	59.5
	Kuwait		199	40.5
Monthly Income	Antalya	2000 TL and less	50	10.2
		2001–4000 TL	104	21.1
		4001–6000 TL	86	17.5
		6001 TL and more	53	10.8
	Kuwait	2000 TL and less	20	4.1
		2001–4000 TL	102	20.6
		4001–6000 TL	38	7.7
		6001 TL and more	39	8.0
Belief	Antalya	Yes	103	21.0
		No	112	22.7
		Partly	78	15.8
	Kuwait	Yes	48	9.8
		No	79	16.0
		Partly	72	14.7

This process is known as Bonferroni correction (Yurtçu, 2020, p.145). Since there are three dependent variables in the study, the significance value of 0,05 was obtained as 0,016 according to the Bonferroni correction.

As seen in Tab. 4, significance levels were obtained according to dependent variables and 0,016 significance value. Cognitive dependent variable varies significantly according to country variable and belief * country interaction. The affective dependent variable differs significantly by country variable. Behavioral dependent variable also shows a significant difference according to belief * monthly income interaction. It is concluded that gender and monthly income variables do not make a significant difference on dependent variables. Tab. 5 shows which groups differ in the country independent variable that creates a significant difference on the dependent variables.

As seen in Tab. 5, although the mean difference of the participants living in Kuwait is higher than Antalya in Behavioral variable this difference was not

Tab. 4: Interaction Between Sub-Dimensions of the LAS and the Source of Variability

Source	**Sub-Dimensions**	**ss**	**df**	**Mean Square**	**F**	**Sig.**	η_p^2
Gender	Cognitive	,044	1	,044	,178	,674	,000
	Affective	,658	1	,658	1,607	,206	,004
	Behavioral	,522	1	,522	1,084	,298	,002
Belief	Cognitive	,724	2	,362	1,469	,231	,007
	Affective	,974	2	,487	1,189	,305	,005
	Behavioral	,633	2	,317	,658	,518	,003
Monthly Income	Cognitive	1,530	3	,510	2,070	,103	,014
	Affective	2,963	3	,988	2,411	,066	,016
	Behavioral	1,144	3	,381	,793	,498	,005
Country/ City	Cognitive	27,122	1	27,122	110,069	**,000***	,198
	Affective	5,574	1	5,574	13,607	**,000***	,030
	Behavioral	,041	1	,041	,086	,770	,000
Gender* Belief	Cognitive	,363	2	,182	,737	,479	,003
	Affective	,089	2	,044	,108	,897	,000
	Behavioral	,703	2	,352	,731	,482	,003
Gender* Monthly Income	Cognitive	,803	3	,268	1,086	,355	,007
	Affective	,720	3	,240	,586	,625	,004
	Behavioral	,377	3	,126	,261	,853	,002
Gender* Country/ City	Cognitive	,385	1	,385	1,564	,212	,004
	Affective	,233	1	,233	,569	,451	,001
	Behavioral	,114	1	,114	,236	,627	,001

Tab. 4: Continued

Source	Sub-Dimensions	ss	df	Mean Square	F	Sig.	η_p^2
Belief * Monthly Income	Cognitive	1,675	6	,279	1,133	,342	,015
	Affective	4,128	6	,688	1,679	,124	,022
	Behavioral	7,069	6	1,178	2,448	**,024***	,032
Belief * Country/ City	Cognitive	1,654	2	,827	3,356	**,036***	,015
	Affective	1,924	2	,962	2,349	,097	,010
	Behavioral	2,132	2	1,066	2,215	,110	,010
Monthly Income * Country/ City	Cognitive	,639	3	,213	,864	,460	,006
	Affective	3,189	3	1,063	2,595	,052	,017
	Behavioral	3,485	3	1,162	2,414	,066	,016
Gender* Belief* Monthly Income	Cognitive	,457	6	,076	,309	,932	,004
	Affective	2,657	6	,443	1,081	,373	,014
	Behavioral	2,946	6	,491	1,020	,411	,014
Gender*Belief * Country/ City	Cognitive	,615	2	,308	1,248	,288	,006
	Affective	,373	2	,187	,456	,634	,002
	Behavioral	,564	2	,282	,586	,557	,003
Gender * Monthly Income * Country/ City	Cognitive	,137	3	,046	,186	,906	,001
	Affective	1,931	3	,644	1,571	,196	,010
	Behavioral	2,052	3	,684	1,422	,236	,009
Belief * Monthly Income * Country/ City	Cognitive	,733	6	,122	,496	,811	,007
	Affective	3,771	6	,629	1,534	,165	,020
	Behavioral	3,262	6	,544	1,130	,344	,015
Gender* Belief * Monthly Income * Country/ City	Cognitive	,095	5	,019	,077	,996	,001
	Affective	3,431	5	,686	1,675	,139	,018
	Behavioral	1,608	5	,322	,668	,648	,007
Error	Cognitive	109,651	445	,246			
	Affective	182,298	445	,410			
	Behavioral	214,135	445	,481			
Total	Cognitive	8788,347	492				
	Affective	8684,49	492				
	Behavioral	7696,32	492				

Tab. 5: Differences of Dependent Variables According to Country

Dependent Variable	I (Country)	J (Country)	Mean Difference (I-J)	Std.Error	Sig.
Cognitive	Antalya	Kuwait	,669	,060	,000
	Kuwait	Antalya	-,669	,060	,000
Affective	Antalya	Kuwait	,291	,078	,000
	Kuwait	Antalya	-,291	,078	,000
Behavioral	Antalya	Kuwait	-,033	,084	,695
	Kuwait	Antalya	,033	,084	,695

found significant. The mean difference of the participants living in Antalya are higher than Kuwait and this difference was found significant in the cognitive and affective variables. In this case, it can be said that Cognitive and Affective variables differ by country.

According to Tab. 6, the participant living in Antalya mostly preferred ball activities while the participants living in Kuwait preferred Far East sport activities. A significant difference was found in the frequency of participation of the participants living in Antalya and Kuwait in all physical activities. The participants living in Antalya mostly preferred eating out while the participants living in Kuwait preferred visiting relatives. A significant difference was found for both Antalya and Kuwait participants in all activities except eating out regarding the frequency of participation in social events. It was observed that the participants living in Antalya mostly preferred reading books and newspapers in terms of cultural and artistic events while the participant living in Kuwait preferred the museum visit. A significant difference was found in the frequency of participation of the participants living in both Antalya and Kuwait in all cultural and artistic activities. Regarding participation in touristic activities, it has been observed that the participants living in Antalya mostly prefer long holidays while the participants living in Kuwait prefer recreation trips. A significant difference was found in the frequency of participation of the participants living in both Antalya and Kuwait in all touristic activities. Except for physical, social, cultural-artistic and touristic activities, it was observed that the participants living in Antalya and Kuwait mostly preferred to listen to music activity. The participants living in Antalya also preferred to watch TV and video activity and the participants living in Kuwait preferred the coffehouse-club activity and shopping activity.

Tab. 6: Chi-Square Values of Participated Leisure Time Activities According to Country

Activity Type		Country/city	N	Participation Frequency			X2	p
				Never	Sometimes	Often		
Physical Activities	**Walking**	Antalya	293	69	184	40	19,729	,000*
		Kuwait	199	18	137	44		
	Jogtrot	Antalya	293	128	135	30	111,914	,000*
		Kuwait	199	23	77	99		
	Aerobic-Yoga	Antalya	293	211	56	26	160,594	,000*
		Kuwait	199	28	129	42		
	Far East Sports	Antalya	293	261	27	5	271,403	,000*
		Kuwait	199	33	65	101		
	Weight Training	Antalya	293	157	96	40	101,884	,000*
		Kuwait	199	19	114	66		
	Racket Sports	Antalya	293	218	65	10	221,728	,000*
		Kuwait	199	22	81	96		
	Ball Activities	Antalya	293	109	126	58	37,702	,000*
		Kuwait	199	25	109	65		
	Outdoor Sports	Antalya	293	74	84	30	132,848	,000*
		Kuwait	199	25	79	95		
	Fishing	Antalya	293	176	97	20	141,388	,000*
		Kuwait	199	24	92	83		

(continued on next page)

Tab. 6: Continued

Activity Type		Country/city	N	Participation Frequency			X2	p
				Never	Sometimes	Often		
Social Activities	**Relatives Visit**	Antalya	293	37	178	78	59,955	,000*
		Kuwait	199	7	71	121		
	Courses	Antalya	293	191	85	17	176,830	,000*
		Kuwait	199	11	141	47		
	Foundation/ Association	Antalya	293	233	51	9	193,168	,000*
	Affairs	Kuwait	199	36	92	71		
	Eating Out	Antalya	293	15	150	128	3.436	,179
		Kuwait	199	14	114	71		
	Picnic	Antalya	293	67	191	35	93,806	,000*
		Kuwait	199	14	84	101		
Cultural-Artistic Activities	**Reading Book/ Newspaper**	Antalya	293	45	134	114	18,104	,000*
		Kuwait	199	12	126	61		
	Cinema	Antalya	293	25	176	92	20,283	,000*
		Kuwait	199	21	79	99		
	Concert	Antalya	293	77	141	75	10,625	,005*
		Kuwait	199	28	109	62		
	Theater	Antalya	293	155	112	26	133,309	,000*
		Kuwait	199	25	73	101		
	Exhibitions/Kermes	Antalya	293	207	72	14	191,478	,000*
		Kuwait	199	18	115	66		
	Museum Visit	Antalya	293	142	117	34	141,608	,000*
		Kuwait	199	20	62	117		

Touristic Activities	**Out of City Trips**	Antalya	293	69	149	75	11,265	,004*
		Kuwait	199	23	119	57		
	Rest Trips	Antalya	293	119	116	58	63,580	,000*
		Kuwait	199	22	85	92		
	Long Holidays	Antalya	293	60	109	124	12,341	,002*
		Kuwait	199	26	105	68		
Other Activities	**Watching TV-Video**	Antalya	293	12	78	203	48,141	,000*
		Kuwait	199	17	107	75		
	Listen to Music	Antalya	293	10	68	215	38,560	,000*
		Kuwait	199	13	95	91		
	Computer Games	Antalya	293	112	112	69	42,924	,000*
		Kuwait	199	23	102	74		
	Gardening	Antalya	293	205	71	17	184,337	,000*
		Kuwait	199	22	92	85		
	Shopping	Antalya	293	42	169	82	10,136	,006*
		Kuwait	199	17	101	81		
	Coffeehouse/Club	Antalya	293	105	131	57	66,465	,000*
		Kuwait	199	14	96	89		

4 Conclusion and Suggestions

In this study which examines the he leisure attitudes of people living in Antalya and Kuwait; attitudes toward leisure time were found difference in country/city variable, belief*monthly income interaction variable and belief*country/city interaction variable according to cities of Antalya and Kuwait. There was no significant difference in gender, belief and monthly income variables.

A significant difference was observed in the cognitive and affective sub-dimensions of the LAS in the country/city variable while a significant difference was not observed in the behavioral sub-dimension. As Ekici (2012) stated, if the cognitive element enables the individual to approach that activity positively what they know about leisure time activities, the individual has a positive awareness and feeling for that activity. According to the result of this research, the awareness and sense levels of the participants living in Antalya and Kuwait for leisure time activities are different. It was seen that the interaction of belief and monthly income caused a significant difference only in the behavioral sub-dimension and the interaction of belief and country/city variables also caused a significant difference in the cognitive dimension. It was examined which groups caused a significant difference in attitudes toward leisure time activities in the country/city variable. The mean difference of the participants living in Antalya was higher than the participants living in Kuwait (in the other words, they have more information about leisure time activities, can analyze, synthesize and evaluate more on these issues). It is important to be informed about individuals' cognitive attitudes as stated by Akgül (2011). People living in Kuwait need to be informed more about leisure time activities. In addition, they should be made aware of the relationship between leisure time and concepts such as health, happiness and business life. Cognitive deficit can also be caused by the lack of awareness about the benefits of participating in leisure time activities. In this sense, increasing public awareness will yield beneficial results. As stated in Al-Moosa (1989), although there are excellent practices for recreation in the country, people need to be persuaded about the quality of services provided in their own country. It was found that the mean difference of the participants living in Antalya was higher than the participants living in Kuwait in the Affective sub-dimension. In the affective field factors such as the individual's feelings, attitude, interest, love, etc. are effective in the affective field (Kablan, 2013). According to the result, it can be said that the participants living in Antalya feel more positive about leisure time activities, they are more interested and have a more positive attitude toward such activities than the participants living in Kuwait. It was observed that, the mean difference of the participants living in Kuwait were higher in the behavioral sub-dimension but this difference was not significant. In this case, it is possible to say that the two cities differ from each other in attitude toward leisure time activities in cognitive and affective sub-dimensions.

Differences were also observed regarding the frequency of participation in leisure time activities of the participants living both in Antalya and Kuwait. It is observed that participants living in Kuwait generally participate in physical activities more often than whose living in Antalya. This result differs from the result of the study conducted by Ramadan, Vuori, Lankenau, Schmid and Pratt (2010). This difference can be explained by the fact that the participation of Kuwaiti people in physical activities may have increased in the nine years between the two studies. The Kuwaiti people continue their traditional life style in physical leisure time activities. Most of these activities are nature activities (riding horse, camel racing, desert safari and desert camping). As a result of this study, it was observed that the participants living in Kuwait participated in nature (outdoor) activities more frequently than the participants living in Antalya. However Antalya has a very rich range of nature sports. At this point, policies should be determined to increase public participation in nature sports. Participants living in Antalya stated that while they frequently participate in ball activities, they do not participate in Far East sports activities. On the contrary, the participants living in Kuwait stated that they frequently participated in Far East sport activities. Regarding social activities, participants living in Antalya preferred to eat outside more frequently, while the participants living in Kuwait preferred the activity of visiting relatives more frequently. It can be said that this situation is a reflection of the fact that the Kuwaiti people also maintain a traditional life style. Except for physical, social, touristic, cultural and artistic activities, the participants living in both cities preferred listening to music activity more frequently. In addition, while the participants living in Kuwait also preferred the coffehouse/club activity, the participants living in Antalya preferred to participate in TV/Video watching activities in accordance with the data of TUIK (2015).

According to this results, these criteria should be taken into account in planning leisure time activities for the people living in Antalya and Kuwait. This study can be a guide for private and public institutions in planning leisure time activities.

The study was carried out only with the data collected from Antalya and Kuwait cities. Similar studies should be repeated in different countries and comparatively on a World scale to have an opinion.

References

Ajzen, I. (2001). Nature and Operations of Attitudes, *Annual Review of Psychology*, 52, ss.27–58.

Ajzen, I. & Fishbein, M. (2005). The Influence of Attitudes on Behavior, D. Albarracín, B. T. Johnson, M. P. Zanna (Eds.). The handbook of attitudes, pp. 173–221, Mahwah, NJ: Lawrence Erlbaum Associates.

Akgül B. M. & Gürbüz, B. (2010). *Bos Zaman Tutum Ölçegi: Geçerlik ve Güvenirlik Çalısması,* 11. Uluslar arası Spor Bilimleri Kongresi, Antalya.

Al-Baho, A. K., Al-Naar, A., Al-Shuaib, H., Panicker, J. K. & Gaber, S. (2016). Levels of Physical Activity among Kuwaiti Adults and Perceived Barriers, *The Open Public Health Journal*, 9, pp. 77–87.

Al Moosa, A. R. A. (1989). Recreation: The Kuwait Model, *Journal Tourism Recreation Model*, 14(2), Published online: 11 Nov 2014, pp. 49–56.

Alp, G. (2019). Otel Çalışanlarının Stres ve Bireysel Performans İlişkisinin İncelenmesi, *Anadolu Üniversitesi Sosyal Bilimler Dergisi*, 19(2), pp. 475–496.

Büyüköztürk, Ş. (2010). *Sosyal bilimler için veri analizi el kitabı.* Ankara: Pegem Akademi.

Chaiklin, H. (2011). Attitudes, Behavior and Social Practice, The *Journal of Sociology & Social Welfare*, 38(1), pp. 31–54.

Cunningham, W. A., Preacher, K. J. ve Banaji, M. R. (2001). Implicit attitude measures: Consistency, stability, and convergent validity. *Psychological science*, 12(2), ss.163–170

Çöllü, E. F. ve Öztürk, Y. E. (2014). Örgütlerde İnançlar-Tutumlar Tutumların Ölçüm Yöntemleri ve Uygulama Örnekleri Bu Yöntemlerin Değerlendirilmesi, *Selçuk Üniversitesi Sosyal Bilimler Meslek Yüksekokulu Dergisi*, 9(1–2), ss.373–404.

Demirci, H. G. (2006). Ticaret Meslek ve Anadolu Ticaret Meslek Liseleri Bilgisayar Programcılığı Bölümü Öğrencilerinin İnternete Yönelik Tutumları ile İnternet Ağ ve Sistemleri Dersindeki Akademik Başarıları Arasındaki İlişki, Çukurova Üniversitesi, Sosyal Bilimler Enstitüsü, Yüksek Lisans Tezi, Adana.

Durmuş, B., Yurtkoru, E. S. ve Çinko, M. (2013). *Sosyal Bilimlerde SPSS'le Veri Analizi*. İstanbul: Beta.

Ekici, T. (2012). Bireysel Ses Eğitimi Dersine Yönelik Tutum Ölçeğinin Geliştirilmesi, GEFAD/GUJGEF, 32(3), ss.557–569.

Güvenç, B. (1972). *İnsan ve Kültür: Antropolojiye Giriş.* Ankara: Ayyıldız Matbaası (Türk Sosyal Bilimler Derneği Yayınları).

Kablan, Z. (2013). *Hedef Belirleme: Bilişsel, Duyuşsal ve Devinişsel Alanlar.* Hasan Şeker (Ed), *Eğitimde Program Geliştirme: Kavramlar, Yaklaşımlar* içinde (ss.1–36). Anı Yayıncılık.

Karaküçük, S., Akgül, B. M., Karaman, M., Ayyıldız, T. (2016). Rekreasyon Bilimi: Türkiye'de Rekreasyonel Durum, Ankara: Gazi kitapevi.

Mahgoub, Y. (2007). Towards Sustainable Desert Eco-Tourism in Kuwait: Learning from the Regional Experince, Open House International, 32(4) December, pp. 75–82.

Mıdık Ö. (2018). *Tutumun Ölçme ve Değerlendirmesi.* Balkan A. (Ed.). Tıp Eğitiminde Ölçme ve Değerlendirme. 1. Baskı. Ankara: Türkiye Klinikleri. ss.19–26.

Paris, C. M. (2012). Failaka Island: A Discussion of the Opportunities and Challenges for the Future of Tourism Development in Kuwait, 3rd Gulf Resort Meeting, Workshop 6: International Tourism Development in the GCC Countries: Opportunities and Challenges, pp. 1–22.

Ragheb M. G. & Beard, J. G. (1982). Measuring Leisure Attitude. *Journal of Leisure Research*, 14 (2), ss.155–167.

Ramadan, J., Vuori, I., Lankenau, B., Schmid, T. & Pratt, M. (2010). Developing a National Physical Activity Plan: the Kuwait Example, *Global Health Promotion*, 17(2), pp. 52–57.

Tabachnick, B. G. & Fidell, L. S. (2001). *Using Multivariate Statistics.* Boston: Ally and Bacon.

Turanlı, M., Taşpınar Cengiz, D. ve Bozkır, Ö. (2012). Faktör Analizi ile Üniversiteye Giriş Sınavlarındaki Başarı Durumuna Göre İllerin Sıralanması, *İstanbul Üniversitesi İktisat Fakültesi Ekonometri ve İstatistik Dergisi*, 17, pp. 45–68.

TÜİK (Türkiye İstatistik Kurumu). (2015). Zaman Kullanım Araştırması 2014–2015, http://www.tuik.gov.tr/PreHaberBultenleri.do?id=18627 accessed: 09.08.2020.

UNDP (2017). Fast Facts United Nations Development Programme, Online Document, https://www.kw.undp.org/content/kuwait/en/home/operations/projects/democratic_governance/institutional-strengthening-and-strategic-development-of-public-.html accessed: 09.08.2020.

Yurtçu, M. (2020). *Çok Değişkenli Varyans ve Kovaryans Analizi Manova ve Mancova.* Cem Oktay Güzeller (Ed.), *Herkes İçin Çok Değişkenli İstatistik* içinde (ss.119–159). Ankara: Maya Akademi.

https://tr.wikipedia.org/wiki/%C4 %B0nsani_Geli%C5 %9Fme_Endeksi#%C3 %87ok_y%C3 %BCksek_geli%C5 %9Fme accessed: 07.08.2020.

https://ticaret.gov.tr/yurtdisi-teskilati/orta-dogu-ve-korfez/kuveyt/ulke-profili/genel-bilgiler accessed: 07.08.2020.

Melda AKBABA

24 Determination of Local Tourists' Travel Motivations

1 Introduction

As travel motivations of local tourists differ from one another, tourism marketers take into consideration the motivations in question while developing tourist products. So as to understand the travel motivations of local tourists, it is of great importance to explore what drives them to consume tourist products.

Rather than referring to the action of leaving and coming back home only, tourism embodies an important structure utilized to identify psychological, social and economic differences among people. From tourist's point of view, tourism is regarded as a response to the needs felt and the values acquired within the framework of temporal, spatial, social and economic parameters. Travel motivation of tourists following a tourism activity activated by personal needs turns into an important parameter in the emergence of expectations. Expectations of tourists define not only their perceptions of tourism product and services but also the perceptions of experience, and travel motivations impact on their perceptions of satisfaction (Gnoth, 1997: 283).

In this vein, the main aim of the current study is to determine travel motivations of local tourists in Turkey.

1.1 Travel Motivation

Due to the increasing competition in the tourism industry in the globalizing world, researches and studies are carried out in order to determine and direct the desires and wishes of the tourists. Travel motivation is one of the topics being studied. The motivations driving tourists to go on holiday and spend their vacation in a certain destination at a certain time refer to the travel motivation (Toksöz and Aras, 2016: 174).

Motivation is defined as a need or state (Aydın and Sezerel, 2017: 119) that drives an individual to certain activities that are likely to satisfy him/her, while travel motivation is described as the reasons prompting individuals to have a certain travel or holiday experience (Harman, 2014: 110). Travel motivation is considered as one of the key elements in the decision-making process of tourists, and it encompasses the answer to the question "Why do people travel?" (Dunne, 2009: 74).

The reasons motivating tourists to travel are researched under the umbrella of the concept of travel motivation, and the researches in question suggest that tourists choose the destinations they will travel in line with specific purposes, needs and wishes, and these choices are directed by various reasons (Aydın and Sezerel, 2017: 119).

The literature involves a great number of studies that examine travel motivations of tourists. For instance;

Jang and Wu (2006) examined the concept of travel motivation in five dimensions: namely, ego-enhancement, self-esteem, knowledge-seeking, relaxation, socialization, and concluded that the main travel motivations of tourists are knowledge-seeking and ego-enhancement.

Lin, Lee and Wang (2012) examined the travel motivation in five dimensions: namely, grow up and endurance, practice and salary, learn and experience, shopping and food, experiences of foreign culture. According to the research results, the main travel motivations of tourists were listed as learn and experience, shopping and food, and experiences of foreign culture.

Li and Cai (2011) investigated the travel motivation in five dimensions: namely, novelty and knowledge, prestigious and luxury experience, self-development, exciting experience, escape and relationship, while Le Serre, Legohérel and Weber (2013) examined it under four dimensions: namely, social travel motivation, relaxation travel motivation, sport/competency travel motivation and intellectual travel motivation.

Yazıcıoğlu and Akbulut (2016) examined the travel motivation in four dimensions as knowledge, relaxation, entertainment and prestige, and determined that the strongest motivations of tourists are knowledge and entertainment.

On the other hand, Sert (2019) examined the travel motivation in four dimensions as socialization, innovation, self-development and health, and reached the conclusion that the strongest motivations of tourists are socialization and novelty-seeking.

The aforementioned researches point out that researchers have different approaches on tourist motivations. In the current study, travel motivations of tourists are examined within the scope of personal/social growth, experiential, relaxation, cultural knowledge, budget travel and independence dimensions in accordance with the scale developed by Paris and Teye (2010).

In this respect, the main objective of this study is to determine the travel motivations of local tourists in Turkey. Furthermore, the study reveals whether travel motivations of local tourists in Turkey vary according to demographical characteristics.

2 Research Methodology

According to 2019 data from Republic of Turkey the Ministry of Culture and Tourism, there are 42,012,998 local tourists traveling in Turkey. This number constitutes the universe of the research. The minimum sample estimation formula developed by Yamane (2001: 116–117) was used to determine the sample diameter. In accordance with this formula, the minimum sample number is determined as 384.

The population of the study consists of local tourists in Turkey. Within the scope of the study, 426 local tourists were reached through the simple random sampling method and the survey utilized as the data collection method in the study. The questionnaire form contains statements aimed at identifying the demographics and measuring the travel motivations of local tourists.

In the study, the travel motivation scale developed by Paris and Teye (2010) was utilized in order to measure the travel motivations of local tourists. The reliability of the scale was determined to be 0.964. The scale is made up of six dimensions in total as personal/social growth, experiential, relaxation, cultural knowledge, budget travel and independence. At the same time, the scale consists of 26 statement with 8 related to personal/social growth, 5 to experiential, 4 to relaxation, 3 to cultural knowledge, 2 to budget travel and 4 to independence. The statements in the scale were determined in accordance with 5-point Likert scale with (1) strongly disagree, (2) disagree, (3) neither agree nor disagree, (4) agree and (5) strongly agree.

The data acquired through the survey were analyzed by statistical software, and descriptive statistics were built in order to identify the travel motivations of local tourists in Turkey who have participated in the survey, and the travel motivations of local tourists which were the most and the least displayed by the tourists were revealed. Furthermore, in order to examine whether the travel motivations of the local tourists who have participated in the survey differ according to demographical characteristics, the t-test and the ANOVA analysis-parametric tests were utilized. The analysis results were evaluated and some recommendations were tried to be proposed.

3 Findings

In the current study which was conducted with the purpose of determining the extent of travel motivations of the local tourists, the demographic characteristics of the tourists who have participated in the survey (Tab. 1), the descriptive statistics on personal/social growth being one dimension of the travel motivations

Tab. 1: Demographic Characteristics of Tourists Participating in the Study

Gender	**f**	**%**
Male	256	60,1
Female	170	39,9
Marital Status	**f**	**%**
Married	206	48,4
Single	220	51,6
Age	**f**	**%**
20 and younger	20	4,7
21–30 age group	142	33,3
31–40 age group	188	44,1
41–50 age group	46	10,8
51 and older	30	7,0
Level of Education	**f**	**%**
Primary School	10	2,3
High School	16	3,8
Associate Degree	84	19,7
Bachelor's Degree	170	39,9
Postgraduate	146	34,3
TOTAL	**426**	**100**

of the local tourists in Turkey (Tab. 2), descriptive statistics on experiential (Tab. 3), descriptive statistics on relaxation (Tab. 4), descriptive statistics on cultural knowledge (Tab. 5), descriptive statistics on budget travel (Tab. 6), descriptive statistics on independence (Tab. 7) and variation of travel motivations in accordance with demographics (Tab. 8) were revealed based on the data collected through surveys.

According to Tab. 1 of the tourists who have participated in the study, 60.1 % are male, 39.9 % female, 48.4 % are married and 51.6 % are single. Looking at the age distribution of tourists participating in the study, 44.1 % are within the 31–40 age group, 33.3 % within the 21–30 age group, 10.8 % within the 41–50 age group, 7.0 % are 51 and older and 4.7 % are 20 and younger. Having considered the education-based distribution of tourists, it was found that the highest level of education is bachelor's degree (39.9 %) while the lowest level is primary school (2.3 %).

The levels of agreement, arithmetic means and standard deviations concerning the 8 statements provided within the scope of the 5-point Likert scale

Tab. 2: Levels of Agreement to Statements Aimed at Identifying the Personal/Social Growth of Tourists

	1	2	3	4	5	x̄	s.d.
	(%)	(%)	(%)	(%)	(%)		
Personal/Social Growth						**2,99**	**0,95**
I travel to use my physical abilities/skills.	13,1	31,9	17,4	27,2	10,3	2,90	1,23
I travel to contribute something to the places I visit	11,7	34,3	15,5	31,5	7,0	2,88	1,18
I travel to challenge my abilities.	16,9	39,4	18,8	16,0	8,9	2,61	1,20
I travel to use my imagination.	11,7	16,9	11,3	39,4	20,7	3,40	1,30
I travel to build friendships with others.	12,7	24,9	16,4	34,3	11,7	3,08	1,25
I travel to gain a sense of belonging.	16,0	18,8	23,0	32,4	9,9	3,01	1,24
I travel to develop close friendships.	11,7	24,9	15,5	35,2	12,7	3,12	1,25
I travel to associate with other travelers.	12,2	26,8	19,7	31,9	9,4	3,00	1,21

Tab. 3: Levels of Agreement to Statements Aimed at Identifying the Experiential of Tourists

	1	2	3	4	5	x̄	s.d.
	(%)	(%)	(%)	(%)	(%)		
Experiential						**3,92**	**1,05**
I travel to experience once in a lifetime activities	9,4	14,1	6,1	41,3	29,1	3,67	1,29
I travel to gain experiences to share with friends and family	6,1	7,0	2,8	39,4	44,6	4,09	1,14
I travel to have a good time with friends.	7,5	5,2	3,3	40,4	43,7	4,08	1,16
I travel to experience excitement	8,0	7,5	3,8	46,5	34,3	3,92	1,18
I travel to attend special events	7,5	7,5	7,5	45,5	31,9	3,87	1,16

Tab. 4: Levels of Agreement to Statements Aimed at Identifying the Relaxation of Tourists

	1	2	3	4	5	x̄	s.d.
	(%)	(%)	(%)	(%)	(%)		
Relaxation						**4,14**	**1,05**
I travel to relax physically.	7,5	6,1	4,2	39,9	42,3	4,03	1,18
I travel to be in a calm atmosphere.	6,1	4,7	4,2	39,9	45,1	4,13	1,10
I travel to relax mentally.	5,6	4,7	2,3	34,3	53,1	4,24	1,09
I travel to avoid hustle.	6,1	4,7	3,8	39,0	46,5	4,15	1,10

developed in order to identify the personal/social growth of the local tourists are presented in Tab. 2. Having analyzed based on average values the statements on personal/social growth of the local tourists, which is one of the dimensions of travel motivation, tourists stated at most that they travel to use their imagination ($\bar{x}$: 3.40) and to develop close friendships ($\bar{x}$: 3.12). In addition, the tourists also travel to build friendships with others ($\bar{x}$:3.08). Another travel motivation of tourists with an mean value of 3.01 travel to gain a sense of belonging.

Taking into consideration the total average value of personal/social growth ($\bar{x}$: 2.99), it can be indicated that the level of personal/social growth being one of the dimensions of travel motivation is the lowest among tourists. Moreover, it can be stressed that the weakest travel motivation displayed by tourists participating in the study is the personal/social growth (Tab. 2).

The levels of agreement, arithmetic means and standard deviations concerning the 5 statements provided within the scope of the 5-point Likert scale developed in order to identify the experiential of the local tourists are presented in Tab. 3. Having analyzed based on average values the statements on experiential of the local tourists, which is one of the dimensions of travel motivation, tourists stated at most that they travel to gain experiences to share with friends and family ($\bar{x}$: 4.09) and to have a good time with friends ($\bar{x}$: 4.08).

In addition, the tourists also travel to experience excitement ($\bar{x}$:3.92). Another travel motivation of tourists with a mean value of 3.87 travel to attend special events.

Looking at the total mean value concerning experiential ($\bar{x}$: 3.92), it can be said that the experiential levels of tourists are above the mean (Tab. 3).

The levels of agreement, arithmetic means and standard deviations concerning the 4 statements provided within the scope of the 5-point Likert scale developed in order to identify the relaxation of the local tourists are presented in Tab. 4. Having analyzed based on average values the statements on relaxation of the local tourists, which is one of the dimensions of travel motivation, tourists

Tab. 5: Levels of Agreement to Statements Aimed at Identifying the Cultural Knowledge of Tourists

	1	2	3	4	5	$\bar{x}$	s.d.
	(%)	(%)	(%)	(%)	(%)		
Cultural Knowledge						**4,04**	**1,04**
I travel to explore other cultures.	6,1	4,2	4,2	32,4	53,1	4,22	1,11
I travel to increase my knowledge.	6,6	4,2	4,2	38,0	46,9	4,15	1,12
I travel to interact with local people.	6,6	10,3	13,1	39,9	30,0	3,77	1,17

Tab. 6: Levels of Agreement to Statements Aimed at Identifying the Budget Travel of Tourists

	1	2	3	4	5	$\bar{x}$	s.d.
	(%)	(%)	(%)	(%)	(%)		
Budget Travel						**3,21**	**1,12**
I travel to travel on a low budget.	8,5	21,6	21,6	30,5	17,8	3,28	1,22
I travel to travel for as long as possible.	10,8	23,0	23,0	26,3	16,9	3,15	1,25

Tab. 7: Levels of Agreement to Statements Aimed at Identifying the Independence of Tourists

	1	2	3	4	5	$\bar{x}$	s.d.
	(%)	(%)	(%)	(%)	(%)		
Independence						**3,60**	**1,13**
I travel to organize one's own journey.	8,5	17,4	16,9	32,4	24,9	3,48	1,26
I travel to get off the beaten track.	10,3	16,9	16,0	31,5	25,4	3,45	1,31
I travel to be free, independent, open-minded.	8,5	12,7	11,7	37,1	30,0	3,68	1,25
I travel to discover myself.	8,9	11,7	8,5	31,9	39,0	3,80	1,30

Tab. 8: Variation of Travel Motivations in Accordance with Demographics

Travel Motivation		**n**	**x̄**	**s.d.**	**F**	**p**
Gender	Male	256	3,57	0,92	4,664	0,752
	Female	170	3,59	0,78		
Marital Status	Married	206	3,53	0,84	0,211	0,272
	Single	220	3,62	0,90		
	20 and younger	20	3,23	1,10	5,628	
	21–30 age group	142	3,60	0,98		
Age	31–40 age group	188	3,69	0,67		**0,000***
	41–50 age group	46	3,61	0,86		
	51 and older	30	2,96	0,99		
	Primary School	10	3,50	1,28	1,688	
	High School	16	3,50	1,26		
Level of Education	Associate Degree	84	3,40	1,17		0,152
	Bachelor's Degree	170	3,57	0,84		
	Postgraduate	146	3,70	0,55		

*Significant at p<0.05

stated at most that they travel to relax mentally (x̄: 4.24) and to avoid hustle (x̄: 4.15). In addition, the tourists also travel to be in a calm atmosphere. (x̄: 4.13).

Another travel motivation of tourists with a mean value of 4.03 travel to relax physically. Taking into consideration the total average value of relaxation (x̄: 4.14), it can be indicated that the level of relaxation being one of the dimensions of travel motivation is the highest among tourists.

Moreover, it can be stressed that the strongest travel motivation displayed by tourists participating in the study is the relaxation (Tab. 4).

The levels of agreement, arithmetic means and standard deviations concerning the 3 statements provided within the scope of the 5-point Likert scale developed in order to identify the cultural knowledge of the local tourists are presented in Tab. 5. Having analyzed based on average values the statements on cultural knowledge of the local tourists, which is one of the dimensions of travel motivation, tourists stated at most that they travel to explore other cultures (x̄: 4.22) and to increase their knowledge. (x̄: 4.15). Taking into consideration the total average value of cultural knowledge (x̄: 4.04), it can be stressed that the other strongest travel motivation displayed by tourists participating in the study is the cultural knowledge (Tab. 5).

The levels of agreement, arithmetic means and standard deviations concerning the 2 statements provided within the scope of the 5-point Likert scale developed in order to identify the budget travel of the local tourists are presented in Tab. 6. Having analyzed based on average values the statements on budget travel of the local tourists, which is one of the dimensions of travel motivation, tourists stated at most that they travel to travel on a low budget ($\bar{x}$: 3.28) and to travel for as long as possible. ($\bar{x}$: 3.15). Looking at the total mean value concerning budget travel ($\bar{x}$: 3.21), it can be said that the budget travel levels of tourists are above the mean.

The levels of agreement, arithmetic means and standard deviations concerning the 4 statements provided within the scope of the 5-point Likert scale developed in order to identify the independence of the local tourists are presented in Tab. 7. Having analyzed based on average values the statements on independence of the local tourists, which is one of the dimensions of travel motivation, tourists stated at most that they travel to discover myself. ($\bar{x}$: 3.80) and to be free, independent, open-minded ($\bar{x}$: 3.68). Looking at the total mean value concerning independence ($\bar{x}$: 3.60), it can be said that the independence levels of tourists are above the mean (Tab. 7).

In order to reveal whether the travel motivations of the tourists who have participated in the study differ in accordance with their demographic characteristics, t-test and ANOVA analysis were carried out (Tab. 8). According to Tab. 8, since $p>0.05$ was observed between the tourists' gender, marital status, level of education and the travel motivations, no statistically significant difference was found. Accordingly, travel motivation does not differ according to the gender, marital status and level of education. However, travel motivation of tourists were found to be $p<0{,}05$ according to the variables of age ($p=0.000$), which points out to a statistically significant difference. Accordingly, travel motivation differs in accordance with the age of the tourists. In this regard, the mean difference between the ages of the participants and travel motivation was found to be the most striking between participants who are 51 and older and in the 31–40 age group. In this scope, it can be expressed that participants who are 31–40 age group display travel motivations more than participants in the 51 and older.

4 Conclusion and Suggestions

Regarded as the smokeless industry, the tourism industry constantly changes in line with the needs and expectations of tourists and the ratio of people traveling around the world increases due to the improving social and economic factors. Tourists' having different characteristics generates different travel motivations and it becomes inevitable to take these motivations into consideration while producing tourism products. In this context, this research aiming to determine the travel motivations of local tourists in Turkey employs a questionnaire.

As a result of the analyses performed in the current study, it was revealed that the strongest motivations of tourists were relaxation and cultural knowledge whereas the weakest motivation was personal/social growth. In addition, certain travel motivations of the respondents such as experiential, budget travel and independence were above average.

In the study, it was also investigated whether travel motivations of local tourists differ by gender, marital status, educational status and age. According to the analysis results, no significant difference was observed between travel motivation and gender, marital status and educational status. On the other hand, travel motivations of tourists differ significantly in line with age. In addition, it was revealed that tourists in the age range of 31–40 had higher travel motivation than those at the age of 51 and above.

In this respect, taking into notice the analysis results put forward in this study which aims to determine travel motivations of local tourists in Turkey, the fact that tourist products are designed in accordance with the need for relaxation and cultural knowledge would increase tourism demand.

References

Aydın, S. and Sezerel, H. (2017). Seyahat motivasyonlarına ilişkin bir yazın incelemesi. *Nevşehir Hacı Bektaş Veli Üniversitesi Sosyal Bilimler Enstitüsü Dergisi*, 7(2), 118–140.

Dunne, G. (2009). *Motivation and decision making in city break travel: The case of Dublin*. VDM Publishing, Saarbrucken, Germany.

Gnoth, J. (1997). Tourism motivation and expectation formation. *Annals of Tourism Research*, 24(2), 283–304.

Harman, S. (2014). Bağımsız hareket eden yerli gezginlerin seyahat motivasyonları üzerine bir araştırma. *Uluslararası Yönetim İktisat ve İşletme Dergisi*. 21(10), 107–128.

Jang, S. C. and Wu, C. M. (2006). Seniors' travel motivation and the influential factors: An examination of Taiwanese seniors. *Tourism Management*, 27, 306–316.

Le Serre D., Legohérel, P. and Weber, K. (2013). Seniors' motivations and perceived risks: A cross-cultural study. *Journal of International Consumer Marketing*, 25(2), 61–79.

Li, M. and Cai, L. A. (2012). The effect of personal values on travel motivation and behavioral intention, *Journal of Travel Research*, 51(4), 473–487.

Lin, Y. H., Lee, Y. C. and Wang. S. C. (2012). Analysis of motivation, travel risk, and travel satisfaction of Taiwan undergraduates on work and travel overseas programmes: Developing measurement scales. *Tourism Management Perspectives*, 2–3, 35–46.

Paris C. M. and Teye, V. (2010). Backpacker motivations: A travel career approach, *Journal of Hospitality Marketing & Management*, 19(3), 244–259.

Sert, A. N. (2019). Üçüncü yaş yerli turistlerin seyahat kısıtları ve motivasyonları üzerine bir araştırma, *Selçuk Üniversitesi Sosyal Bilimler Enstitüsü Dergisi*, 42, 200–211.

Toksöz, D. and Aras, S. (2016). Turistlerin seyahat motivasyonlarında yöresel mutfağın rolü. *Journal of Tourism and Gastronomy Studies*, 4(1), 174–189.

Türkiye Cumhuriyeti Kültür ve Turizm Bakanlığı, (2020). *2019 Yılı Konaklama İstatistikleri Yıllık Bülten Özeti*. http://www.turob.com/Files/dosyalar/pdf/2019Konaklama%C4 %B0statistikleri%C3 %96zet.pdf. (Accessed: 09.06.2020).

Yamane, T. (2001). *Temel Örnekleme Yöntemleri*. (çev. A. Esin, M. A. Bakır, C. Aydın and E. Gürbüzsel) İstanbul, Literatür Yayıncılık.

Yazıcıoğlu, İ. and Akbulut, B. A. (2016). Yabancı turistlerin seyahat motivasyonlarının demografik özellikleri açısından analizi: Ankara örneği. *Journal of Human Sciences*, 13(2), 3231–3241.

Ali KELEŞ & Hüseyin KELEŞ

25 Current Marketing Approaches in Terms of Tourism Destinations: The Sample of Side

1 Introduction

The emergence of globalization enables innovations to reach larger masses by increasing the use of information and technology. All the dynamics in the world are undergoing an inevitable change along with the changing and developing technology. The economic environment, which is considered as one of these dynamics, has started to turn to differences due to the rapid spread of information and technology. In the new economic environment, new markets have emerged where there is intense rivalry, businesses constantly seek for differences and marketing activities differ. The formation of different markets has brought along different marketing concepts, as well. Marketing has existed so far since the day humanity existed and has become an indispensable element of life. Marketing, which has now become a concept that concerns consumers as much as producers, undoubtedly takes place at every stage of life by developing and differentiating. (Özgül, 2008: 27).

The economic, cultural, social and political effects of tourism are quite effective in increasing the importance given to this industry, particularly the economic impacts it has created accelerate the efforts of countries to develop tourism and generate an intense competition environment. Countries have been trying to keep up with the ever-changing and renewed marketing activities in order to get more shares from tourism revenues in this increasingly competitive environment. The primary reasons of the changes encountered in marketing understanding are the developments in information techniques, the prominence of some touristic destinations, changes in the behaviors consumers and changes in terms of demand (Hacıoğlu, 2005: 12–13). As for tourism marketing, this change is observed in the formation of marketing approaches for more specific destinations rather than general or national marketing activities.

According to Philip Kotler (2006: 13), marketing is defined as "the science and art of seeking, creating and delivering value to meet the needs of a target market by getting a certain profit". It is very difficult for businesses to survive without marketing which has an indispensable position for businesses. For this reason, businesses allocate a significant part of their budgets to marketing resources. (Arslan, 2009: 1). However, traditional marketing approaches are no longer sufficient for businesses, and some innovations are also needed. And this leads to the rise of new insights in marketing. While the mass market is targeted in the traditional marketing approaches, businesses are turning to new activities where they can win each customer through new marketing approaches (Arslan and Kendir, 2019). Thanks to the developing technology, customers can access the information they request whenever they want. Customers are no longer satisfied with the product or price offered, but can easily reach the best product at the most affordable price by doing research immediately. The customers' having so many options makes it necessary for businesses to turn to new marketing activities in order to win new customers (Uslu, Ergün and Karabulut, 2020). The changing consumption perceptions of consumers, the emergence of a competitive environment as a result of an increase in rivals, the reactions of the state and social organizations to businesses and the globalization of communication have a great impact on the emergence of new marketing approaches. In addition to these, the new models that have been put forward aim to eliminate the shortcomings of the previous models (Urgancı, 2015: 9). Due to the change of roles in marketing, many factors such as target markets, market positioning, customer demands and needs, and the competitive environment have also been changing (Güneren & Karakuş, 2015; Yayla, 2019).

Destinations have been becoming more and more important in terms of increasing the share of countries in tourism. In general, the countries that want to generate much more income from tourism activities and plan to spread development throughout the country focus on the destinations. In this context, the destinations that want to increase their shares related to increasing tourism movements have been enhancing their marketing activities in order to get ahead of their rivals and become famous in the sector, and putting these thoughts into action via appropriate strategies and policies. (Günaydın, et al., 2015). According to Sarı and Yıldırım (2015); "It has now become necessary to head toward new markets in order to attract more tourists and increase tourism revenues.

In the study, a comprehensive literature review was used as a method. The main objective of this study is; The current marketing approaches of tourism destinations are to increase the awareness and application of the method of new marketing methods, which is an alternative method that will increase the sales of tourism businesses and also enable them to reach their target audiences easily and quickly.

The main objective of the study is to increase the awareness and application of new marketing methods within the academic field and the sector, which have become alternative methods that will increase the sales of tourism businesses, as well as providing easy and fast access to target masses. Besides, it has been aimed to set forth the applicability of these methods in tourism sector. In this context, by giving descriptive information about the specified marketing strategies, their different aspects from traditional marketing strategies have been presented, as well.

2 The Concept of Destination

Destination whose meaning was transferred from English to our language is defined as the place arrived (final target); and it is a word of French origin (https://sozluk.gov.tr/). It is very difficult to define and determine the exact boundaries of the destination. In terms of geographical terms, destination is a piece of land where the perceived variables are similar. It is also defined as a piece of land with similar geological structure, soil, climate, animals and vegetation and some of it was shaped by people (Friedmann and Weaver, 1980, p. 31). Destination is primarily considered as a system that includes inputs and outputs, society, services and business areas (Kocaman, 2012, p. 10).

Cooper and Hall (2008: 112) combined the destination definitions and stated that there are three different destination approaches as settlement or location, sense of place and local place. Destination is defined in these approaches as follows: in the settlement approach, with a geographical view, the territory with definite borders on Earth; in the local place approach a physical environment in which people have their daily life and social relations including their landscape experiences and services; in the sense of place approach, the characteristics that appeal to the visitors' emotions and reveal the difference of the destination.

According to Gunn (1994: 27), destination is defined as the place where an important community with the purpose of travel meet their needs. Although the concept of destination is often understood as a certain area with its physical boundaries, it is not always a piece of land with very precise boundaries due to the values it creates in the minds of the visitors. It is defined as a part of a country or world by communities that are close to each other in terms of language, culture, lifestyle, religion, history. Depending on the changes in tourism market trends, the borders and forms of the destinations may change and the maps showing the destination boundaries are valid only for a certain period of time (Usta, 2008b: 225).

2.1 Tourism Destination

In the tourism literature, the concept of destination is used as a tourist station, tourism destination, touristic destination, tourism region, but all of these concepts describe the place where the holiday experience is realized. It is seen that the definition of touristic destination is frequently used in the literature, but it seems very difficult to make a clear definition that stands out (Martini, Franch and Tomassini, 2003). There are many definitions made for tourism destination. Among these definitions tourism destination is defined as by Coltman (1989) "an area with various natural beauties, features and attractions for the benefit of non-local tourists or excursionists", by Hu and Ritchie (1993), "A package with tourism activities and services, consisting of multidimensional features, and other products that consumers desire", by Buhalis (2000), "Combination of locally offered goods, services and holiday experiences", by Shaw and Morris (2000), "Geographical areas and regions offering versatile tourism products to tourism travelers" and by Tekin and Ardahan (2006), "the place to go or to be sent in real or form perceived in minds". The place expressed as a tourism destination can be a country, as well as a city, a town, an island, etc. (Hall, 1994; 49). In other words, the places as a whole where the tourist leaves his/her place of residence and will go for holiday or business trips can be defined as destination (Kozak, 2005: 77).

According to Atay (2003, 144–145), "a tourism destination is defined as a geographical area which has various touristic values and characteristics and offer all or part of the touristic products that tourists may need during their travels". Tourism destination has a complex structure consisting of tourism attractions, tourism businesses, local people and local administration (Özdemir, 2007, 6).

Tourism destinations are places that attract visitors to stay for a short time and include holiday areas built for a specific purpose, from continents to countries, states, cities, and villages. (Pike, 2004, 3–30). The concept of tourism destination is defined by Usta (2008b: 223) as a mixture of certain elements that must be present together in order to obtain a satisfying holiday experience and which are connected to each other at different levels. Tourism destination is a part of the place that creates a demand on its own with the attractions it creates and meets the secondary requirements of this demand that arise during the main purpose of the visit. For that place, being a destination is related to factors such as having features of which will satisfy one or more of the basic needs of tourism demand such as natural, historical and cultural values; having elements that will meet the needs such as accommodation, nutrition, entertainment and other services; and in addition to these needs, the approach of the local people to tourism, and distance from the tourism market (Usta, 2008b: 224). In addition, in order to define a place as a tourism destination with an economic approach, the economic structure and life of that place must bear the effects of tourism (Olalı, 1990: 85).

3 The Concept of Marketing

Marketing is the management of a process which starts when a good or service is still in the form of an idea and design and including the production of the good and service, determination of prices, delivery to the target audience of which is intended to be marketed, informing the target audience through different channels for purchasing, and requires evaluation after marketing and sales of that good or service realized by collecting feedback (Sadedil, 2016: 5).

Marketing, which is concerned with identifying and meeting human and social needs, provides needs (Kotler and Keller, 2006: 5). Many people consider marketing as just selling and advertising. Contrary to popular belief, sales and advertising are only the visible side of the marketing concept. Kotler and Armstrong say that (2006: 5), "The goods or services you want to sell will be easily sold if you use the concepts of price, distribution and promotion effectively by catching the feelings of understanding the customers' needs, developing products, providing what is desired".

Although there are similar concepts in the definitions related to marketing, the concepts of "value creation", "benefit", "business partners" and "customer relations" have recently come to the fore. Here, the concept of "value" is an important concept for marketing. The value is being able to meet customer needs, which are the focus of marketing, with the lowest costs by analyzing the quality, service and wage policies accurately for the target audience (Mucuk, 2014: 5).

3.1 Tourism Marketing

Tourism marketing, is also defined by the World Tourism Organization (UNWTO) as "it is a management philosophy which aims to conduct research, estimation and make a decision about touristic product, in line with the highest earning goal of a touristic destination or tourism business, with the aim of gaining a good place in the market for the tourism product by taking into account the characteristics of tourism demand, and it is related to this decisions" (İçöz, 2001: 1).

The main factor underlying tourism marketing efforts is to increase sales and increase income. Reaching the most customers at the right time and gaining a place in the market is one of the main goals of businesses operating in the tourism industry. Customer satisfaction comes to the fore in modern marketing.

Tourism marketing is a social and managerial process that involves the creation of products and values and their exchange with others in order to meet the needs and desires of individuals and groups (Kotler, Bowen and Makens, 2014).

İslamoğlu (2010) defines tourism marketing as "the production and pricing, distribution, promotion and positioning of a country and its touristic products in accordance with the needs, wishes, desires and expectations of tourists, in a systematic integrity".

When we look at the figures around the world, according to the data of the United Nations World Tourism Organization (UNWTO, 2019), while the number of tourists traveling in the 1980s was 285 million, the total income was 92 billion dollars, the number of tourists has increased to 698 million in 2000 and the total income has been 475 billion dollars. Especially in the last two decades, the tourism industry data has reached 1.5 million travelers and 1.2 trillion dollars. While Turkey's place in these figures is the 6th on the list of "most tourist visited countries", and 13th on the list of "countries with most tourism income". In the tourism industry where numbers and statistics are growing rapidly, the competition of countries and tourism enterprises serving in these countries has become inevitable.

In line with all these developments, tourism businesses compete with other businesses in the sector in order to gain a solid place in the market and create new markets. Different presentation and sales efforts are required for tourism businesses to take part in the market and in order to meet tourists' needs. This is point where "tourism marketing" activities come into play. Due to the unique characteristics of tourism services, the marketing concepts and techniques of businesses operating in the industry differ from other enterprises in other industries (Akyol 2015: 146).

3.2 Current Marketing Approaches in Tourism

With the worldwide acceptance of the concept of globalization, many innovations in trade, society, politics and culture have been experienced and continue to be experienced. In line with these developments, new marketing strategies are added to the literature. These new marketing strategies and approaches have brought different perspectives to changing and developing marketing activities. Some developments are also taking place in the tourism industry, which has its share of the changing world conjuncture with globalization, and this has led to the need to examine the concept of tourism marketing from different perspectives. In this direction, current marketing approaches, which are closely related to the concepts of tourism and marketing, have been being used frequently in the literature especially in recent years. These approaches are mainly.

Destination Marketing

Destination marketing aims to make a destination identity created in accordance with the characteristics of a place more attractive with effective communication activities for selected target audiences. In the management of destination marketing, since there are real and/or legal entities who live in that destination, know the destination well and have some dreams and expectations about the future of the destination, this type of marketing is a strategy which is able to offer more flexible, fast and creative solutions to differentiate the destination from its competitors and to minimize possible risks. Additionally, since the destination marketing is done on a micro level and on a single location basis, it informs the consumer more about the essence of the product, its basic functions and auxiliary functions and can be more effective on the purchasing decision. With this type of marketing strategy, effects of a negative situation which arises in one region of the country can be prevented for other regions (Yavuz, 2007: 40–41). At the same time, the unique characteristics of the destination can be revealed and used as a competitive advantage. Apart from this, destination marketing is also important in terms of attractions of which are important for the destination.

Post-Modern Marketing

Post-modern marketing is a commodity subject to tourism consumption in society. Tourism is the relationship that arises between this touristic sight and what is looked at, and its object is the target of the tourist's sight. In this respect, visuality is always important in tourism. The tourist consumes images or representational values, and all reality is hidden by these multi-layered symbols. Therefore, the tourist has to consume the symbol, sign or image of the touristic object while not being able to penetrate the underlying reality. Consumption is related to consumption process of signs rather than the satisfaction process of needs. Post-modern tourists are fond of their comfort and taste, in other words they are hedonists. They use consumption to express themselves, to create an identity for themselves. They aim to develop belonging through consumption. While consumption responding to their search for satisfaction, it is important in terms of improving the creativity of the individual and emphasizing his/her abilities (Güneren and Karakuş 2015: 473).

Guerrilla Marketing

The main purpose of this marketing approach is being able to be in big markets with small movements. In this marketing strategy, in which different applications are realized to gain a place in the brain of the customer, product or brand positioning studies are applied in line with this purpose. Effective promotion and marketing practices can be realized by carrying out marketing activities aimed at smaller masses with guerrilla practices. Both free or low-cost guerrilla implementations are important tools especially for small-scale enterprises, and they are powerful tools to convey the message of the business to potential consumers and attract the attention of the consumer (Güripek and Öz 2015: 522).

Niche Marketing

Niche tourism, which has emerged as an alternative to mass tourism in recent years, has its roots in niche marketing. Niche tourism markets are groups of individuals with similar needs, wishes, and interests. For destination managers and planners who see tourism as a dynamic in economic development, the application of niche marketing in tourism is seen as a more profitable, less harmful and most importantly a tool that attracts long-stay tourists. In general, it is an understanding that does not have power in the market, cannot compete in another field and requires expertise. Niche marketing practices provide tourism professionals with many advantages such as lack of competition, facilitated marketing activities, creating a basis for clustering, developing more effective marketing strategies as a result of concentration in a specific industry or geographic area, and putting more profit on cost by means of increased value over against special services (Kılıç, Ön and Yozukmaz, 2015: 307).

Internal Marketing

In order to survive, gain a competitive advantage and grow even more, tourism enterprises of which face with fierce competition in national and international markets, they must meet the needs of both external and internal customers. One of the ways to achieve this is through internal marketing (Aksu and Atay, 2015: 163). According to this marketing strategy, more attention will be paid to customers in a tourism business where employees are satisfied. This will increase customer satisfaction and make the service quality perceived even better. In a tourism business where the problems are minimal, the loyalty of the employees to the business will increase after a while and they will be able to see themselves as a part of the business. Internal marketing is an activity related to the internal structure of the business. All departments of the business must work in a coordinated manner. Marketing of one department by another department is an example of internal marketing.

Social Media Marketing

Developments in information and communication technologies deeply affect the tourism industry. Especially the development of social media-oriented applications and the increase in the number of users of these channels reshape the relationship and communication efforts between consumers and producers, as well as between consumers and consumers in the tourism sector. The main variables that need to be monitored closely are the increasing internet use and the concept of social media with the widespread use of mobile devices. With the opportunities offered by the internet, the development of social media and the diversification of social networks, the increase in the use of mobile devices in direct relation to the internet and social media has created a new generation consumer profile. For this reason, the concept of social media and correspondingly marketing activities carried out on social media are important for businesses operating in the tourism industry as in many other industries (Bayram and Bertan, 2015: 549).

Experiential Marketing

Tourism event is a process based on experiences. During his/her stay in a destination, a tourist constantly experiences different experiences from the place he/she lives in and his/her daily life. These experiences of the tourist in the destination have a direct impact on the tourist's satisfaction. For this reason, rival destinations or tourism businesses must provide tourists with differentiated, unique and unforgettable experiences. This situation makes experiential marketing practices indispensable in tourism (Harman, 2015: 493). The point that can be criticized in this regard is trying to provide services adapted to the country of their origin for foreign country tourists. However, the tourist is curious about the local organizations in the country they visit and acts with the desire to see them.

E-Marketing

The Internet is the largest computer network in the world. E-marketing is an approach which has started to develop in many industries with the widespread use of the Internet, it is also frequently used in the tourism industry. In this industry, the internet enables 7/24 trading. This situation also prompted all tourism businesses, especially hotels and airlines, to open up to foreign countries as well as the domestic market in order to find customers. (Metin, 2015: 94). Within the tourism industry, accommodation businesses, travel businesses, tour operators, food and beverage businesses and entertainment businesses carry out e-marketing activities, but the use of e-marketing is more used in travel businesses and accommodation businesses. Transactions such as purchasing airplane, bus and train tickets in travel businesses and making hotel reservations in accommodation businesses bring the use of e-marketing to the fore.

Interruption Marketing

Consumers can encounter thousands of advertising and marketing tools every day on television, radio, billboards, elevators, public transportation vehicles, postal boxes and e-mail addresses. While these messages about products and services are directed to the consumer, permission and opinion are not received from the consumer. This type of marketing, made in the form of advertising pollution, is called interruption marketing (Dağdaş, 2013: 21).

Side Destination

The destination of Side is located 80 kilometers east of Antalya, 7 kilometers southwest of Manavgat, on a peninsula with a width of 350–400 meters.

Side is the most important maritime tourism destination of Manavgat, which has gained the position of a world-renowned brand of Antalya, with a very high tourism potential. Side and its surroundings make Manavgat a tourism district and attract thousands of tourists every year. Side is an indispensable pearl of tourism of our Mediterranean coast with its unique historical riches, unique nature, blue flag beaches and endless sun. With its ever-increasing tourism potential, destination of Side hosts millions of local and foreign guests every year.

The most important richness that makes the Side immortal is its historical texture and richness. Historical and touristic places have a great importance for the tourism sector. The ancient city of Side is a tourism center rich in these values. Historical artifacts and natural beauties attract the attention of tourists (www.kulturportali.gov.tr).

According to the Ministry of Culture and Tourism, as of 2019, including hotels certified by ministry and municipalities in Turkey has reached a total of 1 million 828 thousand bed capacity. 1 million 200 thousand of this capacity are certified by the Ministry of Culture and Tourism and 628 thousand are certified by the municipalities. 42 % of the bed capacity which has certified by the Ministry is located in Antalya and the surrounding districts. With 2799 accommodation facilities and 628 thousand 660 beds in and around Antalya is the city with the highest bed capacity in Turkey. The destination of Side which has the highest bed capacity of Turkey, shares 30 % of the total bed capacity of Antalya with 529 facilities and 183 thousand 706 beds (https://www.ktb.gov.tr/).

In this study we have conducted, in the light of the information we have obtained from many sector representatives and marketing operators of Side destination, it has been observed that Destination Marketing, Guerrilla Marketing, Experiential Marketing, Interruption Marketing approaches come to the fore. In recent years, it has been determined that Interruption Marketing has come to the fore especially in tourism fairs held in different countries every year. In general, other current marketing approaches are thought to be more oriented toward businesses.

4 Conclusion and Suggestions

The tourism and travel industry has grown rapidly after the 1950s. In general, considering the current situation of tourism, it is possible to say that it is one of the fastest growing sectors. This growth directly or indirectly affects many businesses. Considerable economic development has been achieved by creating employment in all business lines serving in the sector. The place in which the touristic product is presented and consumed and named as tourist station, touristic area, tourism place in the literature has been named as touristic destination in recent years. In the new framework of the emerging competition, traditional marketing methods have fallen behind. Because the consumer can access a lot of different products and information quite fast. For this reason, what the consumers care about and what they consider while making a decision should be examined in depth. In order to be the tourism destination preferred by the consumers in the tourism sector where there is intense competition, it has become a necessity to use different marketing approaches and practices.

Businesses have been trying to determine the most beneficial method by spending great effort and budget so that the marketing strategies they use can be effective to persuade consumers. Technology has also taken its place in the field of marketing as in every field. Consumers while making decisions and purchasing, and manufacturers while marketing their products use technology. In social media, which is the most used tool, content is produced regardless of the subject and type. Creating difference has turned into an obligation and need for tourism businesses rather than a choice. While tourism enterprises can make differences in the products they offer, they should also make innovations in the marketing of these products. With different and new marketing approaches, they can enter different markets, reach potential customers in the existing market and also create customer loyalty among existing customers.

Tourism sector is one of the sectors that is closely affected by the developing technology and it also has to follow the innovations. Traditional marketing methods have become insufficient in the new competition framework. Tourism destinations should turn to strategic, creative, and different marketing methods that can appeal to the emotions of the consumers to attract attention. Otherwise, destinations that do not turn to these methods and avoid innovation will fall behind other destinations in a competitive environment. New marketing approaches will provide economically new inputs for tourism destinations. For example, guerrilla marketing technique is recommended especially for tourism destinations with a limited budget and that want to reach a specific group of target customers. Because, through guerrilla marketing, which has a low cost, a tourism destination is able to attract the attention of consumers and arouse a desire to purchase. The tourism destination managers' sufficient knowledge of different marketing strategies will provide significant benefits while creating their marketing plans. As a result, the widespread use of current marketing approaches in destination marketing is closely related to the benefits of the approaches to the region and knowing how to use them. However, it should not be forgotten that these marketing strategies will differ in every country, every culture, even between destinations, they should be customized to the target, and the marketing approaches should be developed accordingly. This research has been prepared by using secondary data with the aim of evaluating modern marketing approaches that can be used in tourism destinations. The limitation of the study is that the evaluation has been made only for Side destination. In future studies, it is thought that it will be useful to consider the application of current marketing methods specific to different destinations. As a matter of fact, each destination has a number of unique and different features. It is thought that through these studies, the share of our country that will be obtained from tourism sector can be increased.

References

Akyol, C. (2015). Artvin İlinin Mevcut Turizm Potansiyeli ve Niş Pazarlamanın Uygulanabilirliği, *Uluslararası Karadeniz Havzası Halk Bilimi Araştırmaları Dergisi*, 2, 144–172.

Arslan, S. (2009). Turizm Sektöründe Alternatif Bir Pazarlama Stratejisi Olarak Gerilla Pazarlamanın Kullanılması: Konaklama İşletmeleri Üzerine Bir Araştırma, *Yayınlanmamış Yüksek Lisans Tezi*, Gazi Üniversitesi Eğitim Bilimleri Enstitüsü Turizm İşletmeciliği Eğitimi Anabilim Dalı, Ankara.

Arslan, E. and Kendir, S. (2019). Kış Turizmi Potansiyeli Açısından Muş İli'nin SWOT Analizi İle Değerlendirilmesi, *Anemon Muş Alparslan Üniversitesi Sosyal Bilimler Enstitüsü Dergisi*, 7(6), 167–173.

Atay, L. (2003). Turistik Destinasyon Pazarlaması ve Bir Alan Uygulaması, *Yayınlanmamış Doktora Tezi*, Dokuz Eylül Üniversitesi, Sosyal Bilimler Enstitüsü, Turizm işletmeciliği Anabilim Dalı, İzmir.

Coltman, M. M. (1989). *Tourism Marketing*, Van Nostrand Reinhold. New York

Cooper, C. and Hall, C. M. (2008). *Contemporary Tourism and International Approach*. Butterworth-Heinemann, Burlington.

Dağdaş, G. (2013). İşletmelerde Retro Pazarlama Uygulamalarının Müşteri Bağlılığına Etkileri Üzerine Bir Araştırma, *Yayınlanmamış Yüksek Lisans Tezi*, Selçuk Üniversitesi Sosyal Bilimler Enstitüsü, Konya.

Friedmann, J. and Weaver, C. (1980). *Territory and Function, the Evolution of Regional Planning*, California University Press, California.

Gunn, C. A. (1994). *Tourism Planning* (3. Baskı), Taylor & Francis, Washington.

Günaydın, İ., Peker İ., Okan T., Akdu U., (2015). Arap Turistlere Yönelik Hizmet Kalitesinin Analizi Trabzon Üzerine Bir Araştırma, *Gümüşhane Üniversitesi Yayınları*, No: 30.

Güneren, E. and Karakuş, Y. (2015). *Post Modern Pazarlama, Turizm Pazarlamasında Güncel Yaklaşımlar*, Beta Basım Yayım, İstanbul.

Güripek, E. and Öz, M. (2015). *Gerilla Pazarlama, Turizm Pazarlamasında Güncel Yaklaşımlar*, Beta Basım Yayım, İstanbul.

Hacıoğlu, N. (2005). *Turizm Pazarlaması,* 5. Baskı, Nobel Yayın Dağıtım, Ankara.

Hall, C. M. (1994). *Tourism and Politics: Policy, Power and Place*, John Wiley & Sons Publications, Chichester, UK.

Hu, Y. and Ritchie B. J. R. (1993). Measuring destination attractiveness: A contextual approach, *Journal of Travel Research*, 32(3), pp. 25–34.

İçöz, O. (2001). *Turizm İşletmelerinde Pazarlama İlkeler ve Uygulamalar*, Turhan Kitabevi Yayınları, Ankara.

İslamoğlu, H. A. (2010). *Turizm Pazarlaması*, Beta Basım Yayın Dağıtım A.Ş İstanbul.

Kılıç, B., Ön, E. F. and Yozukmaz, N. (2015). Niş Pazarlama." Turizm Pazarlamasında Güncel Yaklaşımlar. Beta Basım Yayım, İstanbul.

Kocaman, S. (2012). Destinasyon Yönetimi Kapsamında Marka Kimliğine Etki Eden Faktörlerin Marka İmajına Etkisi: Alanya Örneği. *Doktora Tezi*, Akdeniz Üniversitesi, Sosyal Bilimler Enstitüsü, Antalya.

Kotler, P. and Armstrong, G. (2006). *Principles of Marketing*, (11.Baskı), Pearson Education Inc., New Jersey.

Kotler, P. and Keller, K. (2006). *Administraçao de Marketing*, Prentice Hall, São Paulo.

Kotler, P., Bowen, J. ve Makens, J. (2014). *Marketing for Hospitality and Tourism*, Pearson New International Edition, USA.

Kozak, N., Akoğlan Kozak, M., & Kozak, M. (2008). *Genel Turizm*, Detay Yayın, Ankara.

Martini U., Franch M., & Tommasini, D. (2003). Hypothesis of Destination Management in the Alps through the Analysis of Regional Tourist Boards, Atti del Convegno: *XII International Symposium on Tourism and Leisure*, Barcelona, 3–4.

Metin, İ. (2015). *Turizm Pazarlamasında Güncel Yaklaşımlar,* Beta Yayınları, İstanbul.

Mucuk İ. (2014). *Pazarlama İlkeleri ve Yönetimi için Örnek Olaylar* (20. Baskım), Türkmen Kitabevi, İstanbul,

Olalı, H. (1990). *Turizm Politikası ve Planlaması*, İşletme İktisadı Enstitüsü, Yayın No: 122, İstanbul.

Özgül, N. (2008). Bütünleşik Pazarlama İletişiminde Gerilla Pazarlama Stratejilerinin Kullanılması, *Yayınlanmamış Yüksek Lisans Tezi,* Ege Üniversitesi sosyal Bilimler Enstitüsü, İzmir.

Özdemir, G. (2007). Destinayon Yönetimi ve Pazarlama Temelleri İzmir İçin Bir Destinasyon Model Önerisi, *Yayınlanmamış Doktora Tezi*, Dokuz Eylül Üniversitesi, Sosyal Bilimler Enstitüsü, İzmir.

Pike, S. (2004). Destination Marketing Organizations Advances, Tourism Research, Elsevier, Butterwortth, UK.

Sadedil Kuruş S. N. (2016). Pazarlama Mesajlarının Etkinliği Açısından Geleneksel Pazarlama Araştırmaları İle Nöropazarlama Araştırmalarının Karşılaştırılması; Sigara Paketleri Üzerindeki Caydırıcı Mesajların, Sigara Kullanma Alışkanlıkları Üzerinde Etkileri, *Yayınlanmamış Doktora Tezi*, Marmara Üniversitesi SBE, İstanbul.

Sarı Y. and Yıldırım, G. (2015). Türkiye Turizmi Bağlamında Ortadoğu Turizmi Üzerine Bir Araştırma, *İşletme Araştırma Dergisi.*

Shaw, M. and Morris S. (2000). *Hospitalty Sales. A Marketing Approach*, John Wiley & Sons Inc. Canada.

Tekin, M. and Ardahan, F. (2006). Bireylerin Tatile Çıkma Sebepleri, Türkiye'nin ve Tatil Yapılan Yörelerin Tatil Destinasyonu Olarak Tercih Edilme Nedenleri, *II. Balıkesir Ulusal Turizm Kongresi*, 20–22 Nisan 2006, Balıkesir.

Urgancı, F. (2015). Gerilla Pazarlama Ve Etik: Nitel Bir Araştırma, Atılım Üniversitesi Sosyal Bilimler Enstitüsü İşletme Anabilim Dalı, *Yayınlanmamış Yüksek Lisans Tezi*, Ankara.

Uslı, A., Ergün, G. S. and Karabulut, A. N. (2020). Otel İşletmelerinde Marka Denkliğinin Müşteri Memnuniyeti Üzerindeki Etkisi: Güven Değişkeninin Aracı Rolü, *Türk Turizm Araştırmaları Dergisi*, 4(3), 2264–2281.

Usta, Ö. (2008). *Turizm, Genel ve Yapısal Yaklaşım*, Detay Yayıncılık, Ankara.

Yavuz, M. C. (2007). Uluslararası Destinasyon Markası Oluşturulmasında Kimlik Geliştirme Süreci: Adana Örneği, *Yayınlanmamış Doktora Tezi*, Çukurova Üniversitesi Sosyal Bilimler Enstitüsü, İşletme Anabilim Dalı, Adana.

Yayla, Ö. (2019). *Factors affecting the tourist purchase behaviour.* eds. Krystev, Efe and Atasoy, in Theory and Practice in Social Sciences, pp: 580–590, Sofia: St. Kliment Ohridski University Pres.

Internet Sources Used

https://sozluk.gov.tr/ Access Date: 29.07.2020.

http://yigm.kulturturizm.gov.tr/TR-201121/isletme-bakanlik-belgeli-tesis-konaklama İstatistikleri.html, Access Date: 29.07.2020.

https://turizmdays.com/tr/news/manavgat-45-milyar-dolar-turizm-geliri-sagladi-4152.html, Access Date: 29.07.2020.

https://www.e-unwto.org/doi/book/10.18111/9789284409570 Access Date: 29.07.2020.

https://www.marketingturkiye.com.tr/haberler/gerilla-pazarlamasi-nedir-ornekleri/, Access Date: 29.07.2020.

https://www.kulturportali.gov.tr/turkiye/antalya/gezilecekyer/side-1, Access Date: 29.07.2020.

Koray ÇAMLICA

26 Cultural Heritage Tourism and Stakeholder Roles in the Scope of Sustainable Tourism

1 Introduction

It is possible to define cultural heritage tourism as the visits made particularly to heritage sites, historical and architectural works, immovable objects with artistic significance, archaeological structures, and the structures built at present by the society to preserve the culture as well as the visits to the intangible cultural heritage elements. It may be stated that cultural heritage tourism covers both the values of the past and today.

Along with the cultural heritage tourism, tourists began to visit sites with cultural properties such as historical places, museums, cultural ruins, festivals, lifestyles, traditions, customs, entertainment, ceremonies, foods and more. Such visits brought out the necessity to take measures to protect the cultural properties. Thus, substantial ruins which shed light to present from the past may be preserved and transferred to future generations. For this reason, it is essential to consider the cultural heritage tourism from the sustainable tourism perspective and to determine the roles of tourism stakeholders. This study has reviewed the roles of the local community, local and senior management, and the tourism establishments at the destination, such as hotels, travel agencies, restaurants in detail within the scope sustainability.

2 Cultural Heritage Tourism

The cultural tourism is a type of tourism with a quite wide scope. When it comes to cultural tourism, the first thing that comes to mind is the tourists being involved in the current culture of a region. In other words, the culture tourists experience the culture in a destination. The culture tourists would like to learn about the lifestyles, history, art, and recreation activities of the habitants in the destinations they visit, and they act accordingly.

The purposes of the cultural tourism include preserving and reanimating, enlivening, sustaining the current cultural values; raising awareness in the local community for the preservation of cultural values; improving and building physical structures based on the culture; and creating destinations prioritizing the local community. On the other hand, there are differences between cultural tourism and the cultural heritage tourism. The cultural tourism focuses on all the cultural properties of a society. The cultural heritage tourism covers the specific tangible and intangible elements that distinguish one destination from others particularly such as heritage sites. For example, the Whistled Language specific to Giresun is listed as intangible cultural heritage, and thus, the visits to the region may be considered within the scope.

Cultural heritage tourism as the visits made particularly to heritage sites, historical and architectural works, immovable objects with artistic significance, archaeological structures and the structures built at the present time by the society to preserve the culture as well as the visits to the intangible cultural heritage elements. It may be stated that the cultural heritage tourism covers both the values of the past and today. In the meantime, the tourist potential interested in this tourism type may be attracted to the region by featuring the cultural values in touristic destinations (Kozak and Bahçe, 2012). Hence, it may provide economical contribution to the destination. Along with the cultural heritage tourism perception, tourists began to visit sites with cultural properties such as historical places, museums, cultural ruins, festivals, lifestyles, traditions, customs, entertainment, ceremonies, foods more. Such visits brought out the necessity to take measures to protect the cultural properties. Thus, substantial ruins which shed light to present from the past may be preserved and transferred to the future generations.

The tourists participating the cultural tourism primarily prefer the heritage site with natural and cultural attractions while choosing a destination. Meanwhile, the tangible and intangible cultural and natural heritage sites in the destinations chosen are attractive elements for the tourists participating in cultural tourism. Besides, perceiving a site as a World Heritage Site makes tourists willing to visit the relevant site, to pay entrance fee, to participate a guided tour or to spend more time to reach the site (Poria et al., 2011).

On the other hand, although the countries consider the preservation, improvement, and management of tangible and intangible cultural elements such as heritage sites or whistled language, they may have insufficient resources. In addition to government resources, there is also a need for support from the local community, local establishments, and local administrations. Being an organization focusing on such matters, UNESCO has started to attach the required importance to the cultural properties around the world by creating heritage sites lists and intangible cultural heritage lists.

"World Heritage Convention" was signed by gathering the international communities by UNESCO in 1972. This Convention intends to take measures to prevent the natural and/or cultural sites valued by the societies and important for

Tab. 1: UNESCO World Heritage Sites in Turkey

Name of the World Heritage Site	Year	Type	Province(s)
Historic Areas of Istanbul	1985	Cultural	Istanbul
Great Mosque and Hospital of Divriği	1985	Cultural	Sivas
Göreme National Park and the Rock Sites of Cappadocia	1985	Cultural-Natural	Nevşehir
Hattusha (Boğazköy) - the Hittite Capital	1986	Cultural	Çorum
Nemrut Dağ	1987	Cultural	Adıyaman - Kâhta
Xanthos-Letoon	1988	Cultural	Antalya - Muğla
Pamukkale-Hierapolis	1988	Cultural-Natural	Denizli
City of Safranbolu	1994	Cultural	Karabük
Archeological Site of Troy	1998	Cultural	Çanakkale
Edirne Selimiye Mosque and Social Complex	2011	Cultural	Edirne
Neolithic Site of Çatalhöyük	2012	Cultural	Konya
Pergamon and its Multi-Layered Cultural Landscape	2014	Cultural	İzmir
Bursa and Cumalıkızık: The Birth of Ottoman Empire	2014	Cultural	Bursa
Diyarbakır Fortress and Hevsel Gardens	2015	Cultural	Diyarbakır
Ephesus	2015	Cultural	İzmir
Archeological Site of Ani	2016	Cultural	Kars
Archeological Site of Aphrodisias	2017	Cultural	Aydın
Göbekli Tepe	2018	Cultural	Şanlıurfa

the history of humanity from being damaged or destroyed by unplanned urbanization, armed terror incidents or various destructions. The Convention made the preservation, improvement, and management of the World Heritage Sites, and thus transferring them to the future generations important.

The regions determined as World Heritage Site are important sites with high-level protection which are frequently visited by culture tourists. The World Heritage Sites are divided into three as natural, cultural, and cultural-natural. Most of the World Heritage Sites possessed by Turkey are listed as cultural heritage sites. UNESCO World Heritage Sites in Turkey which are visited within the scope of cultural heritage tourism are given in Tab. 1.

When the World Heritage Sites around the world are considered, Angkor (Cambodia), Acropolis (Greece), Bagan (Myanmar), Galápagos Islands

(Ecuador), Göreme National Park and the Rock Sites of Cappadocia (Turkey), Great Barrier Reef (Australia), Hampi (India), Iguazu National Park (Brazil and Argentina), Machu Picchu (Peru), Petra (Jordan), Mont Saint-Michel (France), Giza Pyramids (Egypt), Rapa Nui (China), Serengeti National Park (United Republic of Tanzania), Sigiriya (Sri Lanka), Tulum (Mexico), Valetta (Malta), Venice (Italy) and Yellowstone National Park (USA) are seen as the most known and the most frequently visited sites (CNN Travel, 2020).

3 Sustainable Tourism

The concept of sustainability has been developed based on the discussions within the scope of "eco-development" concept addressed in the Report of Global Environment Conference held in Stockholm in 1972, and it is known that man and his interaction with and interrogation of the environment he lives date back to very old times (İncedayı, 2004). Sustainability originally emerged in reaction to the problems caused by industrialization and urbanization in Europe and the United States of America, and it extends back to mid-19 century. The rise of tourism after the Second World War and the developments in the West related to this rise reanimated the philosophy of sustainability (Choi and Sırakaya, 2005). According to Sassi (2006), sustainability is to provide a better quality of life to the present and the future generations. Sassi (2006) indicates that sustainability is not an academic vocation or a professional activity, but an activity which affects anything a person does and his/her life. He mentions that the most important matter is to know the type of relation to have with the global and local environment, and then he discusses how we should attain this desirable relation. In order to put theory into practice, it is necessary to understand the connections between the life activities and the actions we perform (Sassi, 2006). Therefore, this concept offers quality of life to the generations which live and will live.

According to Saarinen (2006), sustainable tourism is defined as a tourism conception which is affordable and which does not damage the resources related to the physical environment and the social structure of the host society on which the future of the tourism depends. Based on this conception, sustainable tourism is required to respect the local culture and the natural environment, to purchase local products and to provide economic return to local families, to attach importance to goods produced by locals, and to protect the endangered flora and fauna, and it is a way of showing respect to the local environment or any destination. While the visitors visit the destination area, they need to be more sustainable to protect and develop their most popular destination for future travel experiences.

"The sustainable tourism is defined as a form of development where the environment with which man interacts or not is preserved without ruining or altering; the cultural integrity, ecological processes, biological diversity and the surviving systems are maintained; and also where all the resources in the region visited are managed in a manner to allow fulfilling the economic, social and aesthetic needs of the locals and the tourists" (Üçer, 2011: 34).

Considering that the principle resource of tourism is the natural, cultural, environmental, and social resources, sustainable tourism is an approach which attempts to reduce the problems which arise due to complex relations between visitors, environment and the areas visited. According to Sarkım (2007: 56), "The development of sustainable tourism is possible through the protection and renewal of natural sites, and the preservation of the environmental balance. If these values, which are the resources of tourism, are not protected, it is clear that tourism will be under threat at local, regional and national levels." Therefore, sustainable tourism is also on the agenda in terms of managing and economically developing a region. For example, if Pamukkale travertine loses its characteristics due to environmental and natural factors, this may lead to a decreased number of or no visitors to come to the region. This would prevent the local establishments from doing business or lead them to put up the shutters and lower the quality of life of the locals economically.

The fundamental content of sustainable tourism consists of stably improving the benefits derived from tourism at regional and local scale without giving harm to the environment, society, history, natural and cultural properties. This improvement represents the dependence between the environmental, economic, and social matters. In other words, sustainable tourism covers a planning process which allows tourism to be integrated with economic developments (Mowforth and Munt, 1998).

Recently, the perspective and supports provided to sustainable tourism in Turkey has improved. For example, in the vision of Tourism Strategy of Turkey 2023, it is mentioned to achieve 2023 goals by adopting a sustainable tourism approach. Another support is the attempt to raise funds for innovative, participatory, sustainable and gender-balanced tourism projects through the "Sustainable Tourism Support Fund" which is the joint project of the Ministry of Culture and Tourism, United Nations Development Program (UNDP) and Anadolu Efes Company. It is considered that this fund will "contribute to increasing cooperation between the stakeholders in sustainable tourism; becoming aware of the impact of sustainable tourism on local economic development and thus increasing the number of initiatives on this matter; allowing the local community to become aware of its own values; raising consciousness for the protection of the cultural and natural heritage; augmenting the number of examples in protection-usage balance; activating the resources for the development of local sustainable tourism; increasing the number of employees in the region, and thus supporting employment" (Gelecek Turizmde, 2020).

According to Choi and Sırakaya (2005), sustainable tourism is the shortest route of development leading to minimizing the unfavorable impacts of tourism. There are many ways to render tourism more sustainable, but there are a few methods to evaluate and control a sustainable tourism system (Assante et al., 2012).

The basic principles of sustainable development provide a basis to sustainable tourism (Sharpley, 2009). When sustainable development is considered, it is seen that three key principles play a role. These are holistic perspective; development and environmental sustainability being a global challenge, the necessity of having the long-term thinking at the forefront and overcoming this challenge despite development being human-centered as well as the necessity for equality within and between the generations; and fair and equal development among all people. When the principles of sustainable tourism are examined in detail, the following results are obtained (Sharpley, 2009).

- The protection and sustainable use of natural, social, and cultural resources are very important. Therefore, tourism should be planned and managed within the framework of environmental limitations and by considering the long-term proper use of the natural and human resources.
- Tourism planning, development and practice should be integrated with national and local sustainable development strategies.
- Tourism needs to support a broad series of economic activities by taking into consideration the environmental costs and benefits.
- The participation of local communities in the planning, development and control of tourism should be promoted with the support of the government and the industry.
- All organizations and individuals should respect the culture, economy, lifestyle, environment, and the political structures in the centers of attraction.
- All stakeholders should be trained on the necessity of development in all areas of tourism.
- Research should be undertaken throughout all stages of tourism development in order to observe the impacts, solve the problems and allow the locals and other to react and to take advantage of the opportunities.
- All agencies, organizations, companies, and individuals should work in cooperation to prevent the potential conflicts involved in the management and development of tourism and to optimize the benefits.

Sustainable tourism is reviewed in terms of the local community, local administration, and visitors, and it is addressed in different dimensions based on the audience. The studies related to the local community (Choi and Murray, 2010) generally discuss sustainable tourism with regards to the aspects such as environmental sustainability, economic sustainability, tourism planning, participation of the community, supporting sustainability, favorable and unfavorable impacts, etc. In studies related to local administration (Ruhanen, 2013), the environmental sustainability, economic sustainability, tourism planning, etc. aspects are discussed in particular. 7 aspects discussed in studies analyzing the visitors (Daud and Rahman, 2011; Nicholas and Thapa, 2010; Yan, 2008) are as follows: Environmental sustainability (Daud and Rahman, 2011; Nicholas and Thapa, 2010), Social sustainability (Daud and Rahman, 2011; Nicholas and Thapa, 2010), Economic sustainability (Nicholas and Thapa, 2010), Supporting sustainability (Yan, 2008), Concern for losing future opportunities (Yan, 2008), Suspicious attitude (Yan, 2008), Temperate sustainable development (Yan, 2008).

4 Destination Stakeholders

Buhalis (2000) breaks tourism stakeholder into five groups consisting of the local community, tour operators, tourists, tourism entrepreneurs and the public sector and governments. When the destinations around the world are reviewed, it is seen that the substantial actors and stakeholders are the public institutions, visitors, university, private sector, local community, and non-governmental organizations.

Mitchell et al. (1997) argue that stakeholders may be business and system oriented. The managers may want to know almost anything about the stakeholders due to the goals such as surviving the business, improving its economy, controlling the loss, taking the opportunities, and being competitive. Based on this, the impact of stakeholders on the businesses and their importance are gathered under headings.

- The businesses and the stakeholders are in a relation.
- The businesses are dependent on the stakeholders, and vice versa.
- The power of stakeholders is above the businesses.
- The power of businesses is above the stakeholders.
- The businesses and stakeholders are interconnected.
- There are contractual relations between the businesses and stakeholders.
- The stakeholders have rights on the business.
- The stakeholders bear some risks.
- The stakeholders have interests in the businesses.

Fig. 1: Destination Stakeholders

Touristic products are important elements that make a destination appealing and attract the tourists. Being able to preserve the appeal of a destination and to market its touristic products, i.e. values in a manner to transfer them to the next generation are very important and interconnected matters. While this model, which is also supported by sustainable tourism, contributes to the development of those who currently live in the touristic destination and who live in the region, it is important for allowing the future generations to benefit from those values and for the continued development. Protecting the cultural, historical, and social values of a site defined as cultural heritage, and to transfer them to future generations by sustaining them should be considered as a matter to be paid attention socially with its financial dimension. Sustaining the local cultural heritage which is the primary factor that is requested and enjoyed by the tourists especially in cultural destinations as well as the elements associated with this may ensure boosting the desire to develop the destination in a sound manner and to protect it.

However, especially in Turkey, it is observed that the local instruments do not work properly for protecting and sustaining the cultural heritage in many cultural destinations. The best method to perform protecting and sustaining activities is to activate the protection motives of the locals by enlivening the governance principles. This may be realized through the cooperation of the local actors or stakeholders. The local stakeholders should cooperate to administer a destination, and they should enhance their common action and strategy development capabilities. The need for this cooperation especially comes forward in developing projects and strategies for the preservation of a destination's cultural values.

The preservation of the cultural heritage in any destination and its transfer to future generations are only possible through the cooperation and common actions of the stakeholders who work in the field and who have impacts. The cultural heritage may also be rapidly consumed or damaged like other products in a destination.

5 Conclusion and Suggestions

If we call all the underlying causes for the visitors to prefer a location as touristic products, the intangible values may also be considered as touristic products like tangible values. It is very likely that this heritage becomes eroded in cases without a proper strategy or sustainable conception at the marketing stage of the tangible and intangible cultural heritage with strong demand. It is possible to develop a sustainable conception and to set the marketing channels on a sound plane in order eliminate such a consequence by establishing a proper cooperation mechanism between the stakeholders. The performance of an accurate analysis of the current situation in a destination, the SWOT (Strengths, Weaknesses, Opportunities and Threats) analysis covering all the factors that represent the value chain, and marketing activities through the correct positioning of the region's cultural heritage in accompany with the experts depend on the availability of such a mechanism. The protection of a cultural heritage of a destination as a result of proper coordination of stakeholders reflected in Fig. 1 would be more realistic than a destination where the actors act unaware of and independent from each other. Public policies and strategies toward keeping the cultural heritage in a destination alive in a sustainable manner may ensure the other actors to be included in the process by being educated with this perception. The studies conducted at the universities and by the experts on the matter and the recipes issued upon analyzing the associated needs of the region may have significant contributions for the elimination of similar problems in the region.

The private sector representatives offering marketing, accommodation and other services in the region constitute another important stakeholder group. Informing the actors in the field regarding the matters such as the requirements of sustainable tourism and the importance of protecting cultural heritage and determining their roles in the protection of the destination and values in advance are prominent for the functioning of the cooperation mechanism. Concomitantly, the cooperation between the stakeholders may serve to protect the cultural heritage as a result of including the organizations such as cooperatives which may involve the voluntary non-governmental organizations, disadvantageous groups and local community in the management system with the same perception and awareness.

The promotion and marketing strategies planned by determining a common agenda and in a manner to observe the protection of cultural heritage moves parallel to the improvement of common action culture of all the actors in the field. Supporting the local community which may be regarded as the most important characteristics of a destination and increasing their income from the tourism sector by growing them stronger in the sector are important requirements as these would contribute to the protection of the cultural heritage. Therefore, strong local community strategy in a touristic destination which is a part of the sustainable tourism may generally contribute to the protection of the cultural heritage. While doing this, raising awareness of the tourists constituting another stakeholder group regarding the destination and the cultural heritage through training is also a product of a common strategy. In cases and destinations lacking stakeholder cooperation, the similar problems are very frequent.

Cappadocia may be one of the most important destinations accommodating the aforementioned matters in its history. It is known that the region which does not have any master plan, or an advanced tourism strategy was severely damaged especially in terms of cultural heritage upon being listed in UNESCO's World Heritage List in 1985. The lack of a mechanism which may allow the stakeholders in the region to cooperate and rendering the existing cooperation weak brought along a gap of management.

References

Assante, L. M., Wen, H. I. & Lottig, K. (2012). An empirical assessment of residents' attitudes for sustainable tourism development: a case study of O'ahu, Hawai'i, *Journal of Sustainability and Green Business*, 1, 1–27.

Buhalis, D. (2000). Marketing the competitive destination of the future. *Tourism Management*, 21(1), 97–116.

Choi, H. & Murray, I. (2010). Resident attitudes toward sustainable community tourism, *Journal of Sustainable Tourism*, 18(4), 575–594.

Choi, H. & Sirakaya, E. (2005). Measuring residents' attitude toward sustainable tourism: Development of sustainable tourism attitude scale, *Journal of Travel Research* 2005(43), 380.

CNN Travel (2020). http://edition.cnn.com/2013/04/28/travel/20-beautiful-unesco-sites/index.html Erişim Tarihi: 10.08.2020.

Daud, N. & Rahman, S. A. (2011). *Tourist Attitudes towards Sustainable Tourism: Empirical Evidence From Malaysian National Park, Taman Negara*, 2010 International Conference on E-business, Management and Economics IPEDR vol.3, IACSIT Press, Hong Kong.

Egresi, I. (Ed.). (2016). *Alternative Tourism in Turkey: Role, Potential Development and Sustainability* (Vol. 121). Springer, Switzerland.

Gelecek Turizmde, (2020). Sürdülebilir Turizm Nedir, https://www.gelecekturizmde.com/surdurulebilir-turizm-nedir/ Access date: 18.07.2020.

İncedayı, D. (2004). Çevresel Duyarlılık Bağlamında Davranış Biçimi Olarak Sürdürülebilirlik, *Mimarlık Dergisi*, Sayı 318, s 39–44.

Kozak, M. A. & Bahçe, A. S. (2012). *Özel İlgi Turizmi*. Ankara: Detay.

Mitchell, R. K., Agle, B. R. & Wood, D. J. (1997). Toward a theory of stakeholder identification and salience: Defining the principle of who and what really counts. *The Academy of Management Review*, 22(4), 853–886.

Mowforth, M. & Munt, I. (1998). *Tourism and Sustainability: New Tourism in the Third World*, London: Routledge Press.

Nicholas, L. & Thapa, B. (2010). Visitor perspectives on sustainable tourism development in the Pitons Management Area World Heritage Site, St. Lucia *Environment, Development and Sustainability*, 12(5), 839–857.

Poria, Y., Reichel, A. & Cohen, R. (2011). World heritage site—Is it an effective brand name? A case study of a religious heritage site. *Journal of Travel Research*, 50(5), 482–495.

Ruhanen, L. (2013). Local government: Facilitator or inhibitor of sustainable tourism development. *Journal of Sustainable Tourism*, 21(1), 80–98.

Saarinen, J. (2006). Traditions of sustainability in tourism studies. *Annals of Tourism Research*, 33(4), 1121–1140.

Sarkım, M. (2007). *Sürdürülebilir turizm kapsamında turistik ürün çeşitlendirme politikaları ve Antalya örneği*, Dokuz Eylül Üniversitesi Sosyal Bilimleri Enstitüsü Doktora Tezi, İzmir.

Sassi, P. (2006). *Strategies for Sustainable Architecture*. New York: Taylor and Francis E-Library.

Sharpley, R. (2009). *Tourism Development and the Environment: Beyond Sustainability*, Earthscan First Published, UK and USA.

Üçer, N. (2011). *Tarihi Dokuların Sürdürülebilirliğinde Turizm Kaynaklı Değişimlerin Etkileri: Kuşadası Dağ Ve Camii-Kebir Mahalleleri Örneği*, Yüksek Lisans Tezi, Dokuz Eylül Üniversitesi, İzmir.

Yan, J. (2008). Is the Chinese Tourist Ready for Sustainable Tourism(s) Attitudes and preferences for sustainable tourism services, Ph. D. Program, Georg-August-University Göttingen, Germany.

Handan ÖZÇELİK BOZKURT

27 Pandemics and Tourism

1 Introduction

Tourism is one of the sectors most affected by many processes and developments. These processes and developments can be exemplified by economic and political crises, natural disasters, terrorist incidents, etc. In those situations people feel extremely vulnerable and their home or area of residence is where they trust most. Therefore, extraordinary developments hinder tourism mobility. For example, after the tsunami that took place in the Indian Ocean in 2004, the number of international tourists coming to Phuket in the first half of 2005 decreased by 67.2 %. Along with these developments, approximately 500 tourism companies were closed and more than 3000 employees lost their jobs (Henderson, 2007). With the outbreak of avian flu in 2005, around 12 million tourists were lost in Asian and Pacific countries (Wilder- Smith, 2006).

Global epidemic diseases such as avian flu, SARS, swine flu, Covid-19 are considered pandemics. Along with these epidemics, many factors such as markets and human psychology are likely to be negatively affected by the change. In such an environment, tourism sector employees may encounter difficulties such as job insecurity, job loss, etc. as well as the depression they experience due to the pandemic. Because the number of visitors decreased significantly during the pandemic period, tourism businesses are also likely to experience organizational downsizing or organizational decline. Tourism is a very important source of power for the country's economy. The current account deficit, which occurs with the decrease in tourism activity, puts the country's economy even more difficult. In 2020, when the Covid-19 epidemic was experienced, the devastating effect of pandemics on the tourism sector came to the fore again. It is important to examine the effects of pandemic periods on the tourism industry. Because it is useful in determining the extent to which the sector employees, those who earn indirectly from tourism (food and beverage producers, furniture companies, oil enterprises, etc.) and the country's economy are affected by this situation.

In this study, the pandemic issue that is closely related to the tourism sector will be examined. In addition, the ongoing Covid-19 pandemic as well as the major pandemics that have been experienced from the modern age to the present will be mentioned and the effects of these pandemics on the tourism sector will be emphasized.

2 Pandemics and Tourism

Infectious diseases are known as epidemics. Pandemics are contagious, epidemic diseases that spread to very large geographies, sometimes to a continent, and even to the whole world, causing diseases and deaths in humans or animals (Aslan, 2020). The main reasons for the rapid increase in infectious diseases in the 21st century are; growing population every year and people getting closer to each other with the increase of transportation possibilities, the spread of industrial foods, uncontrolled food consumption, etc. can be listed as the reasons (Pongsiri et al., 2009; Labonte et al., 2011). As a result of the global change, epidemic and pandemic cases started to increase. According to Potter (2001), there have been about 10 pandemics in the last 300 years. Tourism is one of the sectors most affected by contact, respiratory or sexually transmitted epidemics. In this respect, it is important to have information about the main pandemics that have been experienced and recorded from the past to the present and to examine the effects of these pandemics in the tourism sector.

Spanish Flu and Effects on Tourism

Probably the first pandemic of the modern era is the Spanish flu. It is the most lethal among the pandemics recorded (Holtenius & Gillman, 2014). Although the main place where the disease occurred was the USA, Spain was the first country to report the epidemic, and the name of the pandemic remained Spanish flu. It is estimated that the Spanish flu, which occurred in 1918, infected about 500 million people, and about 21 to 100 million people died from the virus (Johnson & Mueller, 2002; Jeffery & David, 2006).

Measures such as quarantine and travel restrictions in Spanish flu, which lasted about two years (Korstanje, 2011), are estimated to reduce mortality by 50 % (Hatchett et al., 2007). During the pandemic period, a four-month travel restriction was applied (Hospitalitynet.org, 2020). It is thought that this situation sheds light on what needs to be done in later pandemics (Ferguson et al., 2006). The fact that the pandemic occurred during the First World War caused the disease to spread to large masses due to the displacement of the soldiers. However, the emergence of the Spanish flu during the war greatly reduced the impact of the pandemic on international tourism. Because the tourism activity that started at the end of the 19th century came to a halt with the effect of the First World War (www.wishtrip.com, 2020). In the event of a pandemic similar to the Spanish flu in 2020, it is estimated that there will be a decrease of 9 trillion USD in gross domestic product (GDP) on a global basis (McKibbin & Fernando, 2020).

Asian Flu and Effects on Tourism

According to the estimation of the World Health Organization (WHO, 2009), the number of people who lost their lives due to the Asian flu pandemic that emerged in 1957 is around 1 to 4 million people. Rapidly developed vaccine against Asian flu significantly prevented the spread of the disease (Hall, Scott & Gössling, 2020). No source could be found that examines the effects of Asian flu on the tourism industry. However, in one study, it was stated that the Asian flu had a negative impact on the gross domestic product (GDP), income and development level of countries. In addition, it has been stated that Asian flu has negatively affected East Asia, Central Asia, Middle East, Europe, North Africa and South Asia in terms of human and economic effects (Gössling, Scott & Hall, 2020). In addition, this pandemic period, which coincides with the recovery process after World War II, was likely to have negatively affected the tourism sector.

Cholera and Effects on Tourism

Cholera is a pandemic type, the seventh of which emerged in South Asia in 1961 and still exists. Looking at the annual data, it is seen that 1.4 to 4 million new cases emerged and around 143,000 people died due to cholera (Hall et al., 2020). It is a disease with acute and severe diarrhea due to intestinal infection caused by Vibrio cholerae bacteria. Cholera, which was a local feature in India until 1817, later turned into a pandemic and spread to the continents of Asia, Africa, Europe and America. The largest cholera outbreak in Turkey was seen during the 1912–1913 Balkan wars and many soldiers died due to the illness (Yılmaz, 2017; Aslan, 2020). Cholera outbreaks in Zimbabwe (2008–2009), Haiti (since 2010) and Yemen (since 2016) in recent years are the most important (Hall et al., 2020).

Domestic and foreign tourism demands of countries affected by cholera are also negatively affected. Especially when cholera starts to spread in African countries, developed country managers warn their citizens not to visit these countries. This situation negatively affects tourism revenues and causes those who employ in tourism to become unemployed (Kirigia et al., 2009). Reluctant behavior in reporting corona case numbers of countries such as India also prevents access to accurate statistical information. Because countries with cases may want to prevent the tourism sector from being adversely affected by the pandemic (Zuckerman, Rombo & Fisch, 2007). For example, the cost of the pandemic to the country was 770 million dollars due to the food trade embargoes and negative effects on tourism caused by the cholera epidemic that emerged in Peru in 1991 (WHO, 2020).

HIV/AIDS and Effects on Tourism

AIDS/HIV (Human Immunodeficiency Virus) is an infectious disease that causes the human immune system to collapse. It is usually a sexually transmitted disease. It is a global active pandemic and has no cure (Aslan, 2020). A rare HIV/AIDS disease was first described in a group of homosexual men and immigrants from Haiti in the United States in 1981 (Tümer, 2011). According to some sources, the emergence of HIV virus dates back to 1960 (Hall et al., 2020). In fact, it is claimed that the virus first appeared in Cameroon in 1908 (hürriyet.com.tr, 2015). HIV gradually penetrates the immune system, destroying resistance to infections, making the individual vulnerable to illnesses and causing death. According to 2019 data, 38 million people in the world live with HIV virus. In 2019 alone, the number of people infected with HIV was 1.7 million. It is estimated that an average of 7.1 million people live without knowing that they have HIV virus (UNAIDS, 2020).

Tourism is one of the sectors most affected by HIV virüs (Ojha & Pradhan, 2006). The virus indirectly affects the tourism sector employers. Because costs such as travel restrictions, staff training, various analyzes and personnel replacement arise (Gaigbe-Togbe & Weinberger, 2003). In a study conducted in Zimbabwe, it was determined that 8 out of 9 hotels provide training to their employees on HIV virüs (Zengeni & Zengeni, 2012). HIV virus is a pandemic that reduces labor productivity and increases labor costs in the tourism industry (Afsar, 2010). HIV concerns of tourists can prevent them from visiting some tourist destinations. Countries with high incidence such as Kenya and Thailand are examples of these tourist destinations (Forsythe, Hasbun & Lister, 1998).

SARS and Effects on Tourism

SARS virus first appeared in the Guangdong Province of China on November in 2002 (WHO, 2003). SARS virus, thought to be caused by close contact with wild game animals, had a very high human-to-human transmission rate and within three months spread globally (Yang et al, 2020). The virus, which reached large masses in 2003, especially negatively affected the demand for tourism in the Asian continent. In March 2003, when a tourist staying in a hotel in Hong Kong infected many people in the same hotel, SARS virus took a great place in the world press. Seven infected tourists spread to different countries and continents (McKercher & Chon, 2004). Until SARS virus appeared, the concepts of health and tourism aroused the idea of "traveling to be treated" in people, while after this event, tourism had an activity image that increased virus radiation and created concern (Rodriguez-Garcia, 2001).

The World Travel and Tourism Council (WTTC 2003) reported that China, Hong Kong, Singapore, and Vietnam were the countries most affected by this virus and estimated that close to 3 million people lost their jobs and about $ 20 billion in gross domestic product (GDP) loss occurred in these four countries. Even in the countries of the Asian continent where the disease is not seen, the number of tourists has decreased considerably and the number of visitors to the continent has decreased by over 70 % in this period. Worldwide, 8096 people were infected with the SARS virus and 774 people died (WHO, 2006).

Avian Flu and Effects on Tourism

Avian flu (bird flu) is a deadly viral disease of animals. It has been noted that different types of avian flu appeared at different times in the past (Lee & Chen, 2011). The countries where deaths occurred due to avian flu that occurred in China between December 2003 and July 2007 are listed as Indonesia, Vietnam, Egypt, China and Thailand (McAleer et al., 2010). In this pandemic, in which a total of 319 cases were detected, 192 people died due to avian flu. According to the World Health Organization, the possibility of killing avian flu is as high as 59 %, and the epidemic has been seen in about 50 countries (WHO, 2006). Kuo et al. (2008) compared the current tourist loss from SARS virus with the potential tourist loss from avian flu, and predicted that SARS virus would cause a much greater reduction in tourism demand than avian flu. Lee and Chen (2011) found that most of the third age group tourists who visited the Asian continent in 2005 did not cancel their vacations due to avian flu. Researchers estimate that this may be due to the low mortality rates from avian flu.

Considering the data, it is seen that the number of tourists visiting Indonesia, Vietnam, China and Thailand decreased in 2003. After the decline, in 2004 it was observed that these countries exceeded the number of tourists in 2002. Although it is among the five countries with the highest number of cases and deaths, it is surprising that there is no decrease in the number of tourists in Egypt (UNWTO, 2019).

Swine Flu and Effects on Tourism

The swine flu epidemic first appeared in Mexico in March 2009 and then spread to England and other continents. Especially after its emergence in England, swine flu had a great impact in the media. It cannot be said that the world is completely vulnerable to swine flu due to the bird flu that occurred in 2005–2006 and the preparations made by the World Health Organization for the new epidemic (Page, Song & Wu, 2012). The fact that swine flu is a species transmitted by coughing, sneezing, coughing, talking, etc., made people very worried (Haque & Haque, 2018).

The countries most affected by the swine flu pandemic have been the United States, Mexico, Canada, Japan and Spain. USA, Mexico and Japan in 2010; Spain, on the other hand, reached the number of tourists arriving before the epidemic in 2012 (Çeti & Ünlüönen, 2019; WorldBank, 2020). Canada, on the other hand, could not reach the number of tourists in 2008 until 2015 (TIAC-AITC, 2016). Mexico, where the swine flu epidemic first appeared, is estimated to have lost 3.2 billion USD of GDP as well as 2.8 billion USD of tourism income (Hall et al., 2020).

Ebola and Effects on Tourism

It is thought that the Ebola virus type that exists today first appeared in Guinea on December 26, 2013 (Maphanga & Henama, 2019). According to the data of the World Health Organization, a total of 28,652 cases were seen and 11,325 deaths occurred until March 29, 2016. The countries with the highest number of cases are listed as Sierra Leone (14,124), Liberia (10,678), Guinea (3814), Nigeria (20) and Mali (8). 11,324 of the deaths occurred in these five countries. The rest 1 death occurred in the United States of America, where four transmission cases were seen (WHO, 2020).

In 2013, when the Ebola epidemic started, no decrease was observed in the number of tourists in other countries except Guinea. Guinea exceeded the number of tourists arrival before the epidemic in 2016. Guinea exceeded the number of tourists arrival before the epidemic in 2016. However, when it comes to 2014, it is seen that there is a decrease in Sierra Leone, and there are no statistics for Nigeria. It can be said that the tourism sector in Mali and the United States is not affected by the pandemic. The fact that these two countries have relatively fewer cases and deaths than other countries can be seen as the reason for the decrease in the number of tourists (Çeti & Ünlüönen, 2019; WHO, 2020). Ebola vaccine is available today, but it is estimated that the epidemic caused a loss of between 2.8 and 32.6 billion dollars in GDP (Hall et al., 2020).

Dengue Fever and Its Effects on Tourism

Dengue fever is an epidemic originating from a viral vector and spreading globally (Gratz & Knudsen, 1996). The disease is caused by four different types of viruses transmitted by aedes mosquitoes (Gubler, 1998). If the mosquito bites a viral person and the same fly bites another person within 8 to 12 days, the dengue virus is transmitted (Oduber, Ridderstaat & Martens, 2014). Dengue fever occurs worldwide in tropical and subtropical climates, mostly in urban and semi-urban areas. Early diagnosis and use of correct treatment methods reduce the mortality rate due to dengue fever below 1 %. Half of the world's population is at risk due to this pandemic, which has increased in recent years, and approximately 100 to 400 million people are infected with dengue every year. There is no definitive treatment method today (WHO, 2020).

Schwartz et al., (2008) examined the distribution of dengue fever among tourists and observed that this rate was quite high in those who visited South Asia, South-Central Asia, South and the Caribbean. The regions where the virus is common are usually tourist destinations (Wilder-Smith & Schwartz, 2005). It is estimated that the international tourist volume in Asia and Latin America by 2030 will be around 1.8 billion. However, the possibility that these predictions will not materialize due to the dengue epidemic is also considered alınmaktadır (Wilder-Smith, 2012; UN News, 2017). International travel to these areas has increased the spread of dengue fever. With the elimination of pandemic diseases such as dengue fever and ebola, it is estimated that there will be an increase of approximately 10 million dollars in the number of tourists and 120 billion dollars in tourism expenditures (Rosello et al., 2017).

Covid-19 (Coronavirus) and Effects on Tourism

Coronavirus is Coronavirus is an epidemic that destroys weak bodies and drives them to death (Huang, Wang & Li, 2020). In other words, this virus exploits the cells and take them over. It causes death by deforming the cells of patients with weak immune systems (Alpago & Alpago, 2020). The new type of coronavirus was seen in employees of the Chinese Seafood City Market in Wuhan City, Hubei Province, People's Republic of China, and the first cases appeared on December 31, 2019. The disease spread rapidly all over the world due to the human-to-human transmission characteristic of the new type of coronavirus (T.C. Sağlık Bakanlığı, Covid-19 Rehberi, 2020). According to current data, the total number of coronavirus cases in the world is 18,100,204 and the number of deaths from coronavirus is 690,257 (WHO, 2020).

Tourism is one of the leading sectors affected by global and national measures to slow down and stop the spread of the disease, such as travel restrictions, stopping flights, curfews, and suggestions for staying at home (Eryılmaz, 2020). After the epidemic spread to different continents and turned into a pandemic, tourism activities were stopped in many countries. As customers who were anxious at first wanted to cancel their flights, and when cases increased, flights began to be banned. Hotels were closed due to the virus, and travel agencies had to suspend their activities. After the virus was partially controlled, first domestic flights and then international flights were allowed, and hotels started to operate at half capacity and in line with the measures determined by the Ministries of Health. Although these practices vary by country, they are generally implemented in a similar way. In the first four months of 2020, the number of international tourists decreased by 44 % worldwide, and international flights decreased by around 97 % in April. Compared to 2019, this ratio means 180 million international tourists and 195 billion dollars loss in tourism income. The first regions affected in the coronavirus pandemic are Asia and the Pacific with a 51 % reduction in the number of visitors during January-April 2020. The ranking is Europe (-44 %), Middle East (-40 %), America (-36 %) and Africa (-35 %) (UNWTO, 2020). Coronavirus epidemic is still ongoing and it is very difficult to predict what impact it will have on the tourism industry in the future.

3 Conclusion and Suggestions

Pandemics lead people to panic and anxiety, as well as bringing primary needs to the forefront and away from products with high demand flexibility, such as travel. Although tourism has started to take its place among the basic needs today, its hierarchical order is behind. In addition to priority of needs, flight and travel bans applied in pandemic processes negatively affect accommodation businesses, travel agencies, airline companies, people working in the tourism sector and many businesses that earn indirect tourism income (fuel companies, food companies, hotel furniture manufacturers, etc.).

Determining the level of exposure to pandemics sheds light on the current and future developments. When the main pandemics that have been experienced and recorded so far in the study are examined, it is seen that each pandemic has negative effects at different levels, decreases the number of tourists, tourism incomes and causes a decrease in GDP. For example, Canada did not reach the number of tourists in 2008 until 2015 after the swine flu that occurred in 2009. This situation reveals that pandemics have a significant impact on the tourism image. It is important to examine the attempts of Mexico and Japan (WorldBank, 2020), which were able to recover faster in the same pandemic and reached the number of tourists in 2008 in 2010, to destroy the pandemic image or to investigate other factors that made this difference.

Compliance with the tourism measures applied in pandemics, the state's financial support to businesses in this context, and the low number of virus contamination caused by tourism may encourage people to participate in the tourism movement. At the same time, as seen in the Covid-19 pandemic, outbreaks with high spreading power occur and this situation spreads to all areas of life. For this reason, learning to live with the pandemic and shaping life accordingly can reduce the level of pandemic exposure.

It is extremely important to observe the situation of tourism workers who lost their jobs and who are concerned about job insecurity. In addition, it would be useful to seek solutions for tourism businesses experiencing institutional downsizing or organizational decline or bankruptcy. It is possible that the socio-economic problems that these stakeholders will experience until the time when the necessary vaccine or definitive treatment method is found for the ongoing Covid-19 pandemic will turn into different social problems (intra-family conflict, depression, suicide, etc.). At the same time, organizational and individual problems such as professional career regret, low organizational commitment and intention to leave the profession may occur in tourism sector employees. In this respect, it is important for the individual, organizational and social welfare to provide financial and moral support to tourism employees until the day when the transition to a completely normalization process in pandemics is achieved.

References

Afsar, S. M. (2010). *HIV/AIDS & Tourism*. Presentation at the TOI general meeting on 11 November, London. Retrieved on June 2014. Web:http://www.ilo.org/wcmsp5/groups/public/@ed_dialogue/@sector/documents/presentation/wcms_162293.pdf (Accessed: 23.07.2020).

Alpago, H. & Alpago, D. O. (2020). Koronavirüs ve sosyoekonomik sonuçlar. *IBAD Sosyal Bilimler Dergisi*, (8), 99–114.

Aslan, R. (2020). Tarihten günümüze epidemiler, pandemiler ve covid-19, *Ayrıntı Dergisi*, 8(85), 35–41.

Çeti, B. & Ünlüönen, K. (2019). Salgın hastalıklar sebebiyle oluşan krizlerin turizm sektörü üzerindeki etkisinin değerlendirilmesi. *AHBVÜ Turizm Fakültesi Dergisi*, 22(2), 109–128.

Eryılmaz, B. (2020). Türkiye'de faaliyet gösteren otel zincirlerinin yeni tip koronavirüs (covid-19) bilgilendirmeleri, *Turizm Akademik Dergisi*, 7(1), 15–27.

Ferguson, N. M., Cummings, D. A., Fraser, C., Cajka, J. C., Cooley, P. C., & Burke, D. S. (2006). Strategies for mitigating an influenza pandemic. *Nature*, 442(7101), 448–452.

Forsythe, S., Hasbun, J., & Butler de Lister, M. (1998). Protecting paradise: Tourism and AIDS in the Dominican Republic. *Health Policy Plan*, 13(3), 277–286.

Gaigbe-Togbe, V., & Weinberger, M. B. (2004). The social and economic implications of HIV/AIDS. *African Population Studies,* 19(1), 31–59.

Gössling, S., Scott, D., & Hall, C. M. (2020). Pandemics, tourism and global change: A rapid assessment of COVID-19. *Journal of Sustainable Tourism.* Ahead-of-Print, 1–20. https://doi.org/10.1080/09669582.2020.1758708 (Accessed: 02.07.2020).

Gratz, N. G., & Knudsen, A. B. (1996). The rise and spread of dengue, dengue haemorrhagic fever, and its vectors. A historical review (up to 1995). *Geneva: World Health Organization*. Web: https://apps.who.int/iris/handle/10665/64454 (Accessed: 01.08.2020).

Gubler, D. J. (1998). Dengue and dengue hemorrhagic fever. *Clin. Microbiol. Rev.* 11(1), 480–496.

Hall, C. M., Scott, D. & Gössling, S. (2020). Pandemics, transformations and tourism: be careful what you wish for. *Tourism Geographies*, 22(3), 577–598.

Haque, T. H., & Haque, O. (2018). The swine flu and its impacts on tourism in Brunei. *Journal of Hospitality and Tourism Management*, 36(1), 92–101.

Hatchett, R. J., Mecher, C. E., & Lipsitch, M. (2007). Public health interventions and epidemic intensity during the 1918 influenza pandemic. *Proceedings of the National Academy of Sciences*, 104(18), 7582–7587.

Henderson, J. C. (2007). *Tourism Crises: Causes, Consequences and Management*. USA: Butterworth-Heineman.

Holtenius, J., & Gillman, A. (2014). The Spanish flu in Uppsala, clinical and epidemiological impact of the influenza pandemic 1918–1919 on a Swedish county. *Infection Ecology & Epidemiology,* 4(1), Art. 21528.

Hospitality.net (2020). *Unpacking the Impacts of COVID-19 on Tourism and Repackaging the Hotel Service* Web:https://www.hospitalitynet.org/opinion/4098657.html (Accessed: 12.07.2020).

Huang, C., Wang, Y., & Li, X. (2020). Clinical features of patients infected with 2019 novel coronavirus in Wuhan. *China, Lancet*, 395(10223), 497–506.

Hürriyet.com.tr (2015). AIDS Nasıl Ortaya Çıktı? https://www.hurriyet.com.tr/avrupa/aids-nasil-ortaya-cikti-28275375 (Accessed: 23.07.2020).

Jeffery, K. T., & David, M. M. (2006). 1918 influenza: The mother of all pandemics. *Emerging Infectious Diseases*, 12(1), 15–22.

Johnson, N. P., & Mueller, J. (2002). Updating the accounts: global mortality of the 'Spanish' 1918–1920 influenza pandemic. *Bulletin of the History of Medicine*, 76(1), 105–115.

Kirigia, J. M., Sambo, L. G., Yokouide, A., Soumbey-Alley, E., Muthuri, L. K., & Kirigia, D. G. (2009). Economic burden of cholera in the WHO African region. *BMC International Health and Human Rights,* 9(8), 1–14.

Korstanje, M. E. (2011). Swine flu in buenos aires: beyond the principle of resilience. *International Journal of Disaster Resilience in the Built Environment,* 2(1), 59–73.

Kuo, H. I., Chen, C. C., Tseng, W. C., Ju, L. F., & Huang, B. W. (2008). Assessing impacts of SARS and avian flu on international tourism demand to Asia. *Tourism Management*, 29(5), 917–928.

Labonte, R., Mohindra, K., & Schrecker, T. (2011). The growing impact of globalization for health and public health practice. *Annual Review of Public Health, 32*(1), 263–283.

Lee, C., & Chen, C. (2011). The reaction of elderly Asian tourists to avian influenza and SARS. *Tourism Management*, 32(6), 1421–1422.

Maphanga, P. M., & Henama, U. S. (2019). The tourism impact of ebola in Africa: Lessons on crisis management, *African Journal of Hospitality, Tourism and Leisure,* 8(3), 1–13.

McAleer, M., Huang, B., Kuo, H., Chen, C., & Chang, C. (2010). An econometric analysis of SARS and avian flu on international tourist arrivals to Asia. *Environmental Modelling & Software*, 25(1), 100–106.

McKercher, B., & Chon, K. (2004). The over-reaction to SARS and the collapse of Asian tourism. *Annals of Tourism Research,* 31(3), 716–719.

McKibbin, W., & Fernando, R. (2020). *The Global Macroeconomic Impacts of COVID-19: Seven Scenarios* (CAMA Working paper 19/2020). Australian NationalUniversity.Web:https://papers.ssrn.com/sol3/papers.cfm?abstract_id=3547729# (Accessed: 12.07.2020).

Oduber, M., Ridderstaat, J., & Martens, P. (2014). The bilateral relationship between tourism and dengue occurrence: Evidence from Aruba. *Journal of Tourism and Hospitality Management,* 2(6), 223–244.

Ojha, V. P., & Pradhan, B. K. (2006). The Macro-Economic and Sectoral Impacts of HIV and AIDS in India, (Macroeconomics Working Papers 22134), East Asian Bureau of Economic Research. Web:https://ideas.repec.org/p/eab/macroe/22134.html (Accessed:23.07.2020).

Page, S., Song, H., & Wu, D. C. (2012). Assessing the impacts of the global economic crisis and swine flu on inbound tourism demand in the United Kingdom. *Journal of Travel Research*, 51(2), 142–153.

Peiris, J. S., Lai, S. T., Poon, L. L. M., Guan, Y., Yam, L. Y. C, Lim, W., & Yuen, K. Y. (2003). Coronavirus as a possible cause of severe acute respiratory syndrome, *Lancet,* 361 (9366), 1319–1325.

Pongsiri, M. J., Roman, J., Ezenwa, V. O., Goldberg, T. L., Koren, H. S., Newbold, S. C., & Salkeld, D. J. (2009). Biodiversity loss affects global disease ecology. *BioScience*, 59(11), 945–954.

Potter, W. (2001). A history of influenza, *Journal of Applied Microbiology*, 91(1), 572–579.

Rodriguez-Garcia, R. (2001). The health-development link: Travel as a public health ıssue. *Journal of Contemporary Health,* 26(2), 93–112.

Rosello, J., Santana-Gallego, M., & Awans, W. (2017). Infectious disease risk and international tourism demand. *Health Policy and Planning*, 32(1), 538–548.

Schwartz, E., Weld, L. H., Wilder-Smith, A., von Sonnenburg, F., Keystone, J. S., Kain, K. C., & Freedman, D. O. (2008). Seasonality, annual trends, and characteristics of dengue among ill returned travelers, 1997–2006. *Emerging Infectious Diseases*, 14(7), 1081–1088.

T. C. Sağlık Bakanlığı (2020). *Halk Sağlığı Genel Müdürlüğü, Covid-19 (Sars, CoV-2 Enfeksiyonu) Rehberi, Bilim Kurulu Çalışması, 14 Nisan 2020,* https://covid19bilgi.saglik.gov.tr/depo/rehberler/COVID-19_Rehberi.pdf. (Accessed:04.08.2020).

TIAC-AITC (2016). *Annual Report on Canadian Tourism*. https://tiac-aitc.ca/_Library/TIAC_Publications/TIAC_Annual_Report_2016_EN.pdf (Accessed:19.07.2020).

Tümer, A. (2011). *HIV/AIDS Epidemiyolojisi ve Korunma*, http://www.hatam.hacettepe.edu.tr/EpidwebAT2011.pdf (Accessed:23.07.2020).

UNAIDS (2020). *Global HIV & AIDS Statistics—2020 Fact Sheet*. https://www.unaids.org/en/resources/fact-sheet (Accessed: 23.07.2020).

UNWTO (2019). *Dünya Turizm Örgütü (2020). All Countries: Inbound Tourism:Arrivals:1995–2017*. www.eunwto.org/doi/abs/10.5555/unwtotfb0000270019952017201901 (Accessed: 21.07.2020).

UNWTO (2020). *Market Intelligence UNWTO World Tourism Barometer*. https://www.unwto.org/market-intelligence. (Accessed: 04.08.2020).

Yang, Y., Peng, F., Wang, R., Guan, K., Jiang, T., Xu, G., Sun, J., & Chang, C. (2020). The deadly coronaviruses: the 2003 SARS pandemic and the 2020 novel coronavirus epidemic in China. *Journal of Autoimmun*, 109(1), 1–16.

Yılmaz, Ö. (2017). 1847 - 1848 kolera salgını ve osmanlı coğrafyasındaki etkileri. *Avrasya İncelemeleri Dergisi*, 6(1), 23–55.

WHO (2003). *Summary Table of SARS Cases by Country, 1 November 2002–7August 2003* https://www.who.int/csr/sars/country/2003_08_15/en/ (Accessed: 16.07.2020).

WHO (2006). *Cumulative Number of Reported Probable Cases of Severe Acute Respiratory Syndrome (SARS)*. https://www.who.int/csr/sars/country/en/ (Accessed: 17.07.2020).

WHO (2006). *Cumulative Number of Confirmed Human Cases of Avian Influenza A(H5N1) Reported to WHO*. https://www.who.int/influenza/human_animal_interface/H5N1_cumulative_table_archives/en/ (Accessed: 21.07.2020).

WHO (2009). *Past Pandemics*. https://www.euro.who.int/en/health-topics/communicable-diseases/influenza/pandemic-influenza/past-pandemics. (Accessed: 15.07.2020).

WHO (2020). *Global Epidemics and Impact of Cholera*. https://www.who.int/topics/cholera/impact/en/ (Accessed: 28.07.2020).

WHO (2020). *2014–2016 Ebola Outbreak in West Africa*. https://www.cdc.gov/vhf/ebola/history/2014-2016-outbreak/index.html (Accessed: 28.07.2020).

WHO (2020). *Dengue and Severe Dengue*. https://www.who.int/news-room/fact-<sheets/detail/dengue-and-severe-dengue (Accessed: 01.08.2020).

WHO (2020). *WHO Coronavirus Disease (COVID-19) Dashboard*. https://covid19.who.int/ (Accessed: 04.08.2020).

Wilder-Smith, A., & Schwartz, E. (2005). Dengue in travelers. *N Engl J Med,* 353(1), 924–932.

Wilder-Smith, A. (2006). The severe acute respiratory syndrome: Impact on travel and tourism. *Travel Medicine and Infections Disease*, 4(2), 53–60.

Wilder-Smith, A. (2012). Dengue infections in travellers. *Paediatr Int Child Health*, 32(1), 28–32.

WorldBank (2020). *International Tourism Number of Arrivals*. https://data.worldbank.org/indicator/ST.INT.ARVL (Accessed:07.08.2020).

WTTC (2003). Special SARS Analysis: Impact of Travel and Tourism (Hong Kong, China, Singapore and Vietnam reports). *World Travel and Tourism Council,* London.

Zengeni, D. M. F., & Zengeni, N. (2012). Impact of HIV/AIDS to the tourism sector human resources: Case of selected hotels in Harare, *International Journal of Development and Sustainability*, 1(3), 1088–1102.

Zuckerman, J. N., Rombo, L., & Fisch, A. (2007). The true burden and risk of cholera: Implications for prevention and control. *Lancet Infect Dis.* 7(1), 521–530. https://www.wishtrip.com/enterprise/post/what-past-pandemics-can teach-us-about-what-to-expect-for-tourism-recovery Erişim Tarihi: 12.07.2020.

Ediz GÜRİPEK

28 Sirac Turkmen Traditions in the Context of Cultural Heritage Tourism

1 Introduction

Cultural heritage is a sum of cultural values that societies have managed to achieve through the activity of their internal dynamism, nature, and integration with other cultures. Although the scope of cultural heritage is generally understood with its tangible characteristics, when the value of human-based social life is involved it is defined as "Intangible Cultural Heritage". Such values that have been passed down through the generations are supported by the protection programs carried out by UNESCO.

"Siracs", consisting of the Nomadic Turkish tribe living in Zile Provence of Tokat are attached to the Beydilli branch of the Ghuzz Turks. Despite they are still believing in Shamanism, to this date, they had been counted within the Alevite-Bektashi community in Anatolia. The aim of this study is to evaluate the traditions, cultural customs, and rituals of "Siracs" in the context of cultural tourism.

2 Cultural Heritage and Cultural Heritage Tourism

According to the Turkish Language Institution, culture is "the totality of all material and spiritual values created in the historical, social development process and the tools used to create and transmit them to the next generations, showing the measure of human sovereignty over the natural and social environment". It is also defined as; "thought and art specific to a society or the whole of a society's Works. (TDK, 2020).

The concept of culture is also very important in terms of tourism. After the holiday concept known as traditional tourism, it is considered as an important type of tourism with the largest share in tourism. This type of tourism is called cultural tourism. Cultural tourism consists of a wide spectrum. It covers many subjects such as eating and drinking habits, lifestyles, history, understanding of art, clothing, and behavior of a society.

Cultural tourism has become a hope for destinations that do not have a traditional holiday concept of sea, sand, and sun. Many Anatolian cities which are economically underdeveloped Anatolian cities, try to implement policies aimed at developing cultural tourism to get enough share from cultural tourism and to ensure economic vitality in its regions. Cultural heritage tourism is a type of tourism considered within cultural tourism. Although people are curious about different cultures, they want to see and learn different cultural heritage items. Therefore, they participate in cultural tourism.

Although cultural heritage is an old concept, it has become very popular today. In 1972, after the concept of cultural heritage was addressed in the "Convention Concerning the Protection of the World Cultural and Natural Heritage" by UNESCO, it led to the preparation of the "Convention for the Safeguarding of the Intangible Cultural Heritage" in 2003. From 1972 to 2003, concepts such as folklore, popular and traditional culture, intangible heritage, oral and intangible heritage were used by UNESCO in various documents (Oğuz, 2013).

Cultural heritage areas are collected under five headings in UNESCO's contract for the protection of intangible cultural heritage (UNESCO, 2003), these are:

a. oral traditions and expressions, including language as a vehicle of the intangible cultural heritage.
b. performing arts.
c. social practices, rituals, and festive events.
d. knowledge and practices concerning nature and the universe.
e. traditional craftsmanship.

The subject of this study, Sırac Turkmen traditions are evaluated within the scope of social activities, rituals, and festival practices specified under those headings.

3 Sirac Turkmen Culture

The word "Sirac" is not included in Turkish dictionaries. This word, which is referred to as Sirac (light lamp, torch) in Arabic, is used in the meaning of mysticism as; "who does not give secret", "who opens the veil of secrecy in the way of secret or truth", "who is open to the heart, sees the right". It is a local name used for the Beydilli Turkmen tribe. Sirac Turkmens are known to have lived in different cities in Turkey intensively in Tokat, Sivas, Çorum, Amasya, Yozgat (Yilmaz, 2009, p.17). The place where they mostly live is Tokat Zile. There are 18 Sirac villages in the district of Zile province.

Khorasan Region was conquered by Arab Muslims in the 8th century. As a result of Arab pressure, Shia movements were adopted and Turks living in this region were also affected. By the 12th century, Turks living in Turkistan and Khorasan regions gathered around the leader named Ahmet Yesevi. The Sirac

Turkmens in Zile and its surroundings came to life with the settlement of Ahmet Yesevi affiliates in Zile and its surroundings. The spiritual leaders of this community, grandparents, and fathers, are directly descended from Ahmet Yesevi His Holiness (Yılmaz, 2016).

Although the Siracs are considered within the Alevi-Bektashi communities, they are different from the Alevi-Bektashi communities as well as from the "Sunnis". Siracs are a closed society to the outside. They distinguish the Bektashis by calling them "alaca" and they also differ within themselves. The most different group among the ranks are those who have been devoted to Shamanism since the period they lived in Central Asia. Those living in the villages of Karacaoren and Kervansaray in the Zile district of Tokat are especially Sirac Turkmens in this group.

Shamanism is a form of belief that originated in Siberia and Central Asia. It is a set of beliefs and practices in the Tengri belief. Turks who believe in shamanism use the concept of "Kam" instead of the concept of "shaman". It is used as "Tengir" in some regions in Central Asia (Coruhlu, 2000).

Shamanism, which is not a religion but is common among Turkish communities, is defined as a sign of activity that combines the flight of the soul and the underground (Kafesoğlu, 1980, p. 29).

Shaman is defined as a person who acts as an intermediary between spirits and people and is accepted as having the power to heal the sick (İnan, 1972, p. 72). Shaman's ordinary ritual movements (Kamlama) are performed by singing, playing drums, making noise from ironware. It consists of the dance of "Esrime" (ecstatic state) during the pray. Kamlama is called the ability to contact spirits. It is done in two different ways. The first one; spirits enter the shaman's body (or his tambourine) or, on the contrary, the shaman's soul is sent on a journey into the kingdom of spirits. Indeed, the meaning of camouflage consists of hypnotizing those around him and himself, so that the shaman not only compels the audience to believe in his supernatural abilities but also often makes him believe that he and they are delusional during kamlama (Arvas, 2014)

According to the belief, the shaman drives away the spirits, chases some of the undergrounds, and goes up to the heavens to present the sacrifices to the high gods. He establishes a friendship with Gods in the world of darkness, "Ergul" in the sky, or "Erlik" underground and sees them and talks to them. Shaman then finds the souls of the patients, brings them back, and cures the disease (Kafesoğlu, 1980, p. 29). However, the biggest difference from other beliefs is the belief that there is a close relationship between a human being and his dead ancestors (Mandaloğlu, 2011).

In Shamanism, the Tengri, sun, earth, water, ancestors, and hearth/fire ashes are considered sacred. The idea of human and nature unity and harmony lies at the basis of Shamanism, which is believed by Asian people (Izgi, 2012). Because of its similarity with the Alevi-Bektashi beliefs, there are quite a lot of Siracs who have Alevi-Bektashi beliefs and who define themselves as Alevi among the Sirac groups.

There are two types of ceremonies in shamanism, namely ceremonies held on certain days or ceremonies held due to accidental events (Mandaloğlu, 2011). There are differences according to the beliefs, traditions, and customs of various peoples, there is always the number of sacrifices. In addition to the known ritual of sacrificing animals, offering something to a sacred place, a value, dedicating things, tying a cloth to the shaman's drum, sacred trees, offering food to deities made of various materials, pouring alcohol into the fire or throwing it into the fire are also considered bloodless sacrifices. Another form of bloodless sacrifice is animals devoted to the spirits and released into the countryside (Izgi, 2012). Such rituals that take place in shamanism are also included in the Sırac Turkmen culture.

Sirac Turkmen Traditions

Traditions in Sirac Turkmens differ according to various activities. These events are listed as weddings, births, circumcision, sending off soldiers, "Hıdırellez fasting and festival", "Ficenk festival", death and funeral burial rituals. Besides, eating and drinking culture, which is included in the Ordinary Turkmen traditions, is also evaluated in this section (This section is compiled from the information and observations obtained from the interviewees in the study).

Weddings: Due to being a more closed society in the past, Siracs did not establish family unity with other communities. However, they are not as strict today as in the past. "Baglama" is the only instrument used in weddings. This is because other instruments are seen as the devil's work. According to a rumor, at a wedding, the bride was scared by the sound of the drums and zurna, and the bride fell off the horse and was injured. For this reason, drums and zurna are prohibited at their weddings. The groom's side takes sherbet and pastry to the bride's house. The villagers, on the other hand, bring cracked wheat, wheat, and eggs to the wedding house. Tea and coffee are served at weddings. The groom is bathed in a large area by his relatives. The next day, the bride is bathed in the same way. Then "Henna" is applied. Although the duration of the wedding varies depending on the financial situation of the wedding party, the last day of the wedding should be Thursday. The bride must leave the father's house before the sun rises on Thursday morning. For this reason, the lover starts to play and sing at 3 am. This situation is called "dawn tattoo". As other Turks do not know "halay", instead "Semah" is performed in Alevi culture.

Birth: There is a saint or a tilt in almost every Sirac village. Especially women who do not have children make sacrifices (vows) by going to this saint or a place considered sacred. They make wishes by tying generations such as pieces of cloth and cloth cradles to places (The tree, window, etc.) in the holy place. The belief of grandfather in Alevism is also present in Siracs. "Hubyar Dede", "Veli Baba" and "Anşabacı", which are considered important figures among the Sirac Turkmens. In holy places they are also asked to donate children to their grandfathers, who are still living outside of their deposits.

Birth is usually done by experienced women in the village. However, if there is a health problem, you can go to the hospital. Also, those who do not have children are not welcomed to receive treatment due to health problems. Whether or not there is a child is seen as the discretion of Allah (Yılmaz, 2009, p.82).

After the newborn child's navel is cut, it is tied with a rope and kept under the cradle until forty. This is because it is not a traveler. It is then buried in a suitable place. If the child is desired to be educated, he is buried in the school garden.

After birth, forty shirts are worn to prevent forty pressings (mental distress or death). Women and children who have just given birth are not left alone for 40 days. This situation of not leaving it alone is also applied during pregnancy in some of the Sirac groups. According to the belief, if the woman is left alone during this period, her "Alkarisi (Albasti)" can be haunted and kill the mother. For this reason, items (knives, pins, men's clothes, etc.) believed to be afraid of "Alkarisi" are placed in the room or under the pillow. Until the baby comes out forty, it is covered with a yellow gauze to prevent jaundice.

After the baby 40 days old, 40 stones are collected and placed in the strainer, and water is poured through the Stones and to the baby. This is called "Forty Flyer".

Circumcision: Circumcision weddings are not common in Siracs. It is made during the young age of a boy without a ceremony. They state that the reason for this is to prevent the child from being afraid of circumcision.

Military Obligation Ceremony: As in all Turkish communities, military service is sacred in Siracs. Young people who will go to military service collect money among themselves to help the "Cemevi (Alevi Place of Worship)" or build a fountain in their village. As in other ceremonies, drums and flutes are not played and there is no entertainment. Visiting relatives, neighbors, and acquaintances, youngs say goodbye.

Hıdırellez Fast and Feast: This holiday is celebrated after the Hıdırellez fast, which is held for 7 days in February. This fast starts on Friday and ends on Thursday. This fast can be observed on different weeks in February in different Sirac villages. On the feast day, money is collected from the whole village and the unity is sacrificed. No money is received from the poor. Those who are in a good economic situation can also sacrifice for themselves if they wish. On Friday morning, the sacrificed animals are taken to the Cemevi and prayed there. This prayer is called "Kurban Takbirleme". The slaughtered sacrifice is distributed to all villagers (Yılmaz, 2009, p. 88). In some Sirac groups, celebration rituals on Ficenk feast are performed during the Hıdırellez holiday.

Animals, except turkey, goose, and duck, can be sacrificed in Siracs. Those with good economic income prefer sheep as a sacrifice, while those with lower economic status prefer smaller animals such as chicken or roosters.

Apart from the sacrifice of Hıdırellez, there are two more sacrificial cultures in Siracs. One of them is the vow. The person whose wish is accepted slaughtered the sacrifice he made and distributed it to the villager. If a wish is made from a deceased person, the sacrifice is slaughtered by pouring the blood of the animal at his feet. During the Hıdırellez festival, they perform a similar ritual by sacrificing animals in the graves of their dead relatives. Another "Kurban" is the Thanksgiving Sacrifice. This sacrifice happens after the harvest, such as late September and early October.

Ficenk Festival: It consists of the rituals performed by the group defined as "Anşabacılı" only on Saturday and Sunday in the third week of April every year. Ficenk is a cultural event celebrated as the resurrection awakening holiday celebrating the arrival of spring. Although it is similar to the celebrations known as Nevruz, the festival of Hıdırellez and its rituals are quite similar. On Saturday, during the Ficenk celebrations, the youth of the village collect fragrant herbs such as cloves and nettles from seven different mountains. With water brought from seven different sources, these herbs are thrown into a cauldron. Participants in the celebrations throw their personal belongings into the cauldron and the top of the cauldron is covered with a cover called "Cecim". The boiler is then left somewhere in the open area. This event is called "Souring". It is known that the boiler was boiled in the past. However, today boiling is not applied. Young people who will attend Ficenk put stones next to the chimney of their house. In the evening, a big fire is lit in the square and the young people jump over it. The next morning, the stones placed next to the chimneys are removed and the owners of the stones with insects under them are thought to be lucky. In the morning of the Ficenk, a fire is lit in front of the houses and it is jumped over. In this ritual, the Siracs believe that they poured their sins accumulated over a year into the fire and were purified.

Ficenk cauldron is brought to the area where the ficenk celebrations will be held and prayers are read by the devotee or grandfather. The eldest daughters (first daughters) of the families sit next to the ficenk cauldron and sing the mâni. Then Ficenk shooting begins. The hint that is said before each lottery becomes the person whose item is removed and is considered as fortune-telling belonging to him. After the draw is over, the Semah, which is called ascension to the sky, the appeal to the god, returns. Ficenk celebrations are completed by visiting seven different tombs.

Death and Funeral: There are two different beliefs death and after death in the Sırac culture. The first of these is the belief that you will go to heaven or hell after death. However, it is not very common. What is generally accepted by Siracs is the belief that everything ends after death. According to this belief, heaven and hell are in this world. The world is defined as the heaven for those who live well and as hell for those who live badly. It is believed that the soul passed on to another living being, depending on the situation it deserved. In other words, it is believed that the body dies, and the soul is born again. A deceased person is not said to be dead. Instead, expressions such as migrated, walked to the right are used.

His funerals are the ceremonies in which the Shamanism traditions can be seen more clearly. The deceased is placed on the bed in the big room of the house and covered with a rug, which is called "Cecim". The family, relatives, and close friends who gather in the room next to the fire. It is believed that this fire illuminated the person who died by burning fire for 3 days in the place where the body of the deceased was washed. This ritual is called "Kindling". However, the corpses of those who are Dede, Baba, and Sofu are not washed. Since the funeral rituals are different, they do not allow foreigners to attend funerals. For this reason, they hold their funerals early in the morning or after dark in the evening. A bed is laid under the funeral in the grave and the pillow is placed in the grave. The funeral is put in the grave, dressed in the best clothes. Then a quilt is covered. The belongings (reeds, ornaments, personal belongings, etc.) of the deceased person are placed around the bed in the grave and covered with soil. Meanwhile, Dede or Sofu recites sayings and prayers in Turkish. In this community, the Koran is not read, and there are no prayers in the Koran on tombstones. The meal, which is called the "life meal" for the person who died 7 days and 40 days after death, is given to the guests by his family.

Picture 1.1. Sirac Turkmens Dress
Source: http://tokat.gsb.gov.tr/

Sirac Turkmens Dress and Apparels

The clothes of the Ordinc Turkmens also differ. They were still wearing their traditional clothes even when they went out of their villages until 15–20 years ago. However, Siracs has adapted itself to modern life today and started to wear normal clothes. They keep their culture alive by wearing their traditional clothes on special occasions.

The attire of the Sirac Turkmen consists of very flashy and colorful pieces. Women; It consists of "salta (gazeki)", "inner upper", "three loincloths (outer upper)", "apple (hazelnut)", "saddle fez", "cover (fence)", "forehead chin strap", "hair tie", "Yaglik", "belt", "saltshaker", "back tie", "socks", "sandals" and "accessories". The definition of clothes seen in Pictures 1 and 2 are given below.

Salta: The name was given to the velvet fabric Jacket worn by women.
Inner upper: An inner skirt or dress made of etamine or cotton fabric worn by women.
Three chases: It is a garment worn by women with two skirts at the back. It is also known as "three skirt".
Elmalı: It is the most commonly used headcover with apple motifs.
Saddle fez: It is a fez made of felt worn on the head and used with a cover.
Cover: Used in a triangular shape with various motifs and worn over the fez.
Pediment: It is an ornament made of metals such as gold and silver similar to a necklace and attached to the front of the fez.
Chinstrap: It is a beaded thread that prevents the fez from falling and tied from the chin toward the temples.

Hair tie: It is a beaded rope attached to the hair.

Yağlık: It is also known as "Döşlük". A white fabric with edges that goes down to the waist between the neck of women and the lower part of the breasts.

Belt: It is a fabric made of woolen fabric and usually tied to the waist by elderly women.

Saltshaker: It is a bag decorated with beads and embroidery on the front. It is used by crossing the neck.

Back tie: It is a waist tie made of fine beads with a fringed tip.

Socks: They are hand-knitted socks. It takes names such as mirrored, engraved, dark, according to the color and pattern embroidered on it. The colored ones are called "pied".

Sandals: It is a shoe made of buffalo leather with colors such as yellow and red.

Accessories: They are jewelry such as bracelets, necklaces, earrings, amulets, cheeks.

Men wear outfits consisting of "Goynek", "zıvga shalwar", "vest", "workshop", "aba", "zıvga tie", "girdle", "slippers", "variegated socks", "sandals" and "accessories".

Goynek: It is a kind of underwear made of light-colored soft fabric.

Zıvga Shalwar: It is a black outfit worn instead of trousers, with an engraved mesh and woven from mohair.

Vest: It is a garment that is worn under buttoned aba, made of the same fabric as zıvga shalwar, and supported by a U-shaped carved and chain watch.

Workshop: Cotton fabric shirt worn under the vest.

Aba: is a garment worn over a vest, similar to today's jackets made of baggy fabric.

Zıvga tie: Zıvga is a rope decorated with blue beads and with tassels on both ends serving as a belt for shalwar.

Girdle: It is a fabric tied with a striped triangle consisting of yellow, red, and green colors, which is wrapped on a zigga tie and helps to keep the waist tight and upright.

Slippers: It is a fancy and embroidered men's headdress made of cotton fabric.

Variegated socks: It is a sock that extends from wool or mohair to a string of more than one color.

Sandals: Shoes made of buffalo leather and decorated with stitches.

Accessories: These are jewelry such as embroidered handkerchiefs, money pouches, and pocket watches.

Various sewing courses are organized for dolls made up of these clothes and income is earned by selling them.

Food and Beverage Culture

The most distinctive feature of the Sirac cuisine is dishes made with herbs that grow spontaneously in nature. These; town beet, rooted, hibiscus, curled dock, fodder, couscous bread, reddish, madımak, purslane, bridal toe, and sorrel. Dishes made with these herbs are cooked similarly. The herbs that are cleaned and washed before cooking are chopped, Bulgur added, and boiled with water. After it is cooked sufficiently, oil, onion, and spices are added to it and a little more cooked and served.

The culture of making and sharing thin bread, which is on the intangible cultural heritage list, continues in the Siracs. A few people come together, and this bread is made altogether. This unleavened bread, which resembles dough in the region, is called "Iskefe".

4 Conclusion and Suggestions

Sirac Turkmens continue their life with many rituals that are the subject of the study. To perform these rituals every year, Siracs, who live in different cities, come to their villages on the relevant dates and participate in these celebrations. This mobility is also considered a festival-like activity as touristic activity. It provides commercial, economic, and tourism dynamism to the region. Although the Siracs are a closed society, local administrators such as governors, district governors, and mayors, as well as members of the press participate in the events organized today.

The more events and possibilities organized within a destination, the more the destination grows. Especially local administrations that desire the development of tourism activities in the region should support such events and similar activities more and ensure wider participation.

One of the most important motivations of tourists participating in cultural tourism is curiosity. People who are curious about different cultures and communities participate in cultural tourism activities. Festivals such as "Hıdırellez" and "Ficenk" must be evaluated within this scope and wider participation should be ensured. Besides, it is recommended to make an intangible cultural heritage application to UNESCO by the local governments in the region, as in "Nomadic cultural elements", to record and protect these cultures of the Sirac Turkmens.

References

Arvas, A. (2014). Şamanizm. *Uluslararası Türkçe Edebiyat Kültür Eğitim Dergisi.* 3(3), 49–71.

Çoruhlu, Y. (2000). *Türk mitolojisinin anahatları.* İstanbul: Kabalcı Yayınevi.

İnan, A. (1972). *Tarihte ve bugün Şamanizm.* Ankara: Türk Tarih Kurumu.

İzgi, M. C. (2012). Şamanizm ve Şamanlara genel bakış. *Lokman Hekim Dergisi.* 2(1), 31–38.

Mandaloğlu, M. (2011). Türk kültür çerçevesinde Şamanizm ve Şamanlık meselesi. *Türkiye Sosyal Araştırmalar Dergisi.* 15(3), 111–122.

Oğuz, M. Ö. (2013). Terim olarak somut olmayan kültürel miras. *Milli Folklor,* 25 (100), 5–13.

TDK. (2020). *Güncel Türkçe Sözlük,* https://sozluk.gov.tr/ (Accessed: 10.08.2020).

UNESCO. (2003). *Convention for the Safeguarding of the Intangible Cultural Heritage.* https://ich.unesco.org/en/convention (Accessed: 10.08.2020).

Yılmaz, O. (2009). *Sıraçlar (Anşabacılı ve Hubyarlar) Beydilli Alevi Türkmenleri.* Tokat: Veni Vidi Vici Yayınevi.

Yılmaz, O. (2016, Eylül). *Hoca Ahmet Yesevî'yi Pir Kabul Eden Tek Anadolu Alevi Kolu: Zile Sıraç (Beydili) Türkmenleri,* Geçmişten Geleceğe Hoca Ahmet Yesevi Sempozyumu, Bildiriler Kitabı Cilt II. 1087–1092, Istanbul, Turkey. http://tokat.gsb.gov.tr/ (Accessed: 10.08.2020).

Interviewees (March, April, May 2018)

1. Aydoğmuş, Adeli.
2. Bakşi, Temam.
3. Bakşi, Adeli.
4. Baş, Ali.
5. Bayboğa, Adeli.
6. Bektaş, İsmail.
7. Bozdemir, Ali Rıza.
8. Karakurt, Yusuf.
9. Kurt, Ali. 1
10. Kurt, Behçet.
11. Kurt, Beşir.
12. Kurt, Mehmet.
13. Tepecik, Melike.
14. Yılmaz, Döndü.
15. Yılmaz, Orhan.